Classical
and
Modern
Optimization

Advanced Textbooks in Mathematics

Print ISSN: 2059-769X
Online ISSN: 2059-7703

The *Advanced Textbooks in Mathematics* explores important topics for post-graduate students in pure and applied mathematics. Subjects covered within this textbook series cover key fields which appear on MSc, MRes, PhD and other multidisciplinary postgraduate courses which involve mathematics.

Written by senior academics and lecturers recognised for their teaching skills, these textbooks offer a precise, introductory approach to advanced mathematical theories and concepts, including probability theory, statistics and computational methods.

Published

Classical and Modern Optimization
    by Guillaume Carlier

An Introduction to Machine Learning in Quantitative Finance
    by Hao Ni, Xin Dong, Jinsong Zheng and Guangxi Yu

Conformal Maps and Geometry
    by Dmitry Beliaev

*Crowds in Equations: An Introduction to the Microscopic Modeling
of Crowds*
    by Bertrand Maury and Sylvain Faure

Mathematics of Planet Earth: A Primer
    by Jochen Bröcker, Ben Calderhead, Davoud Cheraghi, Colin Cotter,
    Darryl Holm, Tobias Kuna, Beatrice Pelloni, Ted Shepherd and Hilary Weller
    edited by Dan Crisan

Periods and Special Functions in Transcendence
    by Paula B Tretkoff

The Wigner Transform
    by Maurice de Gosson

Advanced Textbooks in Mathematics

Classical and Modern Optimization

Guillaume Carlier
Université Paris Dauphine, France

World Scientific

NEW JERSEY · LONDON · SINGAPORE · BEIJING · SHANGHAI · HONG KONG · TAIPEI · CHENNAI · TOKYO

Published by

World Scientific Publishing Europe Ltd.

57 Shelton Street, Covent Garden, London WC2H 9HE

Head office: 5 Toh Tuck Link, Singapore 596224

USA office: 27 Warren Street, Suite 401-402, Hackensack, NJ 07601

Library of Congress Cataloging-in-Publication Data
Names: Carlier, Guillaume, author.
Title: Classical and modern optimization / Guillaume Carlier, Université Paris Dauphine, France.
Description: Hackensack, New Jersey : World Scientific, [2022] | Series: Advanced textbooks in
    mathematics, 2059-769X | Includes bibliographical references and index.
Identifiers: LCCN 2021028029 (print) | LCCN 2021028030 (ebook) |
    ISBN 9781800610651 (hardcover) | ISBN 9781800610866 (paperback) |
    ISBN 9781800610668 (ebook for institutions) | ISBN 9781800610675 (ebook for individuals)
Subjects: LCSH: Mathematical optimization.
Classification: LCC QA402.5 .C367 2022 (print) | LCC QA402.5 (ebook) | DDC 519.6--dc23
LC record available at https://lccn.loc.gov/2021028029
LC ebook record available at https://lccn.loc.gov/2021028030

British Library Cataloguing-in-Publication Data
A catalogue record for this book is available from the British Library.

For any available supplementary material, please visit
https://www.worldscientific.com/worldscibooks/10.1142/Q0314#t=suppl

Desk Editors: Vishnu Mohan/Michael Beale/Shi Ying Koe

Typeset by Stallion Press
Email: enquiries@stallionpress.com

Printed in Singapore

Preface

The quest for optima, i.e., minima or maxima, is ubiquitous in nature[a] as well as in human decisions. This explains why the development of mathematical optimization has a long history that parallels that of natural sciences. Each technology era raised new optimization problems and witnessed the emergence of new ideas to address them. The recent boom of machine learning techniques and the current interest for large scale data sciences is certainly not an exception. One should however keep in mind that optimization as a mature field of mathematics owes very much to a tradition of great European mathematicians such as Fermat, Lagrange and Euler and to the development of the calculus of variations since the 17th century. Even though finite-dimensional optimization[b] is of chief importance for many practical problems, the calculus of variations is infinite dimensional in essence. Optimization in infinite dimensions, including convex analysis and convex duality theory, developed in particular by Fenchel, Moreau and Rockafellar therefore relies to some extent on functional analysis and some fundamental ideas of Baire, Hilbert and Banach.

Rather than focusing on a particular class of optimization problems (e.g., convex, differentiable, finite dimensional, variational problems in one independent variable, optimization over a Hilbert space...), I have tried to present an overview of classical and more modern ideas and methods. The first three chapters of the book are in some sense devoted to the classical optimization toolbox: topology and functional analysis, differential calculus

---

[a]Think of Fermat's principle in optics which asserts that light travels so as to minimize time.

[b]Often referred to as mathematical programming in the literature.

and convex analysis. Chapter 4 deals with the theory of necessary conditions for differentiable constrained optimization. The other chapters deal with more specialized topics and/or applications. Chapter 5 concerns optimization problems depending on a parameter, a topic, which I thought is important in practice but rarely addressed in textbooks. Convex duality theory, presented in Chapter 6, is a very powerful tool for which several applications are discussed: linear programming, SDP programming and the optimal mass transport problem of Monge and Kantorovich. Chapter 7 is devoted to (some) iterative methods for convex optimization which are of particular importance in numerical optimization algorithms. Chapter 8 presents some optimization problems arising in data sciences. The last chapter is an introduction (hopefully, an invitation?) to the calculus of variations.

It is my sincere hope that this book will be useful to a wide audience, including of course (undergraduate or graduate) students in mathematics but also engineers, data scientists or economists willing to deepen and widen their knowledge of the field. The book contains more than 200 exercises which may also make it useful to anyone teaching a third/fourth-year optimization class. Finally, I have tried to include some results which I think are important but not always easy to find especially in textbooks (e.g., generic uniqueness of extremizers, Ekeland's local surjection theorem, a detailed treatment of first and second-order optimality conditions for constrained minimizers in Banach spaces, the so-called Envelope theorem, the subdifferential of a supremum of convex functions, Birkhoff–von Neumann theorem, Nesterov's accelerated gradient method, convergence of the Douglas–Rachford algorithm, the Moore–Penrose inverse via Tikhonov regularization, sparse solutions of basis pursuit...).

I have benefited from fruitful comments, encouragements and suggestions from many friends and colleagues. I would like to warmly thank in particular, Yves Achdou, Jean-David Benamou, Jérôme Bolte, Lorenzo Brasco, Yann Brenier, Rose-Anne Dana, Ivar Ekeland, Jalal Fadili, Denis Pennequin, Gabriel Peyré, Bruno Nazaret, Filippo Santambrogio, Daniela Vögler and Irène Waldspurger. I am indebted to Laurent Chaminade and Michael Beale at World Scientific Publishing for their wonderful help and infinite patience. Finally, I wish to thank Danielle for her love and support in all circumstances.

About the Author

Guillaume Carlier is Professor of Mathematics at Université Paris Dauphine and a member of the joint research team Mokaplan (INRIA Paris and Dauphine). He has co-authored more than 100 research articles in various fields such as optimal transport, calculus of variations, mathematical economics, partial differential equations, cities modelling, etc. He also belongs to the editorial board of several journals (*Journal de l'Ecole Polytechnique, Applied Mathematics and Optimization, Journal of Mathematical Analysis and Applications, Mathematics and Financial Economics* and *Journal of Dynamics and Games*).

Contents

Preface   v

About the Author   vii

1. Topological and Functional Analytic Preliminaries   1

  1.1   Metric spaces   1

     1.1.1   Completeness and compactness   3

     1.1.2   Continuity, semicontinuity   8

     1.1.3   Baire and Ekeland's theorems   13

  1.2   Normed vector spaces   15

     1.2.1   Finite dimensions   16

     1.2.2   Linear and bilinear maps   17

  1.3   Banach spaces   23

     1.3.1   Definitions and properties   23

     1.3.2   Examples   24

     1.3.3   Ascoli's theorem   27

     1.3.4   Linear maps in Banach Spaces   29

  1.4   Hilbert spaces   30

     1.4.1   Generalities   30

     1.4.2   Projection onto a closed convex set   33

     1.4.3   The dual of a Hilbert space   36

  1.5   Weak convergence   37

  1.6   On the existence and generic uniqueness

     of minimizers   39

     1.6.1   Existence of minimizers: Variations

        on a theme   39

     1.6.2   A generic uniqueness result   41

  1.7   Exercises   42

2. Differential Calculus   **47**

   2.1 First-order differential calculus 47
      2.1.1 Several notions of differentiability 47
      2.1.2 Calculus rules 52
      2.1.3 Mean value inequalities 56
      2.1.4 Partial derivatives 60
      2.1.5 The finite-dimensional case,
           the Jacobian matrix 62
   2.2 Second-order differential calculus 65
      2.2.1 Definitions 65
      2.2.2 Schwarz's symmetry theorem 66
      2.2.3 Second-order partial derivatives 68
      2.2.4 Taylor formula 70
   2.3 The inverse function and implicit function theorems . . . 73
      2.3.1 The inverse function theorem 73
      2.3.2 The implicit function theorem 75
      2.3.3 A local surjection theorem via Ekeland's
           variational principle 77
   2.4 Smooth functions on $\mathbb{R}^d$, regularization, integration
      by parts 78
      2.4.1 Test-functions, mollification 80
      2.4.2 The divergence theorem and other integration
           by parts formulas 83
   2.5 Exercises 88

3. Convexity   **93**

   3.1 Hahn–Banach theorems 93
      3.1.1 The analytic form of Hahn–Banach theorem 93
      3.1.2 Separation of convex sets 94
   3.2 Convex sets 97
      3.2.1 Basic properties 97
      3.2.2 Linear inequalities 98
      3.2.3 Extreme points 101
   3.3 Convex functions 105
      3.3.1 Continuity properties 107
      3.3.2 Differentiable characterizations 109

3.4   The Legendre transform 113
     3.4.1   Basic properties 113
     3.4.2   The biconjugate 118
     3.4.3   Subdifferentiability 120
3.5   Exercises 126

4. Optimality Conditions for Differentiable Optimization   133

4.1   Unconstrained optimization 134
4.2   Equality constraints 136
     4.2.1   Algebraic and topological preliminaries 137
     4.2.2   Lagrange rule in the case of affine constraints . . . 140
     4.2.3   Lagrange rule in the finite-dimensional case 141
     4.2.4   Lagrange rule in Banach spaces 144
     4.2.5   Second-order conditions 145
4.3   Equality and inequality constraints 147
     4.3.1   Karush–Kuhn–Tucker conditions for affine
           constraints 147
     4.3.2   Karush–Kuhn–Tucker conditions in the
           general case 150
4.4   Exercises 153

5. Problems Depending on a Parameter   161

5.1   Setting and examples 161
5.2   Continuous dependence 164
     5.2.1   Notions of continuity for set-valued maps 164
     5.2.2   Semicontinuity of values 167
5.3   Parameter-independent constraints,
     envelope theorems 169
     5.3.1   Differentiability under local uniqueness 170
     5.3.2   Non-smooth cases 172
     5.3.3   The envelope theorem for suprema of convex
           functions 173
5.4   Parameter-dependent constraints 176
     5.4.1   Smoothness of Lagrange points 176
     5.4.2   Multipliers and the marginal price
           of constraints 181

5.5   Discrete-time dynamic programming   182
      5.5.1   Finite horizon   182
      5.5.2   Infinite horizon   183
5.6   Exercises.   187

6. Convex Duality and Applications   195

6.1   Generalities   195
6.2   Convex duality with respect to a perturbation   196
      6.2.1   Setting   196
      6.2.2   A general duality result   198
6.3   Applications   200
      6.3.1   The Fenchel–Rockafellar theorem   200
      6.3.2   Linear programming   203
      6.3.3   Semidefinite programming   205
      6.3.4   Link with KKT and Lagrangian duality   209
6.4   On the optimal transport problem   212
      6.4.1   Kantorovich duality   212
      6.4.2   Characterization of optimal plans   217
      6.4.3   Monge solutions   220
      6.4.4   The discrete case   226
6.5   Exercises.   234

7. Iterative Methods for Convex Minimization   243

7.1   On Newton's method   243
7.2   The gradient method   246
      7.2.1   Convergence of iterates   248
      7.2.2   Convergence of values   249
      7.2.3   Nesterov acceleration   253
7.3   The proximal point method   255
7.4   Splitting methods   262
      7.4.1   Forward–Backward splitting   262
      7.4.2   Douglas–Rachford method   263
      7.4.3   Link with augmented Lagrangian methods   268
7.5   Exercises.   270

8. When Optimization and Data Meet   279

8.1   Principal component analysis   279
      8.1.1   Singular value decomposition   280
      8.1.2   Principal component analysis   282

8.2 Minimization for linear systems 287
    8.2.1 Matrix operator norms and conditioning
        numbers 287
    8.2.2 Least squares and linear regressions 290
    8.2.3 Tikhonov regularization, the Moore–Penrose
        inverse 293
    8.2.4 l^1-penalization and sparse solutions 295
8.3 Classification 300
    8.3.1 Logistic regression 301
    8.3.2 Support-vector machines 303
8.4 Exercises 305

9. An Invitation to the Calculus of Variations   **311**

9.1 Preliminaries 311
    9.1.1 On weak derivatives 313
    9.1.2 Sobolev functions in dimension 1 315
    9.1.3 Sobolev functions in higher dimensions 317
9.2 On integral functionals 319
    9.2.1 Continuity, semicontinuity 319
    9.2.2 The importance of being convex 320
    9.2.3 Differentiability 323
9.3 The direct method 325
    9.3.1 Obstructions to existence 326
    9.3.2 Existence in the separable case 328
    9.3.3 Relaxation 330
9.4 Euler–Lagrange equations and other necessary
    conditions 336
    9.4.1 Euler–Lagrange equations 336
    9.4.2 On existence of minimizers again 340
    9.4.3 Examples 344
9.5 A focus on the case $d = 1$ 348
    9.5.1 Hamiltonian systems 348
    9.5.2 Dynamic programming, Hamilton–Jacobi
        equations 350
    9.5.3 A verification theorem 354
9.6 Exercises 355

Bibliography   363

Index   369

Chapter 1

Topological and Functional Analytic Preliminaries

1.1. Metric spaces

Definition 1.1. Let E be a non-empty set. A distance on E is a map $d :$ $E \times E \to \mathbb{R}_+$ which satisfies the following properties:

1) (symmetry) $d(x, y) = d(y, x)$ for every $(x, y) \in E \times E$;
2) (separation) $d(x, y) = 0 \Leftrightarrow x = y$;
3) (triangle inequality) $d(x, z) \leq d(x, y) + d(y, z)$ for every $(x, y, z) \in E \times E \times E$.

A metric space is a pair (E, d) where E is a non-empty set and d is a distance on E.

A direct consequence of the triangle inequality is that $|d(x, z) - d(y, z)| \leq d(x, y)$, for every $(x, y, z) \in E \times E \times E$. An important class of metric spaces is given by normed spaces.

Definition 1.2. Let E be a real vector space. A norm on E is a map $\| \cdot \|$: $E \to \mathbb{R}_+$ which satisfies the following properties:

1) $\|x\| = 0 \Leftrightarrow x = 0$;
2) $\|x + y\| \leq \|x\| + \|y\|$, $\forall (x, y) \in E^2$;
3) $\|\lambda x\| = |\lambda| \|x\|$, $\forall (\lambda, x) \in \mathbb{R} \times E$.

The distance associated with the norm $\| \cdot \|$ is defined by $d(x, y) := \|x - y\|$, for every $(x, y) \in E \times E$. The absolute value $| \cdot |$ is a norm on $\mathbb{R}$, the *usual* distance of $\mathbb{R}$ is the one induced by the absolute value.

1

Let (E, d) be a metric space. Given $x \in E$ and $r > 0$, the open ball centered at x and of radius r is by definition the set

$$B(x, r) := \{y \in E : d(x, y) < r\},$$

and the closed ball centered at x and of radius $r \geq 0$ is defined by:

$$\overline{B}(x, r) := \{y \in E : d(x, y) \leq r\}.$$

Definition 1.3. Let (E, d) be a metric space and let $A \subset E$. A is said to be bounded if there exist $x \in E$ and $r > 0$ such that $A \subset B(x, r)$.

When $A \subset E$ the diameter of A, $\text{diam}(A)$, is defined by:

$$\text{diam}(A) := \sup\{d(x, y), \ (x, y) \in A^2\}$$

so that A is bounded if and only if $\text{diam}(A) < +\infty$.

Definition 1.4. Let (E, d) be a metric space and let $A \subset E$.

1) A is open if for every $x \in A$, there exists $r > 0$ such that $B(x, r) \subset A$;
2) A is closed if $E \setminus A$ is open;
3) A is a neighborhood of $x \in E$ if there exists $r > 0$ such that $B(x, r) \subset A$.

One directly checks that open (respectively, closed) balls are open (respectively, closed). The family of open sets of (E, d) defines a *topology*, i.e.,

- E and $\emptyset$ are open;
- unions of open sets are open;
- *finite* intersections of open sets are open.

Passing to the complement, we have an analog statement for closed sets:

- E and $\emptyset$ are closed;
- intersections of closed sets are closed;
- *finite* unions of closed sets are closed.

Moreover, the topology of the metric space (E, d) is Hausdorff (or separate points) in the sense that whenever $x \neq y$, x and y admit disjoint neighborhoods.

Definition 1.5. Let (E, d) be a metric space, A be a subset of E and $x \in E$. One says that:

1) x is an interior point of A if there exists $r > 0$ such that $B(x, r) \subset A$;
2) x is a limit point of A if for every $r > 0$, $B(x, r) \cap A \neq \emptyset$;
3) x is a boundary point of A if for every $r > 0$, $B(x, r)$ intersects both A and $E \setminus A$.

The interior of A, denoted $\text{int}(A)$, is the set of interior points of A. The closure of A, denoted $\overline{A}$, is the set of limit points of A. The boundary of A, denoted ∂A, is the set of boundary points of A. Lastly, A is called dense in E if $\overline{A} = E$.

Proposition 1.1. *Let (E, d) be a metric space and let $A \subset E$. Then, we have the following properties:*

1) $\text{int}(A)$ *is open and it is the largest open set contained in A.*
2) $\overline{A}$ *is closed and it is the smallest open set containing A.*
3) A *is dense if and only if for every non-empty open subset U of E, $A \cap U \neq \emptyset$.*

In metric spaces, it is often convenient to express topological properties through sequential properties.

Definition 1.6. Let (E, d) be a metric space and let (x_n) be a sequence of elements of E. Then, $x \in E$ is a limit of (x_n) (which is denoted $x_n \to x$ or $\lim_n x_n = x$) if for every $\varepsilon > 0$, there exists $N \in \mathbb{N}^*$ such that for every $n \geq N$, $d(x_n, x) \leq \varepsilon$. The sequence (x_n) is called convergent if it admits a limit.

Put differently, $x_n \to x$ (in (E, d)) if and only if $d(x_n, x) \to 0$ (in $\mathbb{R}$). Note also that the fact that d separate points straightforwardly implies that if a sequence has a limit, this limit is unique.

Proposition 1.2. *Let (E, d) be a metric space and let $A \subset E$. Then, we have the following properties:*

1) *Let $x \in E$, $x \in \overline{A}$ if and only if x is the limit of a sequence of elements of A;*
2) *A is closed if and only if every convergent sequence (x_n) of elements of A has its limit in A.*

1.1.1. *Completeness and compactness*

Definition 1.7. Let (E, d) be a metric space and $(x_n)_n$ be a sequence of elements of E. The sequence $(x_n)_n$ is a Cauchy sequence if for every $\varepsilon > 0$, there exists $N \in \mathbb{N}^*$ such that for all $(p, q) \in \mathbb{N}^2$ with $p \geq N$ and $q \geq N$ one has $d(x_p, x_q) \leq \varepsilon$.

Note that the previous definition can be expressed as follows: $(x_n)_n$ is a Cauchy sequence if and only if

$$\sup_{p \geq N, \, q \geq N} d(x_p, x_q) \to 0 \quad \text{as } N \to +\infty.$$

Note that Cauchy sequences are bounded. Of course, every convergent sequence is Cauchy but the converse need not be true in general. This leads to the following fundamental definition.

Definition 1.8. The metric space (E, d) is said to be complete if every Cauchy sequence of (E, d) is convergent.

The set of rational numbers $\mathbb{Q}$ (equipped with the usual distance of $\mathbb{R}$) is not complete (the sequence of rationals $x_n := \sum_{k=0}^{n} 1/(k!)$ is Cauchy but does not converge), but $\mathbb{R}$ is (in fact, by construction). The following proposition is a first illustration of the notion completeness.

Proposition 1.3. *Let (E, d) be a complete metric space and let (F_n) be a non-increasing sequence of non-empty closed subsets of E, whose diameter tends to 0. Then there exists $x \in E$ such that $\bigcap_n F_n = \{x\}$ in particular the sequence F_n has non-empty intersection.*

Proof. Let $d_n := \operatorname{diam}(F_n)$ and for every $n \in \mathbb{N}$, pick $x_n \in F_n$. For every pair of integers p and q such that $p, q \geq N$ we then have $d(x_p, x_q) \leq d_N$, and since d_N tends to 0 as $N \to +\infty$, (x_n) is a Cauchy sequence. Since (E, d) is complete, (x_n) converges to a limit x. Since x is the limit of a sequence of elements in F_n and F_n is closed, we have $x \in F_n$, which proves that $x \in \bigcap_n F_n$. Finally, by the fact that $\operatorname{diam}(F_n)$ tends to 0, $\cap_n F_n$ cannot contain any other point than x. $\qquad\square$

Definition 1.9. Let E be a non-empty set and $(x_n)_n$ be a sequence of elements of E. A subsequence (or extraction) of $(x_n)_n$ is a sequence of the form $(x_{\varphi(n)})_n$ where φ is an increasing function from $\mathbb{N}$ to $\mathbb{N}$.

Definition 1.10. Let (E, d) be a metric space and let (x_n) be a sequence of elements of E and $x \in E$. Then, x is called a cluster point of (x_n) if any of the following equivalent assertions holds:

1) (x_n) admits a subsequence that converges to x;
2) $\forall \varepsilon > 0$, $\forall N \in \mathbb{N}$, $\exists n \geq N$ such that $d(x_n, x) \leq \varepsilon$;
3) $\forall \varepsilon > 0$ the set $\{n \in \mathbb{N} : d(x_n, x) \leq \varepsilon\}$ is infinite.

A sequential definition of compact metric spaces then reads as follows.

Definition 1.11. The metric space (E, d) is compact if any sequence of elements of E admits a convergent subsequence. A subset A of E is called compact if any sequence of elements of A admits a subsequence which converges to an element of A.

Proposition 1.4. *Let (E, d) be a metric space and let A be a compact subset of E. Then A is closed and bounded.*

Proof. Let $(x_n)_n \in A^{\mathbb{N}}$ be a convergent sequence and let $x \in E$ denote its limit. Since A is compact, $(x_n)_n$ admits a convergent subsequence in A, but such a subsequence converges to x hence $x \in A$ which shows that A is closed.

If A was not bounded, we could find two sequences $(x_n) \in A^{\mathbb{N}}$ and $(y_n) \in A^{\mathbb{N}}$ such that

$$\lim_n d(x_n, y_n) = +\infty. \tag{1.1}$$

Since A is compact, (x_n) has a convergent subsequence $(x_{\varphi}(n))$ and $(y_{\varphi}(n))$ has a convergent subsequence $(y_{\varphi(\psi(n))})$, so that both $(x_{\varphi(\psi(n))})$ and $(y_{\varphi(\psi(n))})$ converge respectively to some x and y in A. By the triangle inequality, we have

$$d(x_{\varphi(\psi(n))}, y_{\varphi(\psi(n))}) \leq d(x_{\varphi(\psi(n))}, x) + d(x, y) + d(y, y_{\varphi(\psi(n))}) \to d(x, y)$$

which contradicts (1.1). $\qquad\square$

The following proposition is obvious.

Proposition 1.5. *Let (E, d) be a compact metric space and A be a subset of E. Then A is compact if and only if it is closed.*

Note also that compactness is (much) stronger than completeness.

Proposition 1.6. *Every compact metric space is complete.*

Proof. Let (E, d) be a compact metric space and let $(x_n)_n$ be a Cauchy sequence in E. By compactness, (x_n) admits a cluster point $x \in E$. Let us prove that (x_n) converges to x: let $\varepsilon > 0$, since (x_n) is a Cauchy sequence there is an integer N_1 such that for all $p, q \geq N_1$, $d(x_p, x_q) \leq \varepsilon/2$. But since x is a cluster point of (x_n) there exists $N_2 \geq N_1$ such that $d(x_{N_2}, x) \leq \varepsilon/2$. Hence, for every $p \geq N_2$, $d(x_p, x) \leq d(x_p, x_{N_2}) + d(x_{N_2}, x) \leq \varepsilon$; which shows that (x_n) converges to x. $\qquad\square$

Note in passing that in the previous proof, we have shown the following lemma.

Lemma 1.1. *Let (E, d) be a metric space and $(x_n)_n$ be a sequence of elements of E. Then $(x_n)_n$ converges if and only if $(x_n)_n$ is a Cauchy sequence and admits a cluster point.*

Theorem 1.1. *Every bounded and closed subset of $\mathbb{R}$ (equipped with its usual distance) is compact.*

Proof. Let F be a closed and bounded subset of $\mathbb{R}$. Then F is included in a segment $[a, b]$ of $\mathbb{R}$; up to an homothety and a translation, let us further assume that $F \subset [0, 1]$. Let $(x_n)_n \in F^{\mathbb{N}} \subset [0, 1]^{\mathbb{N}}$. We will show that (x_n) admits a Cauchy subsequence by the following induction argument. For every $p \in \mathbb{N}^*$, decompose $[0, 1]$ into 2^p segments of length 2^{-p}:

$$[0, 1] = \bigcup_{k=1}^{2^p} I_k^p, \ I_k^p := [(k-1)2^{-p}, k2^{-p}].$$

For $p = 1$ one of the two segments I_1^1 and I_2^1, denoted J_1, is such that $\{n \in \mathbb{N} : x_n \in J_1\}$ is infinite. Then decompose again J^1 as

$$J_1 = \bigcup_{k \in \{1, \dots, 4\} \ : \ I_k^2 \subset J_1} I_k^2$$

and chose a $k \in \{1, \dots, 4\}$ such that $I_k^2 \subset J_1$ and $J_2 := I_k^2$ is such that $\{n \in \mathbb{N} : x_n \in J_2\}$ is infinite. This way, one inductively builds a sequence of segments $J_1 \supset J_2 \supset \cdots \supset J_p$ such that J_p has length 2^{-p} and for every p, the set $\{n \in \mathbb{N} : x_n \in J_p\}$ is infinite.

Let n_1 be the smallest integer k such that $x_k \in J_1$, n_2 be the smallest integer $k \geq n_1 + 1$ such that $x_k \in J_2, \dots, n_p$ be the smallest integer $k \geq n_{p-1} + 1$ such that $x_k \in J_p$. The sequence $(x_{n_p})_p$ is a subsequence of $(x_n)_n$ and by construction for every $r, s \geq p$, $(x_{n_s}, x_{n_r}) \in J_p$ and since J_p has diameter 2^{-p}, one has $|x_{n_s} - x_{n_r}| \leq 2^{-p}$ so that $(x_{n_p})_p$ is a Cauchy sequence. Since $\mathbb{R}$ is complete, $(x_{n_p})_p$ converges and since F is closed, its limit is in F.

This shows that (x_n) admits a subsequence which converges in F so that F is compact. $\square$

The following important theorem (Bolzano–Weierstrass) gives a characterization of compactness in metric spaces in terms of finite open coverings.

Theorem 1.2 (Bolzano–Weierstrass). *Let (E, d) be a metric space. Then (E, d) is compact if and only if for every family of open sets $(O_i)_{i \in I}$ such that $E = \bigcup_{i \in I} O_i$ there is a finite set $J \subset I$ such that $E = \bigcup_{i \in J} O_i$ (finite covering property).*

Proof. First let us assume that (E, d) has the finite covering property and let us remark that it implies that if F_n is a sequence of non-empty closed subsets of E such that $F_{n+1} \subset F_n$ then $\bigcap_n F_n$ is non-empty (otherwise $O_n = E \setminus F_n$ would be an open covering of E and there would exist a finite covering which would imply that some F_n is empty). Now let $(x_n)_n \in E^{\mathbb{N}}$, for every n, let us set:

$$F_n := \overline{\{x_k, \ k \geq n\}}.$$

We then have $F := \bigcap_n F_n \neq \emptyset$ and it is easy to check that F is the set of cluster points of the sequence (x_n).

Conversely, let us assume that (E, d) is compact and let us prove that it has the finite covering property. Let $(O_i)_{i \in I}$ be a family of open sets such that $E = \bigcup_{i \in I} O_i$.

Claim 1. *There exists $\varepsilon > 0$ such that for every $x \in E$, there is an $i \in I$ such that $B(x, \varepsilon) \subset O_i$.*

If it was not the case (taking $\varepsilon = 1/n$), for every n there would exist some x_n such that $B(x_n, 1/n)$ is not included in any O_i. Since E is compact (taking a subsequence if necessary), we may assume that x_n converges to some $\overline{x}$ that belongs to the open set O_{i_0}, but for n large enough one should have $B(x_n, 1/n) \subset O_{i_0}$ which gives the desired contradiction.

Claim 2. *For every $r > 0$, E can be covered by finitely many open balls of radius r.*

Otherwise, there exists $r > 0$ such that E cannot be covered by finitely many open balls of radius r. We may then choose $x_1 \in E$, $x_2 \notin B(x_1, r)$, $x_3 \notin B(x_1, r) \cup B(x_2, r)$, ..., $x_{n+1} \notin \bigcup_{k=1}^{n} B(x_k, r)$. By construction $d(x_p, x_q) \geq r$ for every $p \neq q$ hence (x_n) does not have any Cauchy subsequence and therefore no cluster point, which yields the contradiction.

Let $\varepsilon > 0$ be as in Claim 1 and take $r = \varepsilon$ in Claim 2. There are points $x_1, \ldots, x_n$ such that $E = \bigcup_{j=1}^{n} B(x_j, \varepsilon)$ and $B(x_j, \varepsilon) \subset O_{i_j}$ for some $i_j \in I$. This proves that the finite family $(O_{i_j})_{j=1,\ldots,n}$ is a covering of E. $\qquad \square$

Passing to the complement, we have an equivalent formulation of Bolzano–Weierstrass' theorem.

Corollary 1.1. *Let (E, d) be a metric space. Then (E, d) is compact if and only if for every family of closed sets $(F_i)_{i \in I}$ such that $\cap_{i \in I} F_i = \emptyset$ there is a finite set $J \subset I$ such that $\bigcap_{i \in J} F_i = \emptyset$.*

Compact metric spaces are *separable*, i.e., admit a countable dense subset.

Proposition 1.7. *Let (E, d) be a compact metric space, Then it is separable in the sense that there exists $(x_n)_n \in E^{\mathbb{N}}$ such that the countable set $\{x_n, \ n \in \mathbb{N}\}$ is dense in E.*

Proof. By the Bolzano–Weierstrass theorem for every $n \in \mathbb{N}^*$ there exists finitely many points $(x_i^n)_{i \in I_n}$ such that E is covered by the union of the open balls $B(x_i^n, 1/n)$. The set $\bigcup_n \{x_i^n, \ i \in I_n\}$ is countable and dense in E by construction. $\qquad \square$

In a compact metric space, we have a convenient criterion for a sequence to converge.

Proposition 1.8. *Let (E, d) be a compact metric space and let $(x_n)_n \in E^{\mathbb{N}}$. Then the sequence (x_n) converges if and only if it possesses a unique cluster point.*

Proof. Of course, if (x_n) converges, it has a unique cluster point. Now assume that some subsequence $(x_{\varphi(n)})$ converges to some $\overline{x}$ but $\overline{x}$ is not the limit of (x_n). This means that there exists $\varepsilon_0 > 0$ such that, for every N, there is some $n \geq N$ such that $d(\overline{x}, x_n) \geq \varepsilon_0$. This implies that (x_n) admits a subsequence $(x_{\psi(n)})$ such that $d(x_{\psi(n)}, \overline{x}) \geq \varepsilon_0$ for all n. By compactness of (E, d), $(x_{\psi(n)})$ possesses a cluster point y and $d(y, \overline{x}) \geq \varepsilon_0 > 0$, which implies that (x_n) has two distinct cluster points. $\qquad \square$

1.1.2. Continuity, semicontinuity

Definition 1.12. Let (E_1, d_1) and (E_2, d_2) be two metric spaces, let $f: E_1 \to E_2$ be a map and $x \in E_1$. One says that f is continuous at x if for every $\varepsilon > 0$, there exists $\delta > 0$ such that $d_1(x,y) \leq \delta \Rightarrow d_2(f(x), f(y)) \leq \varepsilon$. The map f is continuous on E_1 if it is continuous at each point of E_1.

The following characterization of continuity is classic.

Proposition 1.9. *Let (E_1, d_1) and (E_2, d_2) be two metric spaces and let $f: E_1 \to E_2$ be a map. Then the following statements are equivalent:*

1) *f is continuous on E_1.*
2) *For every open subset O of E_2, $f^{-1}(O)$ is an open subset of E_1.*
3) *For every closed subset F of E_2, $f^{-1}(F)$ is a closed subset of E_1.*
4) *For every sequence (x_n) of elements of E_1, one has*

$$\lim_n x_n = x \text{ in } E_1 \Rightarrow \lim_n f(x_n) = f(x) \text{ in } E_2.$$

Proposition 1.10. *Let (E_1, d_1) and (E_2, d_2) be two metric spaces and let $f : E_1 \to E_2$ be continuous. If E_1 is compact, then $f(E_1)$ is a compact subset of E_2.*

Proof. Let $z_n := f(x_n)$ (with $x_n \in E_1$) be a sequence in $f(E_1)$. Since E_1 is compact, (x_n) admits a convergent subsequence $(x_{\varphi(n)})$, f being continuous, the subsequence $z_{\varphi(n)} = f(x_{\varphi(n)})$ converges which shows that $f(E_1)$ is compact. $\square$

Corollary 1.2. *Let (E, d) be a compact metric space and $f : E \to \mathbb{R}$ be continuous, then f achieves its infimum and its supremum on E.*

Proof. By virtue of Proposition 1.10, $f(E)$ is a compact subset of $\mathbb{R}$ hence both its infimum and its supremum are finite and belong to $f(E)$. $\square$

Definition 1.13. Let (E_1, d_1) and (E_2, d_2) be two metric spaces and let $f : E_1 \to E_2$. Then, f is said to be uniformly continuous on E_1 if, for every $\varepsilon > 0$, there exists $\delta > 0$ such that for every $(x, y) \in E_1^2$, $d_1(x, y) \leq \delta \Rightarrow d_2(f(x), f(y)) \leq \varepsilon$.

Clearly, if f is uniformly continuous, it is continuous but the converse is not true (consider $E_1 = E_2 = \mathbb{R}$ and $f(x) = x^2$). A convenient way to express uniform continuity is by the notion of modulus of continuity.

If $f : E_1 \to E_2$, then f is uniformly continuous if and only if the function

$$\omega(\delta) := \sup\{d_2(f(x), f(x')), \ (x, x') \in E_1^2, \ d_1(x, x') \leq \delta\}, \delta \geq 0$$

satisfies

$$\lim_{\delta \to 0^+} \omega(\delta) = 0.$$

Note that by definition ω is non-decreasing, $\omega(0) = 0$ and one has

$$d_2(f(x), f(x')) \leq \omega(d_1(x, x')), \quad \forall (x, x') \in E_1^2.$$

Such a function is called a *modulus of continuity* for f.

Definition 1.14. Let (E_1, d_1) and (E_2, d_2) be two metric spaces, let f: $E_1 \to E_2$ and $k \in \mathbb{R}_+$. Then, f is said to be k-Lipschitz if for every $(x, y) \in E_1 \times E_1$ one has $d_2(f(x), f(y)) \leq kd_1(x, y)$. Finally, f is said to be Lipschitz if it is k-Lipschitz for some $k \geq 0$.

Lipschitz functions are uniformly continuous (with a linear modulus of continuity) but the converse is obviously false (consider $E_1 = E_2 = \mathbb{R}_+$ and $f(x) = \sqrt{x}$). Continuity implies uniform continuity if E_1 is compact.

Theorem 1.3 (Heine's theorem). *Let (E_1, d_1) and (E_2, d_2) be two metric spaces and $f : E_1 \to E_2$ be continuous map. If (E_1, d_1) is compact, then f is uniformly continuous on E_1.*

Proof. We argue by contradiction: if f was not uniformly continuous, there would exist $\varepsilon > 0$, and two sequences (x_n) and (y_n) in E_1, such that $d_1(x_n, y_n)$ tends to 0 as $n \to \infty$ and $d_2(f(x_n), f(y_n)) \geq \varepsilon$ for every n. E_1 being compact, we can extract (not relabeled) convergent subsequences from (x_n) and (y_n) with respective limits x and y. Passing to the limit, $n \to \infty$, we would then have $x = y$ and $d_2(f(x), f(y)) \geq \varepsilon > 0$ which gives the desired contradiction. $\square$

The importance of the notion of completeness is particularly well illustrated by the following simple but extremely useful fixed-point theorem.

Theorem 1.4 (Banach's fixed-point theorem). *Let (E, d) be a complete metric space and f be a contraction of E, i.e., a map from E to E such that for some $k \in (0, 1)$*

$$d(f(x), f(y)) \leq kd(x, y), \quad \forall (x, y) \in E \times E.$$

Then, f admits a unique fixed point: there exists a unique $x \in E$ such that $f(x) = x$. Moreover, for every $x_0 \in E$, if we define inductively the sequence x_n through $x_{n+1} = f(x_n)$, for $n \geq 0$, then x_n converges to x as $n \to +\infty$.

Proof. Let us start with uniqueness: assume that f has two fixed points x_1 and x_2, $f(x_1) = x_1$, $f(x_2) = x_2$, we then have $d(x_1, x_2) = d(f(x_1), f(x_2)) \leq k d(x_1, x_2)$ and since $k < 1$ this yields $d(x_1, x_2) = 0$, i.e., $x_1 = x_2$.

Let $x_0 \in E$, define x_n as in the statement of the theorem and let us prove that (x_n) is a Cauchy sequence. First observe that for any $n \in \mathbb{N}^*$ one has $d(x_{n+1}, x_n) = d(f(x_n), f(x_{n-1})) \leq k d(x_n, x_{n-1})$, iterating the argument thus gives:

$$d(x_{n+1}, x_n) \leq k^n d(x_1, x_0). \tag{1.2}$$

For $q \geq p \geq N$, we then have

$$d(x_p, x_q) \leq d(x_p, x_{p+1}) + \cdots + d(x_{q-1}, x_q) \leq d(x_1, x_0)(k^p + \cdots + k^{q-1})$$

$$\leq d(x_1, x_0) \frac{k^N}{1-k}.$$

Since $k \in (0, 1)$, $k^N \to 0$ as $N \to +\infty$, the previous inequality thus implies that (x_n) is a Cauchy sequence and therefore converges to some limit $x \in E$. Since f is Lipschitz, it is continuous so $x_{n+1} = f(x_n)$ converges to $f(x)$ which gives $x = f(x)$. $\qquad\square$

In optimization, it is useful to consider function $f : E \to \mathbb{R} \cup \{+\infty\}$. We extend the usual order of $\mathbb{R}$ to the set of extended reals $\overline{\mathbb{R}} := \mathbb{R} \cup \{-\infty, +\infty\}$ by setting

$$-\infty < t < +\infty, \quad \forall t \in \mathbb{R}.$$

For $f : E \to \mathbb{R} \cup \{+\infty\}$, we define the domain of f, denoted $\mathrm{dom}(f)$ by

$$\mathrm{dom}(f) := \{x \in E \ : \ f(x) < +\infty\} \tag{1.3}$$

and the epigraph of f by

$$\mathrm{epi}(f) := \{(\lambda, x) \in \mathbb{R} \times E \ : \ f(x) \leq \lambda\}, \tag{1.4}$$

which represents the set of points in $\mathbb{R} \times E$ which lie *above* the graph of f. Remark that if $(\lambda, x) \in \mathrm{epi}(f)$ then $x \in \mathrm{dom}(f)$. We know from Theorem 1.1 that every sequence of reals (t_n) which admits a bounded subsequence possesses a cluster point in $\mathbb{R}$. Now if (t_n) has no bounded subsequence, it necessarily has a subsequence which either tends to $+\infty$ or $-\infty$. In other words, in the set of extended reals $\overline{\mathbb{R}} := \mathbb{R} \cup \{-\infty, +\infty\}$, (t_n) has at least one cluster point. The largest such cluster point

$$\limsup_{n} \{t_k \; : \; k \geq n\} = \inf_{n \in \mathbb{N}} \sup\{t_k \; : \; k \geq n\}$$

is called the lim sup of (t_n) and denoted as

$$\limsup_{n} t_n := \limsup_{n}\{t_k \; : \; k \geq n\}.$$

It is also the supremum[a] of all reals t for which $t_n \geq t$ for infinitely many $n \in \mathbb{N}$. In a similar fashion, the smallest cluster point of (t_n) is called the lim inf of t_n:

$$\liminf_{n} t_n := \liminf_{n}\{t_k \; : \; k \geq n\},$$

it is also the infimum of all reals t for which $t_n \leq t$ for infinitely many $n \in \mathbb{N}$.

Definition 1.15. Let (E, d) be a metric space, let $f : E \to \mathbb{R} \cup \{+\infty\}$ and $x \in E$.

- f is said to be lower-semicontinuous (lsc for short) at x if for every $\lambda \in \mathbb{R}$ such that $f(x) > \lambda$ there is an $r > 0$ such that $f(y) > \lambda$ for every $y \in B(x, r)$.
- f is said to be lsc on E if it is lsc at each point of E.
- In a similar way, $f : E \to \mathbb{R} \cup \{-\infty\}$ is said to be upper-semicontinuous (usc for short) at x if and only if $-f$ is lsc at x and usc on E if it is usc at each point of E.

One easily checks that $f : E \to \mathbb{R}$ is continuous (at a given x or on the whole of E) if and only it is both lsc and usc (at x, at every point of E).

Proposition 1.11. *Let (E, d) be a metric space and $f : E \to \mathbb{R} \cup \{+\infty\}$. Then, the following assertions are equivalent:*

1) *f is lsc on E.*
2) *epi(f) is closed in $\mathbb{R} \times E$.*

---

[a]We will always adopt the convention that $\sup \emptyset = -\infty$ and $\inf \emptyset = +\infty$. This is consistent with the fact that, if a sequence of reals converges to $-\infty$ (respectively, $+\infty$), its limsup (respectively, liminf) is $-\infty$ (respectively, $+\infty$).

3) *For every $x \in E$ and every sequence (x_n) converging to x, one has*

$$\liminf_n f(x_n) \geq f(x).$$

Proof. We shall prove 1) $\Rightarrow$ 2) $\Rightarrow$ 3) $\Rightarrow$ 1). 1) $\Rightarrow$ 2): Let (λ_n, x_n) be a sequence in epi(f) converging to (λ, x), we wish to show $f(x) \leq \lambda$, if $\lambda = +\infty$ there is nothing to prove, we thus assume $\lambda \in \mathbb{R}$. Let us assume by contradiction that $f(x) > \lambda$ and pick $\varepsilon > 0$ such that $f(x) > \lambda + \varepsilon$, then since f is lsc at x and x_n converges to x, we should also have $f(x_n) > \lambda + \varepsilon$ for n large enough, but since λ_n converges to λ we should also have $f(x_n) \geq \lambda_n + \varepsilon/2 > \lambda_n$ for n large enough, contradicting the fact that $(\lambda_n, x_n) \in$ epi(f). 2) $\Rightarrow$ 3): If $\liminf f(x_n) = +\infty$, there is nothing to prove, and taking a subsequence if necessary we may assume that $f(x_n) \in \mathbb{R}$ converges to $\liminf_n f(x_n)$, the sequence $(f(x_n), x_n)$ belongs to epi(f) and converges to $(\liminf_n f(x_n), x)$ which belongs to epi(f) since we have assumed epi(f) is closed, this gives $f(x) \leq \liminf_n f(x_n)$. Finally, 3) $\Rightarrow$ 1), we have to show that for $\lambda \in \mathbb{R}$ the set $\{x \in E : f(x) \leq \lambda\}$ is closed, but this obviously follows from 3). $\qquad \square$

Suprema of continuous functions need not be continuous, but lower-semicontinuity is stable by suprema.

Lemma 1.2. *Let $(f_i)_{i \in I}$ be a family of lsc functions $E \to \mathbb{R} \cup \{+\infty\}$. Then $f := \sup_{i \in I} f_i$ is lsc on E.*

Proof. It is obvious that epi$(f) = \bigcap_{i \in I}$ epi(f_i) and the latter is closed as an intersection of closed sets, lower-semicontinuity of f thus follows from Proposition 1.11. $\qquad \square$

1.1.3. *Baire and Ekeland's theorems*

Another important property of complete metric spaces is as follows.

Theorem 1.5 (Baire's theorem). *Let (E, d) be a complete metric space and (O_n) be a sequence of open and dense subsets of E. Then $\bigcap_n O_n$ is a dense subset of E.*

Proof. Let U be some open set, we have to prove that $\bigcap_n O_n \cap U \neq \emptyset$. First let us fix $x_0 \in E$ and $r_0 > 0$ such that $\overline{B}(x_0, r_0) \subset U$. Since $B(x_0, r_0)$ is open and O_1 is dense there exist x_1 and $0 < r_1 \leq r_0/2$ such that $\overline{B}(x_1, r_1) \subset B(x_0, r_0) \cap O_1$. Inductively, we construct a sequence x_n in E and $r_n > 0$ such that $\overline{B}(x_{n+1}, r_{n+1}) \subset B(x_n, r_n) \cap O_{n+1}$ and $r_{n+1} \leq r_n/2$.

Since x_n is a Cauchy sequence, it converges to some $\bar{x}$. By construction $\bar{x} \in \overline{B}(x_0, r_0) \subset U$ and $\bar{x} \in \overline{B}(x_n, r_n)$ for all n, thus $\bar{x} \in \cap_n O_n$, which completes the proof. $\qquad\square$

Taking complements, we get the equivalent formulation.

Corollary 1.3. *Let (E, d) be a complete metric space and F_n be a sequence of closed subsets of E. If $\mathrm{int}(F_n) = \emptyset$ for all n then $\mathrm{int}(\bigcup_n F_n) = \emptyset$.*

Corollary 1.4. *Let (E, d) be a complete metric space and F_n be a sequence of closed subsets of E, such that $\bigcup_n F_n = E$. Then there exits some n_0 such that F_{n_0} has non-empty interior.*

Another crucial result which illustrates the importance of completeness is the ε-variational principle discovered by Ekeland in 1974 [44].

Theorem 1.6 (Ekeland's variational principle). *Let (E, d) be a complete metric space and $f : E \to \mathbb{R} \cup \{+\infty\}$ be lsc, bounded from below and such that $\mathrm{dom}(f) \neq \emptyset$, let $\varepsilon > 0$ and $x_\varepsilon \in E$ be such that*

$$f(x_\varepsilon) \leq \inf_E f + \varepsilon$$

and let $k > 0$. Then there exists $y_\varepsilon \in E$ such that

$$f(y_\varepsilon) \leq f(x_\varepsilon), \ d(x_\varepsilon, y_\varepsilon) \leq \frac{1}{k} \ and$$

$$f(y_\varepsilon) - k\varepsilon d(x, y_\varepsilon) < f(x), \ \forall x \in E \setminus \{y_\varepsilon\}. \tag{1.5}$$

Proof. Let us set $d' := k\varepsilon d$ and $Y := \{y \in E : f(y) \leq f(x_\varepsilon) - d'(x_\varepsilon, y)\}$. Y is closed because $f(\cdot) + d'(x_\varepsilon, \cdot)$ is lsc. Therefore, (Y, d') is a complete metric space. For any $x \in Y$, set

$$S(x) := \{y \in Y : f(y) \leq f(x) - d'(x, y)\}.$$

Then $S(x)$ is a non-empty (it contains x) and closed (again by lower-semicontinuity) subset of Y. Moreover, if $y \in S(x)$ and $z \in S(y)$, we have

$$f(z) \leq f(y) - d'(y, z) \leq f(x) - d'(x, y) - d'(y, z) \leq f(x) - d'(x, z).$$

This shows that $S(y) \subset S(x)$ whenever $y \in S(x)$. Now let us define $F_0 := Y = S(x_\varepsilon)$ and inductively for $n \geq 1$

$$F_n := S(z_n) \text{ with } z_n \in F_{n-1}, \ f(z_n) \leq \inf_{F_{n-1}} f + \frac{1}{2^n}.$$

If $z \in F_n \subset F_{n-1}$, we thus have

$$d'(z, z_n) \leq f(z_n) - f(z) \leq \frac{1}{2^n}$$

so that $\text{diam}(F_n) \to 0$. Thanks to Proposition 1.3, we deduce that there exists $z \in Y$ such that $\bigcap_n F_n = \{z\}$. Since $z \in S(x_\varepsilon)$ we have

$$f(z) \leq f(x_\varepsilon), \ d'(z, x_\varepsilon) \leq f(x_\varepsilon) - f(z) \leq \varepsilon, \ \text{i.e.,} \ d(z, x_\varepsilon) \leq \frac{1}{k}. \quad (1.6)$$

We also have $S(z) = \{z\}$ (indeed if $y \in S(z)$, since $z \in S(z_n)$ then $y \in S(z_n)$). Hence for every $y \in Y \setminus \{z\}$, we have

$$f(y) > f(z) - d'(z, y). \quad (1.7)$$

Now if $y \in E \setminus Y$, using the fact that $z \in S(x_\varepsilon)$ and the triangle inequality, we get

$$f(y) > f(x_\varepsilon) - d'(y, x_\varepsilon) \geq f(z) - d'(z, x_\varepsilon) - d'(y, x_\varepsilon) \geq f(z) - d'(z, y)$$

so that (1.7) holds for any $y \in E \setminus \{z\}$. This shows that $y_\varepsilon := z$ has all the desired properties in (1.5). $\qquad \square$

1.2. Normed vector spaces

Definition 1.16. A (real) normed vector space is a pair $(E, \|\cdot\|)$ where E is a real vector space and $\|\cdot\|$ is a norm on E.

For instance, on $\mathbb{R}^d$, defining

$$\|x\|_\infty := \max_{i=1,\ldots,d} |x_i|, \ \|x\|_1 := \sum_{i=1}^d |x_i|,$$

for every $x = (x_1, \ldots, x_d) \in \mathbb{R}^d$), both $\|\cdot\|_\infty$ and $\|\cdot\|_1$ are norms on $\mathbb{R}^d$. More generally, for any $p \in [1, +\infty)$, $\|\cdot\|_p$ defined by

$$\|x\|_p := \left(\sum_{i=1}^d |x_i|^p \right)^{1/p}$$

is a norm (for the triangle inequality use Hölder's inequality, see Exercise 1.8); a particular case of interest is $p = 2$, the corresponding norm $\|\cdot\|_2$ being the usual Euclidean norm of $\mathbb{R}^d$.

Definition 1.17. Let E be a real vector space and N_1, N_2 be two norms on E. These norms are said to be equivalent if there are positive constants a and b such that:

$$aN_1(x) \le N_2(x) \le bN_2(x), \quad \forall x \in E.$$

1.2.1. *Finite dimensions*

Theorem 1.7. *Let A be a non-empty subset of $\mathbb{R}^d$. Then A is compact in $(\mathbb{R}^d, \| \cdot \|_\infty)$ if and only if it is closed and bounded. In particular, every bounded sequence of $(\mathbb{R}^d, \| \cdot \|_\infty)$ has a convergent subsequence.*

Proof. We already know from Proposition 1.4 (valid, in any metric space) that compact subsets are closed and bounded. Let F be a closed and bounded subset of $(\mathbb{R}^d, \| \cdot \|_\infty)$. Then there is some $M > 0$ such that $F \subset [-M, M]^d$. Let $(x_n)_n \in F^{\mathbb{N}} \subset ([-M, M]^d)^{\mathbb{N}}$. Thanks to Theorem 1.1, $[-M, M]$ is a compact subset of $\mathbb{R}$. Taking successive extractions for the d coordinates, we deduce that (x_n) admits a cluster point which is in F since F is closed. $\qquad \square$

Theorem 1.8. *If E is finite-dimensional real vector space, all norms on E are equivalent.*

Proof. Let $(e_1, \ldots, e_d)$ be a basis of E and for $x \in E$ denote by $(x_1, \ldots, x_d)$ the coordinates of x in this basis, define the norm $\| \cdot \|_\infty$ on E by $\|x\|_\infty := \max\{|x_i|, \ i = 1, \ldots, d\}$. Let N be another norm on E, we are going to prove that N is equivalent to $\| \cdot \|_\infty$ which will give the result by transitivity of the equivalence between norms. For $x = \sum_{i=1}^d x_i e_i$, we first have

$$N(x) = N\left(\sum_{i=1}^d x_i e_i\right) \le \sum_{i=1}^d |x_i| N(e_i) \le \left(\sum_{i=1}^d N(e_i)\right) \|x\|_\infty.$$

Hence $N(x) \le C\|x\|_\infty$ with $C = \sum_{i=1}^d N(e_i)$. We then also have for every x, y in E:

$$|N(x) - N(y)| \le |N(x - y)| \le C\|x - y\|_\infty \tag{1.8}$$

so that N is continuous for $\| \cdot \|_\infty$.

Let $S := \{x \in E : \|x\|_\infty = 1\}$, S is bounded and closed hence compact for $\|\cdot\|_\infty$ thanks to Theorem 1.7. Since N is continuous, N attains its minimum on S, let then $x_0 \in S$ such that $N(x_0) = \min_S N$. Since $x_0 \in S$, $x_0 \neq 0$ hence $N(x_0) > 0$, set then $\alpha = N(x_0)$. For $x \neq 0$, $x/\|x\|_\infty \in S$ and then:

$$N\left(\frac{x}{\|x\|_\infty}\right) \geq \alpha \Rightarrow \|x\|_\infty \leq \frac{N(x)}{\alpha}.$$

The last inequality is also obviously satisfied for $x = 0$, allowing us to conclude that N and $\|\cdot\|_\infty$ are equivalent. $\square$

Combining Theorems 1.7 and 1.8, we get

Theorem 1.9. *Let $(E, \|\cdot\|)$ be a finite-dimensional normed space, and A be a non-empty subset of E. Then A is compact if and only if it is closed and bounded. In particular, every bounded sequence of $(E, \|\cdot\|)$ has a convergent subsequence.*

1.2.2. *Linear and bilinear maps*

In what follows, $(E, \|\cdot\|_E)$ and $(F, \|\cdot\|_F)$ will denote two normed spaces. We then denote by $\mathcal{L}(E, F)$ (respectively, $\mathcal{L}_c(E, F)$) the space of linear maps (respectively, continuous linear maps) between E and F. For $E = F$, we simply denote $\mathcal{L}(E)$ (respectively, $\mathcal{L}_c(E)$) the space of endomorphisms (respectively, continuous endomorphisms) of E. When $F = \mathbb{R}$, we shall denote by $E' := \mathcal{L}(E, \mathbb{R})$ the space of linear forms on E (algebraic dual) and by $E^* := \mathcal{L}_c(E, \mathbb{R})$ the space of continuous linear forms on E (topological dual).

Theorem 1.10. *Let $f \in \mathcal{L}(E, F)$. Then, the following assertions are equivalent:*

1) *$f \in \mathcal{L}_c(E, F)$.*
2) *f is continuous at a point.*
3) *f is bounded on the closed unit ball of E, $\overline{B}_E(0, 1)$.*
4) *There exists a constant $M \geq 0$ such that $\|f(x)\|_F \leq M\|x\|_E$, $\forall x \in E$.*
5) *f is Lipschitz on E.*

Proof.

1) $\Rightarrow$ 2) The implication is obvious.

2) $\Rightarrow$ 3) Assume that f is continuous at $x_0 \in E$, then there exists $r > 0$ such that for every $x \in \overline{B}_E(x_0, r)$ one has

$$\|f(x) - f(x_0)\|_F \leq 1. \tag{1.9}$$

Let $u \in \overline{B}_E(0, 1)$ since $x_0 + ru \in \overline{B}_E(x_0, r)$. Thanks to (1.9) and the linearity of f, we have

$$\|f(x_0 + ru) - f(x_0)\|_F = \|f(ru)\|_F = r\|f(u)\|_F \leq 1 \tag{1.10}$$

so that $\|f(u)\|_F \leq 1/r$, $\forall u \in \overline{B}_E(0, 1)$.

3) $\Rightarrow$ 4) Assume there exists $M > 0$ such that

$$\|f(u)\|_F \leq M, \quad \forall u \in \overline{B}_E(0, 1). \tag{1.11}$$

Let $x \in E \setminus \{0\}$, since $x/\|x\|_E \in \overline{B}_E(0, 1)$, equation (1.11) yields:

$$\|f(x/\|x\|_E)\|_F = \frac{\|f(x)\|_F}{\|x\|_E} \leq M \Rightarrow \|f(x)\|_F \leq M\|x\|_E. \tag{1.12}$$

and the last inequality is obvious also when $x = 0$.

4) $\Rightarrow$ 5) By linearity, for every $(x, y) \in E \times E$,

$$\|f(x) - f(y)\|_F = \|f(x - y)\|_F \leq M\|x - y\|_E, \tag{1.13}$$

which shows that f is M-Lipschitz on E.

5) $\Rightarrow$ 1) The implication is obvious. $\square$

Example 1.1. Let $E := C^0([-1, 1], \mathbb{R})$ be the space of real-valued continuous functions on $[-1, 1]$. Consider on E, the L^1-norm:

$$\|f\|_1 := \int_{-1}^{1} |f(t)| dt$$

as well as the uniform norm

$$\|f\|_\infty := \max\{|f(t)|, \, t \in [-1, 1]\}.$$

Define then, for all $f \in E$, $T(f) := f(0)$. Clearly $T \in E'$ and for every $f \in E$:

$$|T(f)| \leq \|f\|_\infty.$$

Hence T is continuous on E equipped with $\|\cdot\|_\infty$, but not equipped with $\|\cdot\|_1$. Indeed, consider

$$f_n(t) := \max(0, n(1 - n|t|)), \ t \in [-1, 1], \ n \in \mathbb{N}^*.$$

A direct computation shows that $\|f_n\|_1 = 1$ and $T(f_n) = n \to +\infty$.

When E is finite-dimensional, continuity of linear maps is automatic.

Theorem 1.11. *Let $(E, \|\cdot\|_E)$ and $(F, \|\cdot\|_F)$ be two normed spaces. If E is finite-dimensional, then $\mathcal{L}(E, F) = \mathcal{L}_c(E, F)$.*

Proof. Let $(e_1, \ldots, e_n)$ be a basis of E and equip E with the norm $\|\cdot\|_\infty$ (recall all norms are equivalent on E by Theorem 1.8). Let $f \in \mathcal{L}(E, F)$ and $x \in E$, $x = \sum_i^n x_i e_i$, we then have

$$\|f(x)\|_F = \left\| \sum_{i=1}^n x_i f(e_i) \right\|_F \leq \sum_{i=1}^n |x_i| \|f(e_i)\|_F \leq C\|x\|_E \sum_{i=1}^n \|f(e_i)\|_F,$$

which shows that $f \in \mathcal{L}_c(E, F)$. $\qquad\square$

Spaces of linear maps between normed spaces have a normed space structure themselves.

Proposition 1.12. *On $\mathcal{L}_c(E, F)$:*

$$f \mapsto \|f\|_{\mathcal{L}_c(E,F)} := \sup\{\|f(x)\|_F : \|x\|_E \leq 1\}$$

defines a norm.

Note that if $f \in \mathcal{L}_c(E, F)$ and $x \in E$, one has

$$\|f(x)\|_F \leq \|f\|_{\mathcal{L}_c(E,F)} \|x\|_E. \tag{1.14}$$

As a particular case, the topological E^* of E is equipped with the dual norm

$$\forall f \in E^*, \|f\|_{E^*} := \sup\{|f(x)| : \|x\|_E \leq 1\}. \tag{1.15}$$

Exercise 1.1. *Let E, F and G be three normed spaces $u \in \mathcal{L}_c(E, F)$, $v \in \mathcal{L}_c(F, G)$. Show that*

$$\|v \circ u\|_{\mathcal{L}_c(E,G)} \leq \|v\|_{\mathcal{L}_c(F,G)} \|u\|_{\mathcal{L}_c(E,F)}.$$

Let us now extend the previous results to bilinear maps (this will be useful when dealing with second-order differential calculus in Chapter 2). We leave the proofs, very similar to the ones of the linear case, as an exercise to the reader.

Let E, F and G be three real vector spaces. A bilinear map from $E \times F$ with values in G is a map

$$a : \begin{cases} E \times F \to G, \\ (x, y) \mapsto a(x, y) \end{cases}$$

such that:

- for every $y \in F$, the partial map $x \mapsto a(x, y)$ is linear from E to G;
- for every $x \in E$, the partial map $y \mapsto a(x, y)$ is linear from F to G.

We denote by $\mathcal{L}_2(E \times F, G)$ the (vector space) of bilinear maps from $E \times F$ to G. Here are some examples:

- $E = F = \mathbb{R}^n$, $G = \mathbb{R}$ and $a(x, y) = \sum_{i=1}^{n} x_i y_i$, the usual scalar product on $\mathbb{R}^n$.
- $E = \mathcal{M}_n(\mathbb{R})$ (the space of $n \times n$ matrices with real entries), $F = G = \mathbb{R}^n$ and the map $(A, x) \in E \times F \mapsto Ax$.

Now assume that E, F and G are equipped with norms $\| \cdot \|_E$, $\| \cdot \|_F$, and $\| \cdot \|_G$, respectively. We denote by $\mathcal{L}_{2,c}(E \times F, G)$ the set of continuous bilinear maps from $E \times F$ to G. Note that $\mathcal{L}_{2,c}(E \times F, G)$ is a linear subspace of $\mathcal{L}_2(E \times F, G)$. The following characterization is the bilinear analog of Theorem 1.10 for linear maps:

Theorem 1.12. *Let $(E, \| \cdot \|_E)$, $(F, \| \cdot \|_F)$ and $(G, \| \cdot \|_G)$ be three normed spaces and let $a \in \mathcal{L}_2(E \times F, G)$. Then, the following assertions are equivalent:*

1) $a \in \mathcal{L}_{2,c}(E \times F, G)$.
2) *There exists $M \geq 0$ such that $\|a(x, y)\|_G \leq M \|x\|_E \|y\|_F$, $\forall (x, y) \in E \times F$.*

Again, in finite dimensions, continuity can be taken for granted (simply adapting the proof of Theorem 1.13).

Theorem 1.13. *Let $(E, \| \cdot \|_E)$, $(F, \| \cdot \|_F)$ and $(G, \| \cdot \|_G)$ be three normed spaces. If E and F are finite dimensional, then $\mathcal{L}_2(E \times F, G) = \mathcal{L}_{2,c}(E, F)$.*

This enables one to define, for every $a \in \mathcal{L}_{2,c}(E \times F, G)$,

$$\|a\|_{\mathcal{L}_{2,c}(E \times F, G)} := \sup\{\|a(x, y)\|_G \; : \; \|x\|_E \leq 1, \; \|y\|_F \leq 1\}. \qquad (1.16)$$

It is easy to check that $(\mathcal{L}_{2,c}(E \times F, G), \| \cdot \|_{\mathcal{L}_{2,c}(E \times F, G)})$ is a normed space and, by construction, note that if $a \in \mathcal{L}_{2,c}(E \times F, G)$, one has

$$\|a(x, y)\|_G \leq \|a\|_{\mathcal{L}_{2,c}(E \times F, G)} \|x\|_E \|y\|_F, \; \forall (x, y) \in E \times F. \qquad (1.17)$$

We are going to prove that one can identify $\mathcal{L}(E, \mathcal{L}(F, G))$ (respectively, $\mathcal{L}_c(E, \mathcal{L}_c(F, G))$) to $\mathcal{L}_2(E \times F, G)$ (respectively, $\mathcal{L}_{2,c}(E \times F, G)$). This identification is particularly useful in differential calculus as soon as second derivatives are concerned.

More precisely, let $v \in \mathcal{L}(E, \mathcal{L}(F, G))$ and define for every $(x, y) \in E \times F$:

$$a_v(x, y) := (v(x))(y).$$

One immediately checks that $a_v \in \mathcal{L}_2(E \times F, G)$. Then, let, Φ be defined by

$$\Phi : \begin{cases} \mathcal{L}(E, \mathcal{L}(F, G)) \to \mathcal{L}_2(E \times F, G), \\ v \mapsto a_v. \end{cases}$$

It is clear that Φ is linear (if one really wants to use tedious notations: $\Phi \in \mathcal{L}(\mathcal{L}(E, \mathcal{L}(F, G)), \mathcal{L}_2(E \times F, G))$).

Now let $a \in \mathcal{L}_2(E \times F, G)$. Then, for every $x \in E$, the map $a(x, \cdot)$ belongs to $\mathcal{L}(F, G)$. Moreover, by bilinearity, for every $(x_1, x_2, \lambda) \in E^2 \times F$:

$$a(\lambda x_1 + x_2, \cdot) = \lambda a(x_1, \cdot) + a(x_2, \cdot),$$

which means that:

$$A_a : \begin{cases} E \to \mathcal{L}(F, G), \\ x \mapsto A_a(x) := a(x, \cdot) \end{cases}$$

belongs to $\mathcal{L}(E, \mathcal{L}(F, G))$. Consider now

$$\Psi : \begin{cases} \mathcal{L}_2(E \times F, G) \to \mathcal{L}(E, \mathcal{L}(F, G)), \\ a \mapsto A_a. \end{cases}$$

For $a \in \mathcal{L}_2(E \times F, G)$ and $(x, y) \in E \times F$, one has

$$(\Phi \circ \Psi)(a)(x, y) = (A_a(x))(y) = a(x, y).$$

Similarly, for every $v \in \mathcal{L}(E, \mathcal{L}(F, G))$ and $(x, y) \in E \times F$

$$(((\Psi \circ \Phi)(v))(x))(y) = A_{a_v}(x, y) = (v(x))(y).$$

Hence

$$\Phi \circ \Psi = \text{id} \quad \text{on } \mathcal{L}_2(E \times F, G),$$

$$\Psi \circ \Phi = \text{id} \quad \text{on } \mathcal{L}(E, \mathcal{L}(F, G)).$$

Put differently, Φ is an isomorphism and Ψ is its inverse Φ. The isomorphism Φ enables us to make an (algebraic) identification between $\mathcal{L}(E, \mathcal{L}(F, G))$ and $\mathcal{L}_2(E \times F, G)$. Suppose now, in addition that E, F and G are equipped with respective norms $\| \cdot \|_E$, $\| \cdot \|_F$ and $\| \cdot \|_G$. We then have the following theorem.

Theorem 1.14. *Let Φ and Ψ be defined as above. Then we have the following conditions*

1) *Let $v \in \mathcal{L}(E, \mathcal{L}(F, G))$. We have the following equivalence:*

$$v \in \mathcal{L}_c(E, \mathcal{L}_c(F, G)) \Leftrightarrow \Phi(v) \in \mathcal{L}_{2,c}(E \times F, G).$$

2) *For every $v \in \mathcal{L}_c(E, \mathcal{L}_c(F, G))$,*

$$\|v\|_{\mathcal{L}_c(E, \mathcal{L}_c(F, G))} = \|\Phi(v)\|_{\mathcal{L}_{2,c}(E \times F, G)}.$$

Proof. By construction, $\Phi(v) \in \mathcal{L}_{2,c}(E \times F, G)$ if and only if there exists $M \geq 0$ such that

$$\|v(x)(y)\|_G \leq M \|x\|_E \|y\|_F \quad \forall (x, y) \in E \times F,$$

which is equivalent to

$$\forall x \in E, \ v(x) \in \mathcal{L}_c(F, G) \text{ and } \|v(x)\|_{\mathcal{L}_c(F, G)} \leq M \|x\|_E,$$

i.e., $v \in \mathcal{L}_c(E, \mathcal{L}_c(F, G))$.

Now, let $v \in \mathcal{L}_c(E, \mathcal{L}_c(F, G))$, we have

$$\|\Phi(v)\|_{\mathcal{L}_{2,c}(E \times F, G)} = \sup\{\|(v(x))(y)\|_G : \|x\|_E \leq 1, \ \|y\|_F \leq 1\}$$

$$= \sup\{\|v(x)\|_{\mathcal{L}_c(F, G)} : \|x\|_E \leq 1\} = \|v\|_{\mathcal{L}_c(E, \mathcal{L}_c(F, G))}.$$

$\square$

Theorem 1.14 expresses that not only $\mathcal{L}_c(E, \mathcal{L}_c(F, G)))$ and $\mathcal{L}_{2,c}(E \times F, G)$ are isomorphic, but also isometric.

1.3. Banach spaces

1.3.1. *Definitions and properties*

Definition 1.18. A Banach space is a normed space which is complete for the distance associated to its norm.

An obvious example is given by finite-dimensional spaces.

Theorem 1.15. *Finite-dimensional spaces are Banach spaces.*

Proof. Let E be finite dimensional (recall all norms on E are equivalent thus define the same topology). If $(x_n)_n$ is a Cauchy sequence, it is bounded thus admits a cluster point by Theorem 1.9 and therefore converges by Lemma 1.1. Hence, completeness follows. $\square$

Let $(E, \|\cdot\|)$ be a normed space and $(x_n)_n \in E^{\mathbb{N}}$, the series with general term x_n, denoted $(\sum_n x_n)$, is the sequence formed by the partial sums: $S_n := \sum_{k \leq n} x_k$.

Definition 1.19. Let $(E, \|\cdot\|)$ be a normed space and $(\sum_n x_n)_n$ be a series with values in E, $(\sum_n x_n)_n$ is said to be convergent if the sequence of its partial sums converges in $(E, \|\cdot\|)$. In this case, the limit of the partial sums is denoted $\sum_{n=0}^{+\infty} x_n$. The series $(\sum_n x_n)_n$ is called normally convergent if the series $(\sum_n \|x_n\|)_n$ converges in $\mathbb{R}$.

Recall that the series (with non-negative general term) $(\sum_n \|x_n\|)_n$ converges if and only if its partial sums is a Cauchy sequence, i.e.,

$$\forall \varepsilon > 0, \exists N \in \mathbb{N} \text{ s.t. } \forall p > q \geq N, \quad \sum_{k=q+1}^{p} \|x_k\| \leq \varepsilon, \qquad (1.18)$$

and in this case the sequence of remainders $\sum_{k=n}^{+\infty} \|x_k\|$ tends to 0 as $n \to +\infty$.

Proposition 1.13. *Let $(E, \|\cdot\|)$ be a Banach space and $(\sum_n x_n)$ be a series with values in E. If $(\sum_n x_n)$ is normally convergent, then $(\sum_n x_n)$ converges in E.*

Proof. By completeness, all we have to show is that $S_n := \sum_{k \leq n} x_k$ is Cauchy. But, for $p > q$:

$$\|S_p - S_q\| \leq \sum_{k=q+1}^{p} \|x_k\| \leq \sum_{k=q+1}^{+\infty} \|x_k\| \tag{1.19}$$

and since $(\sum_n x_n)$ is normally convergent, the right-hand side of (1.19) tends to 0 as $q \to +\infty$. Therefore, $(S_n)_n$ is Cauchy and $(\sum_n x_n)$ converges.

$\square$

1.3.2. *Examples*

Let us now consider some examples of infinite-dimensional Banach spaces (see [23] for other examples: Lebesgue spaces, Sobolev spaces, etc.). Let X be a non-empty set, $(E, \| \cdot \|)$ be a Banach space and $B(X, E)$ be the space of bounded functions from X to E:

$$B(X, E) := \left\{ f : X \to E : \sup_{x \in X} \|f(x)\| < +\infty \right\}. \tag{1.20}$$

$B(X, E)$ is obviously a vector space and

$$\|f\|_\infty := \sup_{x \in X} \|f(x)\| \tag{1.21}$$

is a norm, called the uniform norm (or the norm of uniform convergence).

Theorem 1.16. *Let X be a non-empty set and $(E, \|\cdot\|)$ be a Banach space. Then $(B(X, E), \|\cdot\|_\infty)$ is a Banach space.*

Proof. Let $(f_n)_n$ be a Cauchy sequence in $(B(X, E), \|.\|_\infty)$:

$$\forall \varepsilon > 0, \exists N \in \mathbb{N} \text{ such that } \forall p, q \geq N, \ \forall x \in X,$$

$$\|f_p(x) - f_q(x)\| \leq \varepsilon. \tag{1.22}$$

We shall prove that $(f_n)_n$ converges in $(B(X, E), \|.\|_\infty)$ in three steps.

Step 1: identification of a pointwise limit
For fixed $x \in X$, equation (1.22) implies in particular that $(f_n(x))_n \in E^{\mathbb{N}}$ is Cauchy. Since $(E, \| \cdot \|)$ is a Banach space, this sequence converges, we denote by $f(x)$ its limit.

Step 2: $f \in B(X, E)$

By equation (1.22), there exists N such that for all $p, q \geq N$ and every $x \in X$:

$$\|f_p(x) - f_q(x)\| \leq 1. \tag{1.23}$$

For fixed $x \in X$, taking $p = N$ and letting q tend to $+\infty$ in (1.23). Since $f_q(x)$ converges to $f(x)$, we obtain $\|f_N(x) - f(x)\| \leq 1$. Since $x \in X$ is arbitrary in the previous inequality, we deduce

$$\sup_{x \in X} \|f(x) - f_N(x)\| \leq 1 \Rightarrow (f - f_N) \in B(X, E)$$

and since $f_N \in B(X, E)$, we have $f \in B(X, E)$.

Step 3: $(f_n)_n$ **converges to** f **in** $(B(X, E), \|\cdot\|_\infty)$

Let $\varepsilon > 0$, it follows from (1.22) that there exists N such that for all $p, q \geq N$ and every $x \in X$, $\|f_p(x) - f_q(x)\| \leq \varepsilon$. As previously, let us fix x, take $p \geq N$ and let $q \to \infty$, so as to obtain:

$$\|f_p(x) - f(x)\| \leq \varepsilon. \tag{1.24}$$

But since (1.24) holds for every $p \geq N$ and every $x \in X$, we have

$$\forall p \geq N, \ \|f_p - f\|_\infty \leq \varepsilon,$$

which ends the proof. $\qquad\qquad\qquad\qquad\qquad\qquad\qquad\qquad\qquad\quad$ $\square$

Note that in the previous result, the set X is arbitrary. When $X = \mathbb{N}$, $B(\mathbb{N}, E) = l^\infty(E)$ is the space of bounded sequences with values in E. We then have the following corollary.

Corollary 1.5. *Let $(E, \|\cdot\|)$ be a Banach Space. Then $l^\infty(E)$ equipped with the uniform norm is a Banach space.*

Another interesting case is when X is equipped with a distance d, in this case we may wish to consider the space $C_b(X, E)$ consisting of bounded and continuous functions defined on X with values in E, we then have the following theorem.

Theorem 1.17. *Let (X, d) be a metric space and $(E, \|\cdot\|)$ be a Banach space, then $(C_b(X, E), \|\cdot\|_\infty)$ is a Banach space.*

Proof. Thanks to theorem 1.16, it is enough to show that the subspace $C_b(X, E)$ of $B(X, E)$ is closed for $\|\cdot\|_\infty$. Let $(f_n)_n$ be a sequence in $(C_b(X, E), \|\cdot\|_\infty)$, converging to some f in $(B(X, E), \|\cdot\|_\infty)$, we have to show that f is continuous. Let $\varepsilon > 0$ and N be such that for all $n \geq N$:

$$\|f_n - f\|_\infty \leq \frac{\varepsilon}{3}. \tag{1.25}$$

Let $x_0 \in X$, since f_N is continuous at x_0, there exists $\delta > 0$ such that

$$\forall x \in B(x_0, \delta), \quad \|f_N(x_0) - f_N(x)\| \leq \frac{\varepsilon}{3}. \tag{1.26}$$

For $x \in B(x_0, \delta)$, we have

$$\|f(x) - f(x_0)\| \leq \|f(x) - f_N(x)\| + \|f_N(x) - f_N(x_0)\| + \|f_N(x_0) - f(x_0)\|$$
$$\leq \|f - f_N\|_\infty + \varepsilon/3 + \|f - f_N\|_\infty$$
$$\leq \varepsilon \text{ (with (1.25))},$$

which shows that f is continuous at x_0. □

Remark 1.1. If (X, d) is compact, continuous functions defined on X with values in E are bounded so that the space of continuous functions defined on X with values in E, $C(X, E)$, is a Banach space for the uniform norm $\|\cdot\|_\infty$.

Theorem 1.18. *Let E be a normed space and F be a Banach space. Then $(\mathcal{L}_c(E, F), \|\cdot\|_{\mathcal{L}_c(E,F)})$ is a Banach space. In particular (case $F = \mathbb{R}$), $(E^*, \|\cdot\|_{E^*})$ is a Banach space.*

Proof. Let (f_n) be a Cauchy sequence in $(\mathcal{L}_c(E, F), \|\cdot\|_{\mathcal{L}_c(E,F)})$, and let g_n be the restriction of f_n to $\overline{B}_E(0, 1)$. One has $g_n \in C_b(\overline{B}_E(0, 1), F)$ and

$$\|g_n\|_\infty = \|f_n\|_{\mathcal{L}_c(E,F)}. \tag{1.27}$$

By construction of g_n, for every $(p, q) \in \mathbb{N}^2$:

$$\|g_p - g_q\|_\infty = \|f_p - f_q\|_{\mathcal{L}_c(E,F)}. \tag{1.28}$$

This implies that (g_n) is Cauchy in $(C_b(\overline{B}_E(0, 1), F), \|\cdot\|_\infty)$. Thanks to Theorem 1.17, g_n converges to some g in $(C_b(\overline{B}_E(0, 1), F), \|\cdot\|_\infty)$. Let us then define f by $f(0) = 0$ and

$$f(x) = \|x\| g\left(\frac{x}{\|x\|}\right), \quad \forall x \in E \setminus \{0\}. \tag{1.29}$$

Let us observe now that $g = f$ on $\overline{B}_E(0,1)$. Indeed, $g(0) = f(0) = 0$ and for $x \in \overline{B}_E(0,1) \setminus \{0\}$, for every n:

$$\|x\| g_n\left(\frac{x}{\|x\|}\right) = \|x\| f_n\left(\frac{x}{\|x\|}\right) = f_n(x) = g_n(x),$$

which gives $g = f$ on $\overline{B}_E(0,1)$ by letting $n \to \infty$.

Let us now show that f is linear: let $(x_1, x_2, t) \in E \times E \times \mathbb{R}$, for every n, by linearity of f_n. We have[b]

$$0 = f_n(x_1 + tx_2) - f_n(x_1) - tf_n(x_2)$$
$$= \|x_1 + tx_2\| g_n\left(\frac{x_1 + tx_2}{\|x_1 + tx_2\|}\right) - \|x_1\| g_n\left(\frac{x_1}{\|x_1\|}\right) - t\|x_2\| g_n\left(\frac{x_2}{\|x_2\|}\right),$$

letting $n \to \infty$, we get $f(x_1 + tx_2) = f(x_1) + tf(x_2)$. Since $g = f$ on $\overline{B}_E(0,1)$, f is bounded on $\overline{B}_E(0,1)$ and hence $f \in \mathcal{L}_c(E,F)$. Finally, since $g_n = f_n$ and $g = f$ on $\overline{B}_E(0,1)$, we have

$$\|g_n - g\|_\infty = \|f_n - f\|_{\mathcal{L}_c(E,F)}$$

and we can conclude that f_n converges to f in $(\mathcal{L}_c(E,F), \|\cdot\|_{\mathcal{L}_c(E,F)})$. $\quad\square$

We have of course a similar statement for bilinear maps.

Theorem 1.19. *Let E and F be normed spaces and G be a Banach space. Then $(\mathcal{L}_{2,c}(E \times F, G), \|\cdot\|_{\mathcal{L}_{2,c}(E \times F, G)})$ is a Banach space.*

1.3.3. *Ascoli's theorem*

Let (X, d) be a compact metric space and equip $C(X) = C(X, \mathbb{R})$ with the uniform norm. Unless X is finite, $C(X)$ is infinite dimensional, so bounded sequences in $C(X)$ do not have uniformly convergent subsequences in general.

---

[b]We are slightly abusing notations by setting $\|x\| g_n(x/\|x\|) = 0$ for $x = 0$.

Definition 1.20. Let $\mathcal{F}$ be a non-empty subset of $C(X)$. Then $\mathcal{F}$ is called uniformly equicontinuous if, for every $\varepsilon > 0$, there exists $\delta > 0$ such that for every $(x, x', f) \in X^2 \times \mathcal{F}$ one has

$$d(x, x') \leq \delta \Rightarrow |f(x) - f(x')| \leq \varepsilon. \tag{1.30}$$

Note that the uniform equicontinuity of $\mathcal{F}$ can also be expressed as the existence of a common modulus of continuity, ω, for all functions in $\mathcal{F}$:

$$|f(x) - f(x')| \leq \omega(d(x, x')), \ \forall (x, x', f) \in X^2 \times \mathcal{F},$$

with $\omega : \mathbb{R}_+ \to \mathbb{R}_+$ and $\omega(\delta) \to 0$ as $\delta \to 0^+$.

Theorem 1.20 (Ascoli's theorem). *Let $\mathcal{F}$ be a non-empty, bounded and uniformly equicontinuous subset of $C(X)$. Then $\mathcal{F}$ is relatively compact (i.e., has compact closure) for the uniform norm.*

Proof. We have to show that every sequence $(f_n)_n \in \mathcal{F}^{\mathbb{N}}$ has a uniformly convergent subsequence. Since X is compact, it has a countable dense subset which we denote by $(x^k)_k$. For each k, the sequence $(f_n(x^k))$ is bounded hence admits a convergent subsequence. By a standard diagonal extraction argument, there is a subsequence (again denoted f_n) such that $(f_n(x^k))$ converges, for every k, to some limit denoted $g(x^k)$. It follows from the uniform equicontinuity of $\mathcal{F}$ that

$$\forall \varepsilon > 0, \ \exists \delta > 0 \text{ such that } |g(x^k) - g(x^l)| \leq \varepsilon, \ \forall k, l,$$
$$\text{such that } d(x_k, x_l) \leq \delta. \tag{1.31}$$

If $x \in X$ and $(x^{k_n})_n$ converges to x, it follows from (1.31) that $g(x^{k_n})$ is a Cauchy sequence, and hence has some limit. Moreover, again by (1.31), this limit does not depend on the approximating sequence (x^{k_n}). We thus simply denote it by $g(x)$. It also readily follows from (1.31) that g is continuous. Let $\varepsilon > 0$. There is a $\delta > 0$ such that for every n and every x and x' such that $d(x, x') \leq \delta$, one has

$$|f_n(x) - f_n(x')| \leq \varepsilon/3, \ |g(x) - g(x')| \leq \varepsilon/3.$$

By compactness of X, since the family of open balls $(B(x^k, \delta))_{k \in \mathbb{N}}$ covers X, there exists p such that X is covered by $B(x^1, \delta), \ldots, B(x^p, \delta)$. Then, choose N large enough so that for all $n \geq N$ and all $m = 1, \ldots, p$ one has $|f_n(x^m) - g(x^m)| \leq \varepsilon/3$. For $n \geq N$ and $x \in X$, let $m \leq p$ be such that

$d(x, x^m) \leq \delta$. We then have

$$|f_n(x) - g(x)| \leq |f_n(x) - f_n(x^m)| + |f_n(x^m) - g(x^m)| + |g(x^m) - g(x)|$$
$$\leq \varepsilon/3 + \varepsilon/3 + \varepsilon/3 = \varepsilon.$$

This proves that f_n converges uniformly to g. $\qquad\square$

1.3.4. *Linear maps in Banach Spaces*

In Banach spaces, there are additional properties that follow from Baire's theorem. The first one is the Banach–Steinhaus theorem, or principle of uniform boundedness.

Theorem 1.21 (Banach–Steinhaus theorem). *Let E be a Banach space and F be a normed vector space and let $(f_i)_{i \in I}$ be a family of $\mathcal{L}_c(E, F)$. If*

$$\forall x \in E, \quad \sup_{i \in I} \|f_i(x)\|_F < +\infty,$$

then

$$\sup_{i \in I} \|f_i\|_{\mathcal{L}_c(E,F)} < +\infty.$$

Proof. Set $E_n := \{x \in E \ : \ \|f_i(x)\|_F \leq n, \ \forall i \in I\}$, then each E_n is closed and by assumption $\bigcup_n E_n = E$. It then follows from Baire's theorem that E_{n_0} has non-empty interior for some n_0 and then there exist $r > 0$ and $x_0 \in E$ such that

$$\|f_i(x_0 + ru)\|_F \leq n_0, \forall i \in E, \quad \forall u \in \overline{B}_E(0, 1)$$

so that for all $u \in \overline{B}_E(0, 1)$ and all $j \in I$, one has

$$\|f_j(u)\|_F \leq \frac{1}{r} \left(n_0 + \sup_{i \in I} \|f_i(x_0)\|_F \right). \qquad\square$$

Another consequence of Baire's theorem is the open mapping principle.

Theorem 1.22 (Open mapping theorem). *Let E and F be two Banach spaces and $f \in \mathcal{L}_c(E, F)$ be continuous and surjective. Then f is an open mapping in the sense that $f(U)$ is open in F for every U open in E.*

Proof. Due to the linearity of f it is enough to prove that there exists $r_0 > 0$ such that $B_F(0, r_0) \subset f(B_E(0, 1))$. Let $F_n := \overline{nf(B_E(0, 1))}$, since

f is surjective, $F = \bigcup_n F_n$ and it follows from Baire's theorem that there exists n_0 such that F_{n_0} has non-empty interior. There exist then $y_0 \in E$ and $\rho > 0$ such that $B_F(y_0, \rho) \subset \overline{f(B_E(0, n_0))}$. By linearity, we also have $B_F(-y_0, \rho) \subset \overline{f(B_E(0, n_0))}$ and then $B_F(0, \rho) = -y_0 + B_F(y_0, \rho) \subset \overline{f(B_E(0, 2n_0))}$. By homogeneity, we then have $B_F(0, r) \subset \overline{f(B_E(0, 1))}$ with $r = \rho/2n_0$.

Let us now prove that $B_F(0, r) \subset f(\overline{B}_E(0, 2))$. Let $y \in B_F(0, r)$. There exists $x_1 \in B_E(0, 1)$ such that $y - f(x_1) \in B_F(0, r/2)$. Since $B_F(0, r/2) \subset \overline{f(B_E(0, 1/2))}$, there exists $x_2 \in B_E(0, 1/2)$ such that $y - f(x_1) - f(x_2) \in B_F(0, r/4)$. Iterating the argument, we find a sequence $(x_n)_n$ in E such that $\|x_n\|_E \leq 1/2^{n-1}$ and $\|y - f(x_1 + \cdots + x_n)\|_F \leq r/2^n$ for every n. Since $x_1 + \cdots + x_n$ is a Cauchy sequence, it converges to some $x \in \overline{B}_E(0, 2)$ and by continuity $y = f(x)$, which proves that $B_F(0, r) \subset f(\overline{B}_E(0, 2)) \subset f(B_E(0, 5/2))$ and then $B_F(0, r_0) \subset f(B_E(0, 1))$ with $r_0 = 2r/5$. $\qquad\square$

We deduce from the previous theorem the following *automatic continuity* result due to Banach.

Theorem 1.23. *Let* $(E, \| \cdot \|_E)$ *and* $(F, \| \cdot \|_F)$ *be two Banach spaces and* $f \in \mathcal{L}_c(E, F)$. *If* f *is invertible, then* $f^{-1} \in \mathcal{L}_c(F, E)$.

1.4. Hilbert spaces

1.4.1. *Generalities*

Definition 1.21. Let E be a real vector space, an inner product on E is a mapping $\langle \cdot, \cdot \rangle : E \times E \to \mathbb{R}$ which is:

1) *Bilinear:* for every $x \in E$, $y \mapsto \langle x, y \rangle$ is linear and for every $y \in E$, $x \mapsto \langle x, y \rangle$ is linear;
2) *Symmetric:* $\langle x, y \rangle = \langle y, x \rangle$, $\forall (x, y) \in E \times E$;
3) *Positive definite:* $\langle x, x \rangle \geq 0$, $\forall x \in E$ and $\langle x, x \rangle = 0 \Leftrightarrow x = 0$.

The following identities easily follow from bilinearity and symmetry

$$\langle x + y, x + y \rangle = \langle x, x \rangle + \langle y, y \rangle + 2 \langle x, y \rangle,$$

$$\langle x - y, x - y \rangle = \langle x, x \rangle + \langle y, y \rangle - 2 \langle x, y \rangle, \forall (x, y) \in E \times E. \quad (1.32)$$

Definition 1.22. A pre Hilbertian space is a pair $(E, \langle \cdot, \cdot \rangle)$ where E is a vector space and $\langle \cdot, \cdot \rangle$ is an inner product on E.

An inner product enables one to define a norm thanks to the Cauchy–Schwarz inequality.

Proposition 1.14. *Let $(E, \langle \cdot, \cdot \rangle)$ be a pre-Hilbertian space.*
For every $(x, y) \in E \times E$, we have the Cauchy–Schwarz inequality:

$$| \langle x, y \rangle | \leq (\langle x, x \rangle)^{1/2} (\langle y, y \rangle)^{1/2}. \tag{1.33}$$

Moreover, there is equality in (1.33) if and only if x and y are colinear.
Finally, $x \in E \mapsto (\langle x, x \rangle)^{1/2}$ is a norm on E, it is called the norm associated with the inner product $\langle \cdot, \cdot \rangle$.

Proof. If $y = 0$, the inequality is obvious. Therefore, assume $y \neq 0$ and define for $t \in \mathbb{R}$,

$$g(t) := \langle x + ty, x + ty \rangle = t^2 \langle y, y \rangle + 2t \langle x, y \rangle + \langle x, x \rangle.$$

g is a non-degenerate quadratic function of t and $g(t) \geq 0$ for all t. The discriminant of g therefore is non-negative:

$$(\langle x, y \rangle)^2 \leq \langle x, x \rangle \langle y, y \rangle.$$

Taking square roots gives (1.33).

It is obvious that if x and y are colinear, there is equality in (1.33). Conversely, if (1.33) is an equality, the discriminant of g is 0 so that g has a root $t_0 \in \mathbb{R}$, but $g(t_0) = 0$ if and only if $x + t_0 y = 0$ so that x and y are colinear.

Set $\|x\| := (\langle x, x \rangle)^{1/2}$, since $\langle \cdot, \cdot \rangle$ is an inner product $\|x\| = 0$ if and only if $x = 0$. Bilinearity implies that $\|\lambda x\| = |\lambda| \|x\|$. As for the triangle inequality, let $(x, y) \in E \times E$, we have

$$\begin{aligned}
\|x + y\|^2 &= \|x\|^2 + \|y\|^2 + 2 \langle x, y \rangle \\
&\leq \|x\|^2 + \|y\|^2 + 2\|x\| \|y\| \text{ (using Cauchy–Schwarz)} \\
&= (\|x\| + \|y\|)^2,
\end{aligned}$$

and we can conclude that $\| \cdot \|$ is a norm on E. $\qquad \square$

Let $\| \cdot \|$ be the norm associated with $\langle \cdot, \cdot \rangle$. Let us remark that this norm enables one to recover the full inner product via the following *polarization* identity:

$$\langle x, y \rangle = \frac{1}{2}(\|x + y\|^2 - \|x\|^2 - \|y\|^2). \tag{1.34}$$

Let us also mention the *parallelogram* identity:

$$\|x + y\|^2 + \|x - y\|^2 = 2(\|x\|^2 + \|y\|^2). \tag{1.35}$$

Remark 1.2. It follows from Cauchy–Schwarz inequality that, for every $x \in E$, the linear form $y \mapsto \langle x, y \rangle$ is continuous on E (for the norm associated with $\langle \cdot, \cdot \rangle$). Cauchy–Schwarz inequality also gives the continuity of $\langle \cdot, \cdot \rangle$ as a bilinear form.

Definition 1.23. Let $(E, \langle \cdot, \cdot \rangle)$ be a pre-Hilbertian space; $(E, \langle \cdot, \cdot \rangle)$ is a Hilbert space if it is complete for the norm associated with $\langle \cdot, \cdot \rangle$.

Example 1.2. Take $E = \mathbb{R}^d$ and let $A = [a_{ij}]$ be a symmetric positive definite matrix. Then

$$\langle x, y \rangle_A := x^T A y = \sum_{1 \le i,j \le d} a_{ij} x_i y_j$$

is an inner product on $\mathbb{R}^d$ (the usual scalar product corresponds to the case where A is the identity matrix). As completeness being automatic in finite dimensions, $(\mathbb{R}^d, \langle \cdot, \cdot \rangle)$ is a Hilbert space.

For $E = \mathbb{R}^d$, we shall denote by $x \cdot y := x^T y = \sum_{i=1}^d x_i y_i$ the usual inner product of x and y in $\mathbb{R}^d$ (which corresponds to the case where A is the identity matrix in the example above). The corresponding norm $x \mapsto \sqrt{x \cdot x}$ will simply be denoted $|\cdot|$ in most of this book.

Example 1.3. The space l^2 of real sequences (x_n) such that $\sum |x_n|^2 < +\infty$ equipped with:

$$\langle x, y \rangle := \sum_{n \ge 0} x_n y_n$$

is a Hilbert space.

Example 1.4. The Lebesgue space, $L^2((0,1), \mathbb{R})$ equipped with:

$$(f, g) \mapsto \int_0^1 f(t) g(t) dt$$

is a Hilbert space, but $C([0,1], \mathbb{R})$ equipped with the same norm is only pre-Hilbertian.

Given a pre-Hilbertian space $(E, \langle \cdot, \cdot \rangle)$, two vectors u and v are called orthogonal if $\langle u, v \rangle = 0$. For $A \subset E$, the orthogonal of A is defined by:

$$A^\perp := \{x \in H : \langle x, y \rangle = 0, \, \forall y \in A\}.$$

Note that $A^\perp$ is a closed linear subspace of E.

1.4.2. *Projection onto a closed convex set*

Definition 1.24. Let E be a real vector space. A subset C of E is convex if, for every x and y in C and every $\lambda \in [0, 1]$, $\lambda x + (1 - \lambda)y \in C$.

Basic examples of convex sets are subspaces, half-spaces, balls etc. Let us also remark that intersections or convex sets are convex. By an obvious induction argument, it is easy to see that if C is convex then, for every $p \in \mathbb{N}^*$, every $x_1, \ldots, x_p \in C^p$ and every $\lambda_1, \ldots, \lambda_p$ such that each $\lambda_i \geq 0$ and $\sum_{i=1}^p \lambda_i = 1$, one has

$$\sum_{i=1}^p \lambda_i x_i \in C.$$

Any vector that can be written in the form $\sum_{i=1}^p \lambda_i x_i$ with non-negative weights λ_i that sum to 1 is called a *convex combination* of the vectors x_i.

Theorem 1.24. *Let $(H, \langle \cdot, \cdot \rangle)$ be a Hilbert space and C be a non-empty closed convex subset of H. For every $x \in H$, there exists a unique element of C called projection of x onto C and denoted $\mathrm{proj}_C(x)$ such that:*

$$\|x - \mathrm{proj}_C(x)\| = \inf\{\|x - y\|, y \in C\}.$$

Moreover, $\mathrm{proj}_C(x)$ is characterized by $\mathrm{proj}_C(x) \in C$ and the variational inequality:

$$\langle x - \mathrm{proj}_C(x), y - \mathrm{proj}_C(x) \rangle \leq 0, \quad \forall y \in C. \tag{1.36}$$

Proof. Let us denote $d^2(x, C)$ the squared distance between x and C:

$$d^2(x, C) := \inf\{\|x - y\|^2, \, y \in C\}.$$

Let us recall the following parallelogram identity:

$$\left\|\frac{u - v}{2}\right\|^2 + \left\|\frac{u + v}{2}\right\|^2 = \frac{1}{2}(\|u^2\| + \|v\|^2), \quad \forall (u, v) \in H^2. \tag{1.37}$$

Uniqueness: Suppose y_1 and y_2 belong to C and satisfy:

$$\|x - y_1\|^2 = \|x - y_2\|^2 = d^2(x, C).$$ (1.38)

We have $\frac{(y_1 + y_2)}{2} \in C$ since C is convex, and then

$$\left\| x - \frac{(y_1 + y_2)}{2} \right\|^2 \geq d^2(x, C).$$ (1.39)

Applying (1.37) to $u = (x - y_1)$ and $v = (x - y_2)$ and using (1.38) and (1.39), we get:

$$d^2(x, C) = \frac{1}{2}(\|x - y_1\|^2 + \|x - y_2\|^2)$$

$$= \left\| x - \frac{(y_1 + y_2)}{2} \right\|^2 + \left\| \frac{(y_1 - y_2)}{2} \right\|^2 \geq d^2(x, C) + \left\| \frac{(y_1 - y_2)}{2} \right\|^2$$

(1.40)

so that $y_1 = y_2$.

Existence: For $n \in \mathbb{N}^*$, let $y_n \in C$ be such that:

$$\|x - y_n\|^2 \leq d^2(x, C) + \frac{1}{n^2}.$$ (1.41)

Identity (1.37) applied to $u = (x - y_p)$ and $v = (x - y_q)$ gives

$$\frac{1}{2}(\|x - y_p\|^2 + \|x - y_q\|^2) = \left\| x - \frac{(y_p + y_q)}{2} \right\|^2 + \left\| \frac{(y_p - y_q)}{2} \right\|^2.$$ (1.42)

Since $\frac{(y_p - y_q)}{2} \in C$, we have

$$\left\| x - \frac{(y_p + y_q)}{2} \right\|^2 \geq d^2(x, C).$$ (1.43)

From (1.41), (1.42) and (1.43), we thus get

$$\|(y_p - y_q)\|^2 \leq \frac{1}{2p^2} + \frac{1}{2q^2}.$$ (1.44)

It follows from (1.44) that (y_n) is a Cauchy sequence thus converges to some limit denoted $\text{proj}_C(x)$. Since C is closed, $\text{proj}_C(x) \in C$ and passing to the limit in (1.41) yields $\|x - \text{proj}_C(x)\|^2 = d^2(x, C)$.

Variational characterization: Let $y \in C$ and $t \in [0,1]$. Since $(1-t)\mathrm{proj}_C(x) + ty \in C$, we have

$$\|x - \mathrm{proj}_C(x)\|^2 \leq \|x - ((1-t)\mathrm{proj}_C(x) + ty)\|^2$$
$$= \|x - \mathrm{proj}_C(x) - t(y - \mathrm{proj}_C(x))\|^2$$
$$= \|x - \mathrm{proj}_C(x)\|^2 + t^2\|y - \mathrm{proj}_C(x))\|^2$$
$$- 2t\langle x - \mathrm{proj}_C(x), y - \mathrm{proj}_C(x)\rangle.$$

Dividing by t and letting t go to 0^+ we deduce that $\mathrm{proj}_C(x)$ satisfies (1.36).

Conversely, assume that $z \in C$ satisfies:

$$\langle x - z, y - z\rangle \leq 0, \quad \forall y \in C. \tag{1.45}$$

Let $y \in C$. Then we have

$$\|x - y\|^2 = \|x - z\|^2 + \|z - y\|^2 + 2\langle x - z, z - y\rangle \geq \|x - z\|^2,$$

which proves that $z = \mathrm{proj}_C(x)$. $\qquad\square$

Proposition 1.15. *Under the same assumptions and notations as in Theorem 1.24, for all $(x,y) \in H^2$, one has:*

$$\langle x - y, \mathrm{proj}_C(x) - \mathrm{proj}_C(y)\rangle \geq 0,$$
$$\|\mathrm{proj}_C(x) - \mathrm{proj}_C(y)\| \leq \|x - y\|. \tag{1.46}$$

In particular, proj_C is 1-Lipschitz continuous.

Proof. Using the variational inequalities characterizing $\mathrm{proj}_C(x)$ and $\mathrm{proj}_C(y)$, we have

$$\langle x - \mathrm{proj}_C(x), \mathrm{proj}_C(y) - \mathrm{proj}_C(x)\rangle \leq 0,$$
$$\langle y - \mathrm{proj}_C(y), \mathrm{proj}_C(x) - \mathrm{proj}_C(y)\rangle \leq 0.$$

Summing these inequalities and using Cauchy–Schwarz inequality yields

$$\|\mathrm{proj}_C(x) - \mathrm{proj}_C(y)\|^2 \leq \langle \mathrm{proj}_C(x) - \mathrm{proj}_C(y), x - y\rangle$$
$$\leq \|\mathrm{proj}_C(x) - \mathrm{proj}_C(y)\|\|x - y\|,$$

which proves (1.46). $\qquad\square$

An important special case is when C is a closed subspace of H. In this case, proj_C is linear (and continuous thanks to (1.46)); it is the orthogonal projection on C.

Proposition 1.16. *Let C be a closed subspace of the Hilbert space $(H, \langle \cdot, \cdot \rangle)$. Defining proj_C as in Theorem 1.24, for $x \in H$, $\text{proj}_C(x)$ is characterized by*

$$\text{proj}_C(x) \in C, \ (x - \text{proj}_C(x)) \in C^\perp.$$

Moreover, $\text{proj}_C \in \mathcal{L}_c(H, C)$ and proj_C is called the orthogonal projection on C.

Proof. If $z \in C$ and $x - z \in C^\perp$, then for every $y \in C$ we have $\langle x - z, y - z \rangle = 0$ so that z satisfies (1.36). Conversely, equation (1.36) implies that $\langle x - \text{proj}_C(x), y - \text{proj}_C(x) \rangle \leq 0$ for all $y \in C$. Taking $y = 2\text{proj}_C(x)$ and $y = \text{proj}_C(x)/2$, we get

$$\langle x - \text{proj}_C(x), \text{proj}_C(x) \rangle = 0$$

and then $\langle x - \text{proj}_C(x), y \rangle \leq 0$ for all $y \in C$. Since C is a subspace, we deduce that $(x - \text{proj}_C(x)) \in C^\perp$. Finally, it remains to prove that proj_C is linear. For x_1, x_2 in H and $t \in \mathbb{R}$, set $x = x_1 + tx_2$ and $z = \text{proj}_C(x_1) + t\text{proj}_C(x_2)$. We have $z \in C$ and $(x - z) \in C^\perp$ so that $z = \text{proj}_C(x)$. $\square$

1.4.3. *The dual of a Hilbert space*

An important consequence of the projection theorem is that one can identify a Hilbert space to its dual:

Theorem 1.25 (Riesz' representation theorem). *Let $(H, \langle \cdot, \cdot \rangle)$ be a Hilbert space and $f \in H^*$. There exists a unique $x \in H$ which represents f such that:*

$$f(u) = \langle x, u \rangle, \quad \forall u \in H. \tag{1.47}$$

Proof. Uniqueness is obvious: indeed if both x_1 and x_2 represent f then $f(x_1 - x_2) = \langle x_1, x_1 - x_2 \rangle = \langle x_2, x_1 - x_2 \rangle$ which implies $\|x_1 - x_2\| = 0$. If $f = 0$, obviously $x = 0$ represents f, let us then suppose $f \neq 0$. In this case $F := \ker(f)$ is a closed hyperplane of H. Let $x_0 \in H$ such that $f(x_0) = 1$. By Proposition 1.16, we can define y_0, the orthogonal projection of x_0 on F.

We then have $x_0 \neq y_0$ since $x_0 \notin F$ and y_0 is characterized by

$$y_0 \in F = \ker(f), \text{ and } (x_0 - y_0) \in F^\perp. \tag{1.48}$$

In particular, since $\langle x_0 - y_0, y_0 \rangle = 0$, we have

$$\langle x_0 - y_0, x_0 \rangle = \|x_0 - y_0\|^2 \neq 0. \tag{1.49}$$

Let us then define:

$$x := \frac{x_0 - y_0}{\|x_0 - y_0\|^2} = \frac{x_0 - y_0}{\langle x_0 - y_0, x_0 \rangle}. \tag{1.50}$$

Thanks to (1.48), $x \in F^\perp$ so that for $u \in F$ we have $f(u) = \langle x, u \rangle = 0$. Finally

$$\langle x, x_0 \rangle = \frac{\langle x_0 - y_0, x_0 \rangle}{\langle x_0 - y_0, x_0 \rangle} = 1 = f(x_0),$$

which enables to conclude that (1.47) holds since $F \oplus \mathbb{R}x_0 = H$. □

Remark 1.3. Denoting by x_f the solution of:

$$\langle x, u \rangle = f(u), \quad \forall u \in H.$$

Let us call T the map $f \in H^* \mapsto T(f) = x_f \in H$. It is easy to see that $T \in \mathcal{L}_c(H, H^*)$ and T is an isomorphism. Finally, T is an isometry (we leave the easy proof of this fact as an exercise to the reader):

$$\|T(f)\| = \|f\|_{H^*}, \quad \forall f \in H^*.$$

1.5. Weak convergence

In finite-dimensional spaces, closed balls are compact, but this is *always* false in infinite-dimensional Banach spaces (see Exercise 1.14). Compactness being an essential property (for the existence of minimizers, for instance), one naturally has to consider another topology than the norm (also called strong) topology, for which there are less open sets but more compact sets. Given a normed vector space $(E, \|\cdot\|)$, the weak topology is the coarsest topology for which $x \in E \mapsto f(x)$ is continuous for every $f \in E^*$, this topology (which coincides with the strong one in finite dimensions) is coarser than the strong one in the sense that it has less open sets. We refer to [23] for a detailed study of the weak topology. In the present book, we shall mainly use the more pedestrian concept of weak convergence.

Definition 1.25. Let E be a Banach space and E^* denote its dual. The sequence (x_n) of elements of E is said to weakly converge to x, which is denoted $x_n \rightharpoonup x$, if for every $f \in E^*$ one has $f(x_n) \to f(x)$ as $n \to \infty$.

Of course, strong convergence implies weak convergence. We shall see in Chapter 3 that, as a consequence of Hahn–Banach's theorem, the norm can be represented in a dual way

$$\|x\|_E = \sup\{f(x),\ f \in E^*,\ \|f\|_{E^*} \le 1\}, \quad \forall x \in E. \tag{1.51}$$

This implies that if a subsequence converges weakly, its weak limit is unique. We also deduce from the Banach–Steinhaus theorem that weakly convergent sequences are bounded. Finally, the dual representation of the norm (1.51) shows that the norm is a supremum of continuous linear forms; the norm is therefore (sequentially) weakly lsc in the sense that

$$x_n \rightharpoonup x \Rightarrow \liminf_n \|x_n\|_E \ge \|x\|_E.$$

We will also see in Chapter 3 that, more generally, convex functions which are lsc for the norm topology are also lsc for the weak topology. The chief advantage of using weak convergence is that bounded sequences admit weakly convergent subsequences in *reflexive* Banach spaces which we now define.

Let us first recall that E can be imbedded into E^{**} (the topological bidual, i.e., $E^* = (E^*)^*$) through $J : E \to E^{**}$ defined by

$$J(x)(f) := f(x), \quad \forall x \in E,\ \forall f \in E^*.$$

It directly follows from (1.51) that J is an isometry between E and E^{**} (i.e., $\|J(x)\|_{E^{**}} = \|x\|_E$); in particular it is continuous.

Definition 1.26. The normed space E is said to be reflexive if J is surjective.

In other words, E is reflexive when E^{**} can be identified with E. Hilbert spaces are reflexive (from Riesz's theorem, the dual of a Hilbert space can already be identified with its topological dual). For any $p \in (1, \infty)$, the spaces l^p, L^p and $W^{1,p}(\Omega)$ are reflexive. On the contrary, l^1, L^1, $W^{1,1}$, l^∞, L^∞, $W^{1,\infty}$, spaces of continuous functions, spaces of measures are not reflexive. We again refer to Brezis' book [23] for more details.

Theorem 1.26 (Kakutani). *Let E be a reflexive Banach space and (x_n) be a bounded sequence in E. Then (x_n) admits a weakly convergent subsequence.*

1.6. On the existence and generic uniqueness of minimizers

1.6.1. *Existence of minimizers: Variations on a theme*

The purpose of this book is to study optimization problems. That is, given, a non-empty set E and a function $f : E \to \mathbb{R}$, or more generally $f : E \to \mathbb{R} \cup \{+\infty\}$, we are interested in solving

$$\inf_{x \in E} f(x). \tag{1.52}$$

The function f is called the cost or the minimization criterion, the fact that f may take the value $+\infty$ is a trivial but useful trick to take into account constraints. It may indeed well be the case in applications that not all points of E are admissible for the minimization problem and then it is natural to set the value of the cost to $+\infty$ for those points which are not admissible. Writing a generic optimization problem in the form of the minimization (1.52) is without loss of generality since changing f into $-f$ in (1.52) leads to the maximization of E. A solution of (1.52) is a point $x^* \in E$ such that $f(x^*) \leq f(x)$, for every $x \in E$ and a first question to address when facing an optimization problem is the existence issue: when can the existence of such an x^* be taken for granted? A first basic classical result in this direction is the following theorem.

Theorem 1.27 (Weierstrass' theorem). *Let (E, d) be a compact metric space and let $f : E \to \mathbb{R}$ be continuous. Then (1.52) admits at least one solution.*

Proof. By continuity of f and compactness of E, $f(E)$ is a compact subset of $\mathbb{R}$. It therefore contains its infimum: $\inf\{f(x),\ x \in E\} \in f(E)$. $\qquad\square$

One can slightly weaken the assumptions and in particular relax the continuity assumption into a lower semicontinuity assumption and allow f to take the value $+\infty$.

Theorem 1.28. *Let (E, d) be a metric space, $f : E \to \mathbb{R} \cup \{+\infty\}$ be lower semicontinuous. Assume also that there exists $x_0 \in E$ such that*

$$f(x_0) < +\infty \ \text{and} \ \{x \in E : f(x) \leq f(x_0)\} \ \text{is compact.}$$

Then (1.52) admits at least one solution.

Proof. Clearly, the infimum of f over E coincides with that on f on the compact set $F := \{x \in E : f(x) \leq f(x_0)\}$. Let $(y_n)_n$ be a minimizing sequence for f on F:

$$\lim_n f(y_n) = \inf_F f = \inf_E f.$$

Since F is compact, there exists a (not relabeled) subsequence of $(y_n)_n$ which converges to some point $x^* \in F$. Since f is lsc, we have

$$f(x^*) \leq \liminf_n f(y_n) \leq \inf_E f$$

so that x^* solves (1.52).   □

In finite dimensions, we therefore have the following corollary.

Corollary 1.6. *Let $f : \mathbb{R}^d \to \mathbb{R} \cup \{+\infty\}$ be lower semicontinuous. Assume also that there exists $x_0 \in E$ such that $f(x_0) < +\infty$ and*

$$\lim_{\|x\| \to +\infty} f(x) = +\infty.$$

Then (1.52) admits at least one solution.

For instance, in the calculus of variations one minimizes over spaces of functions which are infinite-dimensional, but, in infinite-dimensional Banach spaces, sets with non-empty interior never are relatively compact for the norm topology (see Exercise 1.14). Theorem 1.28 is therefore of little use if one uses the strong topology in infinite dimensions. However, in reflexive Banach spaces, bounded sequences admit weakly convergent subsequences (Kakutani's theorem) so that we have the following theorem.

Theorem 1.29. *Let $(E, \|\cdot\|)$ be a reflexive Banach space, $f : E \to \mathbb{R} \cup \{+\infty\}$ be weakly sequentially lsc, i.e., such that*

$$x_n \rightharpoonup x \Rightarrow \liminf_n f(x_n) \geq f(x). \tag{1.53}$$

Assume also that there exists $x_0 \in E$ such that $f(x_0) < +\infty$ and

$$\lim_{\|x\| \to +\infty} f(x) = +\infty.$$

Then (1.52) admits at least one solution.

Proof. Let (y_n) be a minimizing sequence for f. We can assume that $f(y_n) \leq f(x_0)$. Hence, (y_n) is bounded. It therefore admits a weakly convergent subsequence (again denoted (y_n)), Denoting by x^* the weak limit of $(y_n)_n$, thanks to (1.53) we have $f(x^*) \leq \liminf_n f(y_n) = \inf_E f$ so that x^* solves (1.52). $\qquad \square$

Of course, being weakly sequentially lsc is much more demanding than being lsc for the strong topology, we will see however (see Chapter 3) that for convex functions, strong lower semicontinuity implies weak sequential lower semicontinuity.

1.6.2. *A generic uniqueness result*

Once the existence of a minimizer for (1.52) is known, one has to wonder if such a minimizer is unique. Of course, it is not always the case (unless E is a singleton!) but one may intuitively think that multiplicity of minimizers is a rather exceptional phenomenon. The aim of this section is to clarify this intuition. To fix ideas, let us work in the compact setting. Let (E, d) be a compact metric space and equip $C(E, \mathbb{R})$ with the topology of uniform convergence. We already know that every $f \in C(E, \mathbb{R})$ attains its minimum. we are interested in the unique minimum attainment property, i.e., in the set:

$$U(E) := \{f \in C(E, \mathbb{R}) \ : \ f \text{ has a unique minimizer on } E\} \qquad (1.54)$$

and wonder how large this set is. The next obvious observations give two partial answers from which no clear conclusion can be drawn

- $U(E)$ is dense in $C(E, \mathbb{R})$ (indeed, if $f \in C(E, \mathbb{R})$ and $\overline{x} \in E$ is such that $f(\overline{x}) = \min_E f$ then the sequence $f_n := f + \frac{1}{n} d(\cdot, \overline{x})$ admits $\overline{x}$ as unique minimizer and converges to f);
- if (E, d) has no isolated point (i.e., for every $x \in E$ and every $r > 0$, $B(x, r) \setminus \{x\} \neq \emptyset$), then $U(E)$ has an empty interior (indeed if $f \in C(E, \mathbb{R})$, the sequence $f_n := \max(\min_E f + \frac{1}{n}, f)$ converges to f as $n \to \infty$ and has several minimum points).

The following theorem however enables us to conclude that the unique minimum attainment property is generic in the sense of Baire.

Theorem 1.30. *Let (E, d) be a compact metric space. Then the set $U(E)$ defined in (1.54) is a dense G_δ (i.e., a countable intersection of open dense subsets) of $C(E, \mathbb{R})$.*

Proof. Of course if E is a singleton there is nothing to prove we may therefore assume that $\text{diam}(E) > 0$. For $\varepsilon \in (0, \text{diam}(E))$ define for every

$f \in C(E, \mathbb{R})$,

$$F_\varepsilon(f) := \min\{\max(f(x_1), f(x_2))\,(x_1, x_2) \in E^2,\, d(x_1, x_2) \geq \varepsilon\}$$

and observe that F_ε is 1-Lipschitz: $|F_\varepsilon(f) - F_\varepsilon(g)| \leq \|f - g\|_\infty$ for every $(f, g) \in C(E, \mathbb{R})$. Likewise, $f \in C(E, \mathbb{R}) \mapsto \min_E f$ is 1-Lipschitz so that the set

$$U_\varepsilon := \{f \in C(E, \mathbb{R}) \,:\, F_\varepsilon(f) > \min_E f\}$$

is open. We now claim that for every $\varepsilon > 0$, U_ε is dense in $C(E, \mathbb{R})$. Indeed, let $f \in C(E, \mathbb{R})$ and $\overline{x} \in E$ be such that $f(\overline{x}) = \min_E f$, and define for every $\lambda > 0$, $f_\lambda := f + \lambda d(\cdot, \overline{x})$. Then by construction $\min_E f = \min_E f_\lambda$. Let now $(x_1, x_2) \in E^2$ with $d(x_1, x_2) \geq \varepsilon$. By the triangle inequality, we have $\max(d(x_1, \overline{x}), d(x_2, \overline{x})) \geq \frac{\varepsilon}{2}$ and since $f(x_i) \geq f(\overline{x}) = \min_E f_\lambda$, we have

$$\max(f_\lambda(x_1), f_\lambda(x_2)) \geq \min_E f_\lambda + \frac{\lambda \varepsilon}{2}$$

hence $F_\varepsilon(f_\lambda) > \min_E f_\lambda$, i.e., $f_\lambda \in U_\varepsilon$. Since f_λ converges to f as $\lambda \to 0$, the density of U_ε follows. Since $U(E) = \bigcap_{k \in \mathbb{N}^*} U_{\frac{1}{k}}$, it follows from the completeness of $C(E, \mathbb{R})$ and Baire's theorem that $U(E)$ is a dense G_δ. $\square$

1.7. Exercises

Exercise 1.2. *Let (E, d) be a metric space. (E, d) is called precompact if for every $r > 0$, E can be covered by finitely many balls of radius r. Prove that (E, d) is compact if and only if it is precompact and complete.*

Exercise 1.3 (Dini's theorems). *Let (X, d) be a compact metric space, let (f_n) be a sequence of functions from X to $\mathbb{R}$ and $f : X \to \mathbb{R}$.*

1) *Assume that f_n and f are continuous, such that $f_n(x) \leq f_{n+1}(x)$ for every $x \in X$ and that (f_n) converges pointwise to f (i.e., $f_n(x)$ converges to $f(x)$ for every $x \in X$). Prove that (f_n) converges uniformly on X to f.*
2) *Assume that if $X = [a, b]$, such that each function f_n is non-decreasing on $[a, b]$, that (f_n) converges pointwise to f and that f is continuous. Prove that (f_n) converges uniformly on $[a, b]$ to f.*

Exercise 1.4. *Let (X, d) be a compact metric space and let (f_n) be a sequence of lsc functions from X to $\mathbb{R}$. Show that if (f_n) converges uniformly on X to some function f, then f is lsc.*

Exercise 1.5. Let $E := [0,1]^{\mathbb{N}}$ given $x = (x_n)_{n \in \mathbb{N}}$ and $y = (y_n)_{n \in \mathbb{N}}$ in E. Define

$$d(x,y) := \sum_{n \in \mathbb{N}} \frac{|x_n - y_n|}{2^n}.$$

1) Show that d is well-defined and that (E,d) is a metric space.
2) Prove that (E,d) is complete.
3) Prove that (E,d) is compact.

Exercise 1.6. Let F be a finite-dimensional linear subspace of a normed vector space E. Prove that F is closed.

Exercise 1.7. Show that there exists a unique real-valued function f defined on $[0,1]$ such that, for every $x \in [0,1]$,

$$f(x) = \sup_{y \in [0,1]} \left\{ \cos(x^2 - y^{12}) + \frac{1}{2} f(y) \right\}.$$

Prove that f is Lipschitz.

Exercise 1.8. Let $p \in (1, +\infty)$ and p' denote its conjugate exponent, i.e., $1/p + 1/p' = 1$,

1) Let $(a,b) \in \mathbb{R}_+ \times \mathbb{R}_+$ show that $ab \leq \frac{a^p}{p} + \frac{a^{p'}}{p'}$. (Hint: use the convexity of $t \in \mathbb{R} \mapsto \exp(t)$.)
2) Let $(a_1, \ldots, a_d)$, and $(b_1, \ldots, b_d)$ be in $\mathbb{R}^d$. Prove Hölder's inequality:

$$\sum_{i=1}^{d} |a_i b_i| \leq \left(\sum_{i=1}^{d} |a_i|^p \right)^{1/p} \left(\sum_{i=1}^{d} |b_i|^{p'} \right)^{1/p'}. \tag{1.55}$$

(Hint: set $S := (\sum_{i=1}^{d} |a_i|^p)^{1/p}$, $T := \sum_{i=1}^{d} |b_i|^{p'})^{1/p'})$, assume $S > 0$, $T > 0$ and apply the previous question to $|a_i|/S$, $|b_i|/T$.)
3) Let $(x_1, \ldots, x_d)$, and $(y_1, \ldots, y_d)$ be in $\mathbb{R}^d$. Prove Minkowski's inequality:

$$\left(\sum_{i=1}^{d} |x_i + y_i|^p \right)^{1/p} \leq \left(\sum_{i=1}^{d} |x_i|^p \right)^{1/p} + \left(\sum_{i=1}^{n} |y_i|^p \right)^{1/p}, \tag{1.56}$$

and conclude.

Exercise 1.9. Let $(E, \|\cdot\|)$ be a real normed space and let f be a linear form on E. Show that $f \in E^*$ if and only if $\ker(f)$ is closed.

Exercise 1.10. *Let E be a Banach space, $u \in \mathcal{L}_c(E)$ with $\|u\|_{\mathcal{L}_c(E)} < 1$. Prove that $(\sum_n (-1)^n u^n)$ converges in $\mathcal{L}_c(E)$ to a limit v. Identify $(\mathrm{id}+u)\circ v$ and conclude.*

Exercise 1.11. *Let E be a Banach space and let $\mathrm{GL}_c(E)$ be the set of invertible elements of $\mathcal{L}_c(E)$. Show that if $\|u - \mathrm{id}\|_{\mathcal{L}_c(E)} < 1$ then $u \in \mathrm{GL}_c(E)$. Show that $\mathrm{GL}_c(E)$ is open and that the map $u \mapsto u^{-1}$ is continuous on $\mathrm{GL}_c(E)$.*

Exercise 1.12. *Let E be a vector space, N_1 and N_2 be two norms on E such that E is complete for both norms and there exists $M > 0$ such that*

$$N_1(x) \leq M N_2(x), \quad \forall x \in E.$$

Prove that N_1 and N_2 are equivalent.

Exercise 1.13. *Let $(H, \langle \cdot, \cdot \rangle)$ be a Hilbert space and $g : H \to H$ (a priori non linear) such that there exist $\alpha > 0$ and $M > 0$ such that for all $(x, y) \in H^2$, one has*

$$\langle g(x) - g(y), x - y \rangle \geq \alpha \|x - y\|^2, \quad \|g(x) - g(y)\| \leq M\|x - y\|.$$

Let $y_0 \in H$ and $\rho > 0$. Define for every $x \in H$

$$T_\rho(x) := x - \rho(g(x) - y_0).$$

1) *Show that for every $(x_1, x_2) \in H^2$, one has*

$$\|T_\rho(x_1) - T_\rho(x_2)\|^2 \leq (1 + M^2\rho^2 - 2\alpha\rho)\|x_1 - x_2\|^2,$$

and deduce that for $\rho > 0$ well-chosen, T_ρ is a contraction of H.
2) *Prove that g is a bijection.*
3) *Prove that the inverse of g is Lipschitz.*

Exercise 1.14. *Let E be an infinite-dimensional Banach space. Assume that the closed unit ball $\overline{B}$ of E can be covered by finitely many balls of radius $1/2$, denote by $y_1, \ldots, y_N$ the centers of such balls. Show that the vector space spanned by $\{y_1, \ldots, y_N\}$ is dense and conclude.*

Exercise 1.15. *Consider $\mathbb{R}^d$ equipped with its usual euclidean structure, and define $C := (\mathbb{R}_+)^d$.*

1) *Show that C is closed and convex, then recall the variational characterization of proj_C, the projection onto C.*

2) *Show that for every* $x = (x_1, \ldots, x_d)$, $\mathrm{proj}_C(x) = (x_1^+, \ldots, x_d^+)$ *($t^+ :=$* $\max(t, 0)$ *denotes the positive part of* t*).*

Exercise 1.16. *Let* $\mathcal{M}_d(\mathbb{R})$ *be the space of* $d \times d$ *matrices with real entries. For* A *and* B *in* $\mathcal{M}_d(\mathbb{R})$ *set*

$$\langle A, B \rangle := \mathrm{tr}(AB^T),$$

(where B^T *stands for the transpose of* B*, i.e.,* $(B^T)_{ij} = B_{ji}$*).*

1) *Show that* $(\mathcal{M}_d(\mathbb{R}), \langle \cdot, \cdot \rangle)$ *is a Hilbert space.*
2) *Show that the orthogonal of the subspace* $\mathcal{S}_d(\mathbb{R})$ *of* $\mathcal{M}_d(\mathbb{R})$ *is the subspace of skew-symmetric matrices (recall that* A *is skew-symmetric when* $A^T = -A$*).*

Exercise 1.17. *For* x *and* y *in* $\mathbb{R}^d$*, let* $x \cdot y := \sum_{i=1}^d x_i y_i$ *denote the usual scalar product of* x *and* y*. Denote by* $\mathcal{S}_d(\mathbb{R})$ *the space of symmetric matrices and recall that any* A *in* $\mathcal{S}_d(\mathbb{R})$ *admits an orthogonal basis of eigenvectors (see Exercise 4.13 for an optimization-based proof). We denote by* $\mathcal{S}_d^+(\mathbb{R})$ *and* $\mathcal{S}_d^{++}(\mathbb{R})$*, respectively, the set of positive semidefinite matrices and the set of positive definite matrices, i.e,*

$$\mathcal{S}_d^+(\mathbb{R}) := \{A \in \mathcal{S}_d(\mathbb{R}) \; : \; Ax \cdot x \geq 0, \; \forall x \in \mathbb{R}^d\}$$

and

$$\mathcal{S}_d^{++}(\mathbb{R}) := \{A \in \mathcal{S}_d(\mathbb{R}) \; : \; Ax \cdot x > 0, \; \forall x \in \mathbb{R}^d \setminus \{0\}\}.$$

1) *Let* $A \in \mathcal{S}_d(\mathbb{R})$*. Show that* $A \in \mathcal{S}_d^+(\mathbb{R})$ *if and only if its eigenvalues are non-negative.*
2) *Show that* $\mathcal{S}_d^+(\mathbb{R})$ *is convex and closed and that* $\mathcal{S}_d^{++}(\mathbb{R})$ *is convex and open (in* $\mathcal{S}_d(\mathbb{R})$*).*
3) *Show that every* $A \in \mathcal{S}_d^+(\mathbb{R})$ *admits a square root (i.e., a matrix whose square is* A*) in* $\mathcal{S}_d^+(\mathbb{R})$*.*
4) *Show that every* $A \in \mathcal{S}_d^+(\mathbb{R})$ *admits a unique square root in* $\mathcal{S}_d^+(\mathbb{R})$*.*
5) *Equip* $\mathcal{S}_d(\mathbb{R})$ *with the scalar product of Exercise 1.16. For* $A \in \mathcal{S}_d(\mathbb{R})$*, diagonalize* A *in an orthogonal basis, i.e, write* A *as* $A = U^T \Delta U$ *with* U *orthogonal (i.e.,* $U^T U = I_d$*,* I_d *being the identity matrix) and* $\Delta = \mathrm{diag}(\lambda_1, \ldots, \lambda_d)$ *diagonal, what is the projection of* A *onto* $\mathcal{S}_d^+(\mathbb{R})$*?*

Exercise 1.18. *Let M_1 and M_2 be in $\mathcal{S}_d^{++}(\mathbb{R})$. Show that*

$$\operatorname{tr}(M_1^{-1}M_2) + \operatorname{tr}(M_2^{-1}M_1) \geq 2\operatorname{tr}(I_d) = 2d$$

and identify the equality case. (Hint: use the scalar product of Exercise 1.16, and the squared norms of the matrices $M_2^{1/2}M_1^{-1/2}$ and $M_1^{1/2}M_2^{-1/2}$ where $M_i^{1/2}$ denotes the positive semidefinite square root of M_i, see Exercise 1.17.)

Exercise 1.19. *Let $(H, \langle \cdot, \cdot \rangle)$ be a Hilbert space and B denote its closed unit ball.*

1) *Define and characterize the projection onto B, $p_B(x)$.*
2) *Compute $p_B(x)$ for every $x \in H$.*

Exercise 1.20. *Define $f_n(t) = t^n$ for every $t \in [0,1]$ and every $n \in \mathbb{N}^*$. Show that (f_n) has no weakly convergent subsequence in $(C([0,1], \| \cdot \|_\infty)$. (Hint: show that if f_n converges weakly to f, then it converges pointwise to f.) Deduce that $C([0,1])$ is not reflexive.*

Exercise 1.21. *Let E be a Hilbert space and $(x_n)_n$ be a bounded sequence in E. Show that $(x_n)_n$ weakly converges if and only if it admits a unique weak cluster point (that is the weak limit of some subsequence).*

Exercise 1.22. *Let E be a Banach space and E^* be its topological dual. The weak-* topology of E^* is by definition the coarsest topology on E^* making $p \in E^* \mapsto p(x)$ continuous for every $x \in E$. A sequence $(p_n)_n$ in E^* converges weakly-* to p, which is denoted $p_n \overset{*}{\rightharpoonup} p$ if $p_n(x) \to p(x)$, for every $x \in E$. Establish the following properties of the weak $*$ convergence:*

1) *If (p_n) converges weakly-*, its limit is unique.*
2) *If $p_n \to p$ (i.e., $\|p_n - p\|_{E^*} \to 0$), then $p_n \overset{*}{\rightharpoonup} p$.*
3) *If $p_n \rightharpoonup p$ for the weak topology of E^*, then $p_n \overset{*}{\rightharpoonup} p$.*
4) *Every sequence which is weakly-* convergent in E^* is bounded.*
5) *If $p_n \overset{*}{\rightharpoonup} p$, then $\|p\|_{E^*} \leq \liminf_n \|p_n\|_{E^*}$.*
6) *If $p_n \overset{*}{\rightharpoonup} p$ and $x_n \to x$ in E (strongly), then $p_n(x_n) \to p(x)$.*

Chapter 2

Differential Calculus

2.1. First-order differential calculus

2.1.1. *Several notions of differentiability*

In what follows, we will be given two real normed spaces $(E, \|\cdot\|_E)$ and $(F, \|\cdot\|_F)$, Ω a non-empty open subset of E and f a map defined on Ω with values in F. Since Ω is open, for every $x \in \Omega$ and every $h \in E$, $x + th \in \Omega$ for $t \in \mathbb{R}$ small enough in absolute value. Let us start with the most basic notion of differentiability: differentiability at x in the direction h.

Definition 2.1. Let $x \in \Omega$ and $h \in E$. f is differentiable at x in the direction h if the following limit exists (in the sense of the topology of $(F, \|\cdot\|_F)$):

$$\lim_{t \to 0,\ t \neq 0} \frac{1}{t}(f(x + th) - f(x)).$$

If this limit exists, it is called the directional derivative of f at x in the direction h and denoted $Df(x; h)$.

Note that f is differentiable at x in the direction h if and only if both limits:

$$\lim_{t \to 0^+} \frac{1}{t}(f(x + th) - f(x)) \quad \text{and} \quad \lim_{t \to 0^-} \frac{1}{t}(f(x + th) - f(x))$$

exist and are equal. This naturally leads to the next definition:

Definition 2.2. Let $x \in \Omega$ and $h \in E$. f is right-differentiable at x in the direction h if the following limit exists:

$$\lim_{t \to 0^+} \frac{1}{t}(f(x + th) - f(x)).$$

47

If this limit exists, it is called the right-directional derivative of f at x in the direction h and denoted $D^+f(x;h)$.

Observing that if f is right-differentiable at x in the direction $-h$ then

$$D^+f(x;-h) = -\lim_{t\to 0^-}\frac{1}{t}(f(x+th)-f(x)).$$

We deduce that f is differentiable at x in the direction h if and only if it is right-differentiable at x in both directions h and $-h$ and

$$D^+f(x;-h) = -D^+f(x;h).$$

An easy exercise is as follows:

Exercise 2.1. *Let f be as above, $x \in \Omega$ and $h \in E$. Show that:*

1) $Df(x;0)$ *exists and equals 0.*
2) *If f is right-differentiable at x in the direction h, then for every $\lambda \geq 0$, it is also right-differentiable at x in the direction λh and*

$$D^+f(x;\lambda h) = \lambda D^+f(x;h).$$

3) *If f is differentiable at x in the direction h, then for every $\lambda \in \mathbb{R}$, it is also differentiable at x in the direction λh and*

$$Df(x;\lambda h) = \lambda Df(x;h).$$

Definition 2.3. Let $x \in \Omega$. f is Gateaux-differentiable at x if

- for every $h \in E$, f is differentiable at x in the direction h;
- the map $h \in E \mapsto Df(x;h) \in F$ is linear and continuous, hence can be written as $Df(x;h) = D_G f(x)(h)$ with $D_G f(x) \in \mathcal{L}_c(E,F)$, $D_G f(x)$ is called the Gateaux derivative (or Gateaux differential) of f and x.

Finally, f is Gateaux-differentiable on Ω if it is Gateaux-differentiable at each point of Ω.

It is important to understand that if f is Gateaux-differentiable on Ω, its Gateaux derivative $D_G f$ is a map defined on Ω which takes values in $\mathcal{L}_c(E,F)$. Let us also emphasize that the previous definition depends on the norms on E and F; however, the choice of equivalent norms give the same definition for the Gateaux derivative, in particular in finite dimensions, the choice of particular norms is irrelevant in Definition 2.3. If E is finite-dimensional, since $\mathcal{L}_c(E,F) = \mathcal{L}(E,F)$ (see Theorem 1.13), the continuity requirement in Definition 2.3 can be omitted since it automatically follows from the linearity of $D_G f(x)$.

Remark 2.1. Gateaux-differentiability is a rather weak notion, in particular it does not imply continuity. To see this, consider the behavior near 0 of the function $f : \mathbb{R}^2 \to \mathbb{R}$ given by

$$f(x, y) = \begin{cases} 1 & \text{if } y = x^2 \text{ and } x \neq 0, \\ 0 & \text{otherwise.} \end{cases}$$

It is discontinuous at 0 but is Gateaux-differentiable at 0 with $D_G f(0) = 0$.

Remark 2.2. The fact that f admits directional derivatives at x in every direction is not enough to guarantee that it is Gateaux-differentiable at x. As an exercise, we advise the reader to study near 0 the behavior of $f : \mathbb{R}^2 \to \mathbb{R}$ given by

$$f(x, y) = \begin{cases} 0 & \text{if } (x, y) = (0, 0), \\ \dfrac{x^3}{x^2 + |y|} & \text{otherwise.} \end{cases}$$

The previous remarks explain why one needs a stronger notion of differentiability, called Fréchet-differentiability.

Definition 2.4. Let $x \in \Omega$. f is Fréchet-differentiable (in the sequel, we will simply say differentiable) at x if there exist $L \in \mathcal{L}_c(E, F)$ and a map ε defined in a neighborhood of 0 in E and taking its values in F such that

$$f(x + h) = f(x) + L(h) + \|h\|_E \cdot \varepsilon(h) \text{ with } \lim_{h \to 0} \|\varepsilon(h)\|_F = 0. \quad (2.1)$$

In a more detailed way, (2.1) can be expressed as follows: for every $\varepsilon > 0$, there exists $\delta > 0$ such that, for $h \in E$, one has

$$\|h\|_E \leq \delta \Rightarrow \|f(x + h) - f(x) - L(h)\|_F \leq \varepsilon \|h\|_E.$$

Remark 2.3. If f is (Fréchet) differentiable at x, then it is obviously continuous at x.

It is often convenient to write (2.1) in a more concise way, using Landau's notation:

$$f(x + h) = f(x) + L(h) + o(h), \quad (2.2)$$

where $o(h)$ designates a function that tends to 0 (in F) *faster* than h as h tends to 0 (in E), that is

$$\lim_{h \to 0,\ h \neq 0} \frac{\|o(h)\|_F}{\|h\|_E} = 0. \tag{2.3}$$

Uniqueness of the derivative is given by the following lemma.

Lemma 2.1. *Let $x \in \Omega$. If $L_1 \in \mathcal{L}_c(E, F)$ and $L_2 \in \mathcal{L}_c(E, F)$ are such that*

$$f(x + h) = f(x) + L_1(h) + o(h) = f(x) + L_2(h) + o(h),$$

then $L_1 = L_2$.

Proof. Since $(L_1 - L_2)h = o(h)$, for every $\varepsilon > 0$, there is a $\delta > 0$ such that for every $h \in B(0, \delta)$

$$\|(L_1 - L_2)(h)\|_F \leq \varepsilon \|h\|_E,$$

hence

$$\|L_1 - L_2\|_{\mathcal{L}_c(E,F)} \leq \varepsilon$$

letting $\varepsilon > 0$ tend to 0 gives the desired result. $\qquad \square$

Exercise 2.2. *Consider the map $T : C([-1, 1], \mathbb{R}) \to \mathbb{R}$ defined by $T(f) = f(0)^2$ for every $f \in C([-1, 1], \mathbb{R})$. Show that T admits directional derivatives in all directions. Is T Gateaux-differentiable when E is equipped with the uniform norm? Same question with the L^1-norm? Same questions for Fréchet-differentiability?*

Lemma 2.1 enables us to define unambiguously the Fréchet derivative.

Definition 2.5. Let $x \in \Omega$ and f differentiable at x. The derivative of f at x, denoted $f'(x)$, is the unique element of $\mathcal{L}_c(E, F)$ such that

$$f(x + h) = f(x) + f'(x)(h) + o(h).$$

f is differentiable on Ω if it is differentiable at each point of $x \in \Omega$.

Of course, if f is differentiable at x, it is Gateaux-differentiable at x, and both notions coincide:

$$f'(x) = D_G f(x), \ Df(x; h) = f'(x)(h), \ \forall h \in E.$$

Definition 2.6. f is of class C^1 (or continuously differentiable) on Ω (which we denote $f \in C^1(\Omega, F)$) if f is differentiable on Ω and f' is continuous, i.e., $f' \in C(\Omega, \mathcal{L}_c(E, F))$.

Example 2.1. Let $f : \mathcal{M}_n(\mathbb{R}) \to \mathcal{M}_n(\mathbb{R})$ be defined by $f(A) = A \times A = A^2$ for every $A \in \mathcal{M}_n(\mathbb{R})$. Let us show that f is of class C^1. Since all norms are equivalent on $\mathcal{M}_n(\mathbb{R})$, let us choose a convenient one. First fix a norm $|\cdot|$ on $\mathcal{M}_{n,1}(\mathbb{R}) \simeq \mathbb{R}^n$ and define the corresponding *operator* norm:

$$\|A\| := \sup\{|AX|, \ X \in \mathcal{M}_{n,1}(\mathbb{R}), \ |X| \le 1\}.$$

This norm is indeed convenient because it satisfies

$$\|AB\| \le \|A\| \, \|B\|. \tag{2.4}$$

Let A and H be in $\mathcal{M}_n(\mathbb{R})$. We then have

$$f(A + H) = f(A) + AH + HA + H^2.$$

Thanks to (2.4), $H^2 = o(H)$ and since $H \mapsto AH + HA$ is linear, we deduce that f is Fréchet-differentiable and

$$f'(A)(H) = AH + HA, \ \forall A \in \mathcal{M}_n(\mathbb{R}), \ \forall H \in \mathcal{M}_n(\mathbb{R}).$$

Now let B be another matrix. We have

$$(f'(A) - f'(B))(H) = (A - B)H + H(B - A)$$

so that, using (2.4) again and the triangle inequality, we deduce

$$\|(f'(A) - f'(B))(H)\| \le 2\|A - B\| \, \|H\|$$

or, equivalently

$$\|(f'(A) - f'(B))(H)\|_{\mathcal{L}(\mathcal{M}_n(\mathbb{R})} := \sup\{\|(f'(A) - f'(B))(H)\|, \ \|H\| \le 1\}$$
$$\le 2\|A - B\|,$$

which means that f' is 2-Lipschitz, in particular f is of class C^1.

A simple criterion for being C^1 is given by the next result that we shall prove in Section 2.1.3.

Theorem 2.1. *If f is Gateaux-differentiable on Ω and $D_G f \in C(\Omega, \mathcal{L}_c(E, F))$, then f is of class C^1 on Ω.*

Theorem 2.1 is very useful, it can be used in practice as follows to show that a function f is of class C^1:

- **Step 1:** Show that f admits directional derivatives and compute $Df(x; h)$ for $(x, h) \in \Omega \times E$.

- **Step 2:** Show that $h \mapsto Df(x; h)$ is linear and continuous so that f is Gateaux-differentiable at x and $Df(x; h) = D_G f(x)(h)$.
- **Step 3:** Prove that $D_G f : x \mapsto D_G f(x)$ is a continuous map from Ω to $\mathcal{L}_c(E, F)$. This amounts to show that for every $x \in \Omega$ and every $\varepsilon > 0$ there exists δ such that for every $y \in \Omega$, if $\|x - y\|_E \le \delta$ then $\|D_G f(x) - D_G f(y)\|_{\mathcal{L}_c(E,F)} \le \varepsilon$ which is the same as

$$\|D_G f(x)(h) - D_G f(y)(h)\|_F \le \varepsilon \|h\|, \ \forall h \in E. \tag{2.5}$$

Regarding invertible maps, we have the following definitions.

Definition 2.7. Let Ω be an open subset of E, Ω' be an open subset of F and f be a bijection between Ω and Ω'. One says that:

- f is a homeomorphism between Ω and Ω' if $f \in C(\Omega, \Omega')$ and $f^{-1} \in C(\Omega', \Omega)$,
- f is a C^1-diffeomorphism between Ω and Ω' if $f \in C^1(\Omega, \Omega')$ and $f^{-1} \in C^1(\Omega', \Omega)$.

When E is a Hilbert space and $F = \mathbb{R}$, the Gateaux derivative at a given point being an element of E^*, it can be identified thanks to Riesz' representation theorem (Theorem 1.25) to an element of E, called the *gradient*.

Definition 2.8. Let $(E, \langle \cdot, \cdot \rangle)$ be a Hilbert space, Ω be an open subset of E, $f \colon \Omega \to \mathbb{R}$ and $x \in \Omega$. If f is Gateaux-differentiable at x, the gradient of f at x, denoted $\nabla f(x)$, is the unique element of E such that

$$D_G f(x)(h) = \langle \nabla f(x), h \rangle \text{ for every } h \in E.$$

In particular, if $(E, \langle \cdot, \cdot \rangle)$ is a Hilbert space, $x \mapsto \langle x, x \rangle = \|x\|^2$ is of class C^1 with $\nabla(\|.\|^2)(x) = 2x$ for every $x \in E$.

2.1.2. Calculus rules

Let us start with some obvious observations.

Proposition 2.1. Let $f : \Omega \to F$ and $g : \Omega \to F$ and $x \in \Omega$.

1) If f is constant in a neighborhood of x, then f is differentiable at x and $f'(x) = 0$.

2) *If f is differentiable at x, then, for every $\alpha \in \mathbb{R}$, αf is differentiable at x and*

$$(\alpha f)'(x) = \alpha f'(x).$$

3) *If f and g are differentiable at $x \in \Omega$, then so is $f + g$ and*

$$(f + g)'(x) = f'(x) + g'(x).$$

4) *If $L \in \mathcal{L}_c(E, F)$, then $L \in C^1(E, F)$ and $L'(z) = L$ for every $z \in E$.*

For continuous bilinear maps, we have the following (we leave the easy proof as an exercise to the reader).

Proposition 2.2. *Let E, F and G be three normed spaces, and $a \in \mathcal{L}_{2,c}(E \times F, G)$. Then $a \in C^1(E \times F, G)$ and for every $(x, y) \in E \times F$ and $(h, k) \in E \times F$, one has*

$$a'(x, y)(h, k) = a(x, k) + a(h, y).$$

Proposition 2.3. *Let E, $F_1, \ldots, F_p$ be $(p + 1)$ normed spaces, Ω be an open subset of E, for $i = 1, \ldots, p$, let $f_i : \Omega \to F_i$. For $x \in \Omega$, then define:*

$$f(x) = (f_1(x), \ldots, f_p(x)) \in F := \prod_{i=1}^{p} F_i.$$

Then f is differentiable at $x \in \Omega$ if and only if f_i is differentiable at $x \in \Omega$ for $i = 1, \ldots, p$ and in this case:

$$f'(x)(h) = (f_1'(x)(h), \ldots, f_p'(x)(h)) \text{ for every } h \in E.$$

In a more concise way, this can be rewritten as

$$f' = (f_1, \ldots, f_p)' = (f_1', \ldots, f_p').$$

Differentiating a composite function can done by the chain rule.

Theorem 2.2 (Chain rule). *Let E, F and G be three normed spaces, Ω be an open subset of E, U be an open subset of F, $f : \Omega \to F$, $g : U \to G$ and $x \in \Omega$. If f is differentiable at x, $f(x) \in U$ and g is differentiable at $f(x)$ then $g \circ f$ is differentiable at x and*

$$(g \circ f)'(x) = g'(f(x)) \circ f'(x).$$

Proof. Since g is differentiable at $f(x)$, we have

$$g(f(x) + k) = g(f(x)) + g'(f(x))k + \|k\|_F \, \varepsilon_2(k),$$

$$\text{with } \lim_{k \to 0} \|\varepsilon_2(k)\|_G = 0.$$

If we apply the previous to $k = f(x + h) - f(x)$, which thanks to the differentiability of f at x can be expressed as

$$f(x + h) - f(x) = f'(x)h + \|h\|_E \, \varepsilon_1(h), \text{ with } \lim_{h \to 0} \|\varepsilon_1(h)\|_F = 0,$$

we get

$$g \circ f(x + h) = g(f(x)) + (g'(f(x)) \circ f'(x))(h) + R(h)$$

with

$$R(h) = \|h\|_E g'(f(x))\varepsilon_1(h) + \|f'(x)h + \|h\|_E \, \varepsilon_1(h)\|_F$$
$$\times \varepsilon_2(f'(x)h + \|h\|_E \, \varepsilon_1(h)),$$

which is easily seen to be an $o(h)$. $\qquad\square$

In particular, if $(E, \langle \cdot, \cdot \rangle)$ is a Hilbert space, the chain rule implies that $x \mapsto \sqrt{\langle x, x \rangle} = \|x\|$ is of class C^1 on $E \setminus \{0\}$ with

$$\nabla(\|\cdot\|)(x) = \frac{x}{\|x\|}, \ \forall x \in E \setminus \{0\}. \tag{2.6}$$

Corollary 2.1. *If, in addition to the assumptions of Theorem 2.2, we assume that f is of class C^1 on Ω, that $f(\Omega) \subset U$ and that g is of class C^1, then $g \circ f$ is of class C^1 on Ω.*

Concerning the differentiability of a product (scalar function multiplying a vector function), we have the following proposition.

Proposition 2.4. *Let E and F be two normed spaces, Ω be an open subset of E, $f : \Omega \to F$, $u : \Omega \to \mathbb{R}$ and $x \in \Omega$. If f and u are differentiable at x, then $u \cdot f$ is differentiable at x and*

$$(u \cdot f)'(x)(h) = (u'(x)(h)) \cdot f(x) + u(x) \cdot f'(x)(h) \text{ for every } h \in E.$$

Finally, concerning the differentiability of an inverse map, we have the following result. Note that it requires the source and target spaces to be Banach spaces, because its proof relies on Banach's theorem (Theorem 1.23).

Theorem 2.3. *Let E and F be two Banach spaces, Ω be an open subset of E, U be an open subset of F, and f be a homeomorphism between Ω and U (we denote its inverse by $f^{-1} : U \to \Omega$) and $x \in \Omega$. If f is differentiable at x and if $f'(x)$ is invertible, then f^{-1} is differentiable at $f(x)$ and*

$$(f^{-1})'(f(x)) = [f'(x)]^{-1}.$$

Proof. Set $y = f(x)$ our aim is to show that for $k \in F$ small enough $h_k := f^{-1}(y + k) - x = f^{-1}(y + k) - f^{-1}(y)$ can be written as

$$h_k = [f'(x)]^{-1}(k) + o(k). \tag{2.7}$$

First, we observe that, since f^{-1} is continuous at y,

$$\lim_{k \to 0} h_k = 0. \tag{2.8}$$

Then since $f(x + h_k) = y + k$ and f is differentiable at y we have

$$k = f(x + h_k) - f(x) = f'(x)h_k + \|h_k\| \, \varepsilon(h_k), \quad \text{with } \lim_{h \to 0} \varepsilon(h) = 0.$$

Recall that from Banach's theorem (Theorem 1.23), $f'(x)^{-1}$ is continuous, then rewrite the previous identity as

$$[f'(x)]^{-1}(k) = h_k + \|h_k\|\widetilde{\varepsilon}(h_k) \text{ with } \widetilde{\varepsilon}(h) :$$
$$= [f'(x)]^{-1}\varepsilon(h) \to 0 \text{ as } h \to 0.$$

Now, from (2.8), there exists $\delta > 0$ such that whenever $\|k\| \leq \delta$, one has $\|\widetilde{\varepsilon}(h_k)\| \leq 1/2$. for such a k we thus have

$$\|h_k\| \leq \|[f'(x)]^{-1}(k)\| + \frac{1}{2}\|h_k\|,$$

hence

$$\|h_k\| \leq 2\|[f'(x)]^{-1}\| \, \|k\| \tag{2.9}$$

and then, with

$$h_k = [f'(x)]^{-1}(k) - \|h_k\|\widetilde{\varepsilon}(h_k),$$

we can conclude that (2.7) holds because $\|h_k\|\widetilde{\varepsilon}(h_k) = o(k)$ thanks to (2.9) and the fact that $\widetilde{\varepsilon}(h_k) \to 0$ as $k \to 0$. $\qquad \square$

Exercise 2.3. *Let us review some classic examples.*

1) *Given A an $n \times n$ matrix (and denoting by A^T its transpose) and identifying $\mathbb{R}^n$ with $n \times 1$ matrices, define*

$$f_1(x) := Ax \cdot x = \sum_{1 \leq i,j \leq n} A_{ij} x_j x_i, \ \forall x \in \mathbb{R}^n$$

show that f_1 is of class C^1 and

$$\nabla f_1(x) = Ax + A^T x.$$

2) *Denoting by $\| \cdot \|$ the usual Euclidean norm:*

$$f_2(x) = \|Ax - b\|^2,$$

show that f_2 is of class C^1 and

$$\nabla f_2(x) = 2A^T A x - 2A^T b.$$

3) *Show that $\| \cdot \|$ is of class C^1 on $\mathbb{R}^n \setminus \{0\}$ and*

$$\nabla (\| \cdot \|)(x) = \frac{x}{\|x\|}, \forall x \neq 0.$$

4) *Given $\alpha \in \mathbb{R}$ set $g_\alpha(x) := \|x\|^\alpha$, show that g_α is of class C^1 on $\mathbb{R}^n \setminus \{0\}$ and compute its gradient.*

2.1.3. *Mean value inequalities*

Let us start with some reminders about real-valued functions of a real variable and in the first-place Rolle's classic theorem:

Theorem 2.4 (Rolle's theorem). *Let $a < b$, $f \in C([a,b], \mathbb{R})$. If $f(a) = f(b)$ and f is differentiable on (a,b), then there exists $c \in (a,b)$ such that $f'(c) = 0$.*

The mean value theorem then reads as follows.

Theorem 2.5 (Mean value theorem I). *Let $a < b$, $f \in C([a,b], \mathbb{R})$. If f is differentiable on (a,b), there exists $c \in (a,b)$ such that*

$$f'(c) = \frac{f(b) - f(a)}{b - a}.$$

Given a vector space E and $(a, b) \in E^2$ recall that:

$$[a, b] = \{ta + (1 - t)b, \, t \in [0, 1]\} \text{ and } (a, b) = \{ta + (1 - t)b, \, t \in (0, 1)\}.$$

One straightforwardly deduces from Theorem 2.5, a mean value formula for *real-valued* functions of a vector-valued variable.

Theorem 2.6 (Mean value theorem II). *Let E be a normed vector space, Ω be an open subset of E, $a \neq b$ be two distinct points in Ω such that $[a, b] \subset \Omega$ and f be a real-valued function defined on Ω. If f is continuous on $[a, b]$ and the directional derivative $Df(x; b - a)$ exists for every $x \in (a, b)$, then there exists $c \in (a, b)$, such that*

$$f(b) - f(a) = Df(c; b - a).$$

Proof. Define for every $t \in [0, 1]$, $g(t) := f(a + t(b - a))$, then g is continuous on $[0, 1]$ and for $t \in (0, 1)$ one has

$$\lim_{h \to 0, \, h \neq 0} \frac{g(t + h) - g(t)}{h}$$

$$= \lim_{h \to 0, \, h \neq 0} \frac{f(a + (t + h)(b - a)) - f(a + t(b - a))}{h}$$

so that g is differentiable on $(0, 1)$ with:

$$g'(t) = Df(a + t(b - a); b - a).$$

Applying the mean value equality given by Theorem 2.5 to g, we can find a $t \in (0, 1)$ such that $g(1) - g(0) = g'(t)$ setting $c = a + t(b - a)$, this means

$$f(b) - f(a) = Df(c; b - a). \qquad \square$$

Remark 2.4. Consider $f \colon \mathbb{R} \to \mathbb{R}^2$ defined by $f(t) := (\cos(t), \sin(t))$ for every $t \in \mathbb{R}$. By periodicity $f(2\pi) - f(0) = 0$ but $f'(t) = (-\sin(t), \cos(t))$ always have unit Euclidean norm hence never vanishes in particular there is no $c \in (0, 2\pi)$ such that $f(2\pi) - f(0) = 2\pi f'(c)$. This simple example shows that a mean value equality can be expected only in the case of real-valued functions. The best one can hope for in the case of vector-valued functions is a mean value *inequality*.

For functions with values in a normed vector space, the mean value inequality theorem reads as follows.

Theorem 2.7 (Mean value inequality). *Let E and F be real normed vector spaces, Ω be an open subset of E, $f : \Omega \to F$, $(a, b) \in \Omega^2$ with $a \neq b$ such that $[a, b] \subset \Omega$, assume that f is continuous on $[a, b]$. If the directional derivative $Df(x; b - a)$ exists for each $x \in (a, b)$, then there exists $c \in (a, b)$ such that*

$$\|f(b) - f(a)\| \leq \|Df(c; b - a)\|. \tag{2.10}$$

Proof. It follows from Hahn–Banach's theorem (Theorem 3.1) (see Corollary 3.2 in Chapter 3) that there exists $p \in F^*$ such that

$$\|p\|_{F^*} = 1 \text{ and } p(f(b) - f(a)) = \|f(b) - f(a)\|. \tag{2.11}$$

Let us then define the real-valued function $t \in [0, 1] \mapsto g(t) := p(f(a + t(b - a))$. Since g fulfills the assumptions of Theorem 2.5, there is a $t \in (0, 1)$ such that

$$p(f(b) - f(a)) = g(1) - g(0) = g'(t) = p(Df(c; b - a)), \tag{2.12}$$

where we have set $c = a + t(b - a) \in (a, b)$. Together with (2.11), this yields

$$\|f(b) - f(a)\| \leq \|p\|_{E^*} \|Df(c; b - a)\| \leq \|Df(c; b - a)\|. \qquad \square$$

If f is Gateaux-differentiable on Ω (so that it is continuous on $[a, b]$), since $\|Df(c; b - a)\| \leq \|D_G f(c)\| \|b - a\|$, Theorem 2.7 implies that there is a $c \in (a, b)$ such that

$$\|f(b) - f(a)\| \leq \|D_G f(c)(b - a)\| \leq \|D_G f(c)\| \|b - a\|. \tag{2.13}$$

Let us now state some corollaries.

Corollary 2.2. *Let E and F be real normed vector spaces, Ω be an open subset of E, $f : \Omega \to F$, $(a, b) \in \Omega^2$ with $a \neq b$ such that $[a, b] \subset \Omega$. If the directional derivative $Df(x; b - a)$ exists for each $x \in (a, b)$, then*

$$\|f(b) - f(a)\| \leq \sup_{c \in (a, b)} \|Df(c; b - a)\|, \tag{2.14}$$

(allowing the supremum in the right-hand side to be $+\infty$).

If f is Gateaux-differentiable on Ω, Corollary 2.2 implies

$$\|f(b) - f(a)\| \leq \sup_{c \in (a, b)} \|D_G f(c)\| \|b - a\|. \tag{2.15}$$

An obvious consequence is the following sufficient condition for a non-linear map to be Lipschitz.

Corollary 2.3. *Let E and F be real normed vector spaces, Ω be an open convex subset of E, $f : \Omega \to F$. If f is Gateaux-differentiable on Ω and $\|D_G f(x)\| \leq k$ for every $x \in \Omega$, then f is k-Lipschitz on Ω, i.e.,*

$$\|f(b) - f(a)\| \leq k\|b - a\|, \ \forall (a,b) \in \Omega^2. \tag{2.16}$$

Another useful mean value inequality is given by the following corollary.

Corollary 2.4. *Let E and F be real normed vector spaces, Ω be an open subset of E, $f : \Omega \to F$, $(a,b) \in \Omega^2$ with $a \neq b$ such that $[a,b] \subset \Omega$. If f is Gateaux-differentiable on Ω, then for every $z \in \Omega$ one has*

$$\|f(b) - f(a) - D_G f(z)(b-a)\| \leq \sup_{c \in (a,b)} \|D_G f(c) - D_G f(z)\| \|b-a\|.$$

In particular,

$$\|f(b) - f(a) - D_G f(a)(b-a)\| \leq \sup_{c \in (a,b)} \|D_G f(c) - D_G f(a)\| \|b-a\|$$

(allowing the supremum in the right-hand side to be $+\infty$).

Proof. Apply Corollary 2.2 to $x \mapsto f(x) - D_G f(z)(x)$. $\qquad\square$

We are now in position to prove Theorem 2.1.

Theorem 2.8. *If f is Gateaux-differentiable on Ω and $D_G f \in C(\Omega, \mathcal{L}_c(E,F))$, then f is of class C^1 on Ω.*

Proof. If we prove that for every $x \in \Omega$ and $\varepsilon > 0$, there exists $\delta_\varepsilon > 0$ such that for every $h \in E$ such that $\|h\| \leq \delta_\varepsilon$, there holds:

$$\|f(x+h) - f(x) - D_G f(x)(h)\| \leq \varepsilon \|h\|.$$

We will have established that f is (Fréchet) differentiable with $f'(x) = D_G f(x)$ for each x so that f is of class C^1 on Ω since $D_G f \in C(\Omega, \mathcal{L}_c(E,F))$. Let $r > 0$ be such that $B(x,r) \subset \Omega$, $\delta \in (0,r)$ and $h \in B(0,\delta)$. Then $[x, x+h] \subset B(x,\delta)$, hence thanks to Corollary 2.2, we have

$$\|f(x+h) - f(x) - D_G f(x)(h)\| \leq \sup_{c \in [x, x+h]} \|D_G f(c) - D_G f(x)\| \|h\|$$

$$\leq \sup_{c \in B(x,\delta)} \|D_G f(c) - D_G f(x)\| \|h\|.$$

Since $D_G f \in C(\Omega, \mathcal{L}_c(E, F))$, there exists $\delta_\varepsilon \in (0, r)$ such that

$$\sup_{c \in B(x, \delta_\varepsilon)} \|D_G f(c) - D_G f(x)\| \leq \varepsilon,$$

which gives the desired result.   □

2.1.4. *Partial derivatives*

Let us now focus on the case where the space E is a product of normed vector spaces, $E = E_1 \times \cdots \times E_p$ each E_k being equipped with a norm N_k so that a natural choice is to equip E with the norm:

$$N : x = (x_1, \ldots, x_p) \mapsto N(x) := N_1(x_1) + \cdots + N_p(x_p)$$

(or any equivalent norm).

In what follows, we will assume that Ω is an open subset of E which (say) will have a product structure, i.e., $\Omega = \prod_{k=1}^{p} \Omega_k$ with Ω_k an open subset of E_k. F will be another normed vector space and a map $f : \Omega \to F$.

Definition 2.9. Let $x = (x_1, \ldots, x_p) \in \Omega$ and $k \in \{1, \ldots, p\}$. f is said to admit a partial derivative with respect to its kth variable at x if the kth partial map:

$$y \in \Omega_k \mapsto f(x_1, \ldots, x_{k-1}, y, x_{k+1}, x_p) \in F$$

is differentiable at x_k. In this case, the kth partial derivative of f with respect to its kth variable at x, denoted $\partial_k f(x)$ (or also sometimes as $\partial_{x_k} f(x)$, $\frac{\partial f}{\partial x_k}(x)$, $D_k f(x)$, $f_{x_k}(x)$, $f'_{x_k}(x)$), is the derivative of the previous kth partial map at x_k.

Remark 2.5. The notion of partial derivative is tightly related to the notion of directional derivative. Indeed, it is easy to see that if f admits a partial derivative with respect to its kth variable at x then for every $h_k \in E_k$, f is differentiable at x in the direction $(0, \ldots, 0, h_k, 0, \ldots, 0)$ and one has

$$Df(x; \underbrace{(0, \ldots, 0, h_k, 0, \ldots, 0)}_{k}) = \partial_k f(x)(h_k).$$

Note that the notion of partial differentiability we adopted above is that of Fréchet and that $\partial_k f(x) \in \mathcal{L}_c(E_k, F)$. The connection between derivative and partial derivatives is given by the following (whose elementary proof is left to the reader as an exercise).

Proposition 2.5. *Let* $x = (x_1, \ldots, x_p) \in \Omega$. *If* f *is differentiable at* x, *then, for every* $k \in \{1, \ldots, p\}$, f *admits a partial derivative with respect to its kth variable at* x *and:*

$$\partial_k f(x)(h_k) = f'(x)(\underbrace{(0, \ldots, 0, h_k, 0, \ldots, 0)}_{k}), \quad \forall h_k \in E_k, \qquad (2.17)$$

and for every $h = (h_1, \ldots, h_p) \in E$:

$$f'(x)(h) = \sum_{k=1}^{p} \partial_k f(x)(h_k). \qquad (2.18)$$

Being differentiable implies admitting partial derivatives and the converse is not true in general (see the example in Remark 2.2). However (and this is a very useful result in practice), if f admits partial derivatives and if these partial derivatives are continuous with respect to x then f is in fact of class C^1.

Theorem 2.9. *If, for every* $x \in \Omega$ *and every* $k \in \{1, \ldots, p\}$, f *admits a partial derivative with respect to its kth variable at* x *and* $x \mapsto \partial_k f(x)$ *is continuous (as a map from* Ω *to* $\mathcal{L}_c(E_k, F)$*) then* f *is of class* C^1 *on* Ω *and (2.18) holds.*

Proof. Let us prove the result for $p = 2$ (the general case can then easily be treated by induction, we leave the details to the reader). Let $x = (x_1, x_2) \in \Omega$ and $\varepsilon > 0$, it is enough to show that there is a $\delta_\varepsilon > 0$ such that whenever $(h_1, h_2) \in E_1 \times E_2$ satisfy $\|h_1\| + \|h_2\| \leq \delta_\varepsilon$, one has:

$$\|f(x_1 + h_1, x_2 + h_2) - f(x) - \partial_1 f(x)(h_1)$$
$$- \partial_2 f(x)(h_2)\| \leq \varepsilon(\|h_1\| + \|h_2\|).$$

Let us first remark that

$$f(x_1 + h_1, x_2 + h_2) - f(x) - \partial_1 f(x)(h_1) - \partial_2 f(x)(h_2)$$
$$= (f(x_1 + h_1, x_2 + h_2) - f(x_1, x_2 + h_2) - \partial_1 f(x)(h_1))$$
$$+ (f(x_1, x_2 + h_2) - f(x_1, x_2) - \partial_2 f(x)(h_2)).$$

Then observe that for $y \in \Omega$, f is differentiable at y in directions $(h_1, 0)$ and $(0, h_2)$ with:

$$Df(y; (h_1, 0)) = \partial_1 f(y)(h_1) \text{ and } Df(y; (0, h_2)) = \partial_2 f(y)(h_2).$$

It therefore follows from Theorem 2.7 that there exists $t_1 \in (0,1)$ such that

$$\|f(x_1 + h_1, x_2 + h_2) - f(x_1, x_2 + h_2) - \partial_1 f(x)(h_1)\|$$
$$\leq \|(\partial_1 f(x_1 + t_1 h_1, x_2 + h_2) - \partial_1 f(x))(h_1)\|.$$

In a similar way, we can find $t_2 \in (0,1)$ such that

$$\|f(x_1, x_2 + h_2) - f(x_1, x_2) - \partial_2 f(x)(h_2)\|$$
$$\leq \|(\partial_2 f(x_1, x_2 + t_2 h_2) - \partial_2 f(x))(h_2))\|.$$

Since $\partial_1 f$ and $\partial_2 f$ are continuous, there is a $\delta_\varepsilon > 0$ such that $B(x, \delta_\varepsilon) \subset \Omega$ and for every $y \in B(x, \delta_\varepsilon)$, one has

$$\max(\|\partial_1 f(y) - \partial_1 f(x)\|, \|\partial_2 f(y) - \partial_2 f(x)\|) \leq \varepsilon.$$

If $\|h_1\| + \|h_2\| \leq \delta_\varepsilon$, the points $(x_1 + t_1 h_1, x_2 + h_2)$ and $(x_1, x_2 + t_2 h_2)$ belong to $B(x, \delta_\varepsilon)$ so that:

$$\|f(x_1 + h_1, x_2 + h_2) - f(x) - \partial_1 f(x)(h_1)$$
$$- \partial_2 f(x)(h_2)\| \leq \varepsilon(\|h_1\| + \|h_2\|). \qquad \square$$

2.1.5. *The finite-dimensional case, the Jacobian matrix*

In this section, we focus on the case where both E and F are finite dimensional. In this case since the derivative of f at x is a linear map between E and F, it can be represented by a matrix which represents this linear map in a given choice of bases for E and F. Throughout this section, we assume that $E = \mathbb{R}^n$, $F = \mathbb{R}^p$, Ω is an open subset of $E = \mathbb{R}^n$ and $f : \Omega \mapsto F = \mathbb{R}^p$ is a map: Elements of $\mathbb{R}^n$ will be represented (as usual in matrix calculus) as their column vector of coordinates in the canonical basis:

$$x = \begin{pmatrix} x_1 \\ \vdots \\ x_n \end{pmatrix} \text{ likewise } f, \text{ is written as } f(x) = \begin{pmatrix} f_1(x) \\ \vdots \\ f_p(x) \end{pmatrix},$$

the real-valued map f_i being the ith component of f. We already know that f is differentiable at $x \in \Omega$ if and only if its each component is. In this case, each f_i admits a partial derivative with respect to its jth argument (note that $\partial_j f_i$ as a linear map from $\mathbb{R}$ to $\mathbb{R}$ can and should be identified with a real). Thanks to formula (2.18) for every $j \in \{1, p\}$ and $h \in \mathbb{R}^n$ one has

$$f_i'(x)(h) = \sum_{j=1}^{n} \partial_j f_i(x) h_j,$$

hence

$$f'(x)(h) = \begin{pmatrix} f_1'(x)(h) \\ \vdots \\ f_p'(x)(h) \end{pmatrix} = \begin{pmatrix} \sum_{j=1}^{n} \partial_j f_1(x) h_j \\ \vdots \\ \sum_{j=1}^{n} \partial_j f_i(x) h_j \end{pmatrix},$$

which can be conveniently be rewritten in matrix form as:

$$f'(x)(h) = \begin{pmatrix} \partial_1 f_1(x) & \cdots & \partial_n f_1(x) \\ \vdots & \ddots & \vdots \\ \partial_1 f_p(x) & \cdots & \partial_n f_p(x) \end{pmatrix} \begin{pmatrix} h_1 \\ \vdots \\ h_n \end{pmatrix}.$$

In other words, the linear map $f'(x) \in \mathcal{L}(\mathbb{R}^n, \mathbb{R}^p)$ is represented in the canonical bases of $\mathbb{R}^n$ and $\mathbb{R}^p$ by the $p \times n$ matrix with entries $\partial_j f_i(x)$. This matrix is called the Jacobian matrix of f at x and denoted $Jf(x)$:

$$Jf(x) := \begin{pmatrix} \partial_1 f_1(x) & \cdots & \partial_n f_1(x) \\ \vdots & \ddots & \vdots \\ \partial_1 f_p(x) & \cdots & \partial_n f_p(x) \end{pmatrix}$$

so, in matrix form, the differential of f at x is given by

$$f'(x)(h) = Jf(x)h, \text{ for every } h \in \mathbb{R}^n.$$

Note that differential calculus rules (composition, inverse, etc.) can easily be written in matrix form as well. If f is differentiable at x and g is a map defined in neighborhood (in $\mathbb{R}^p$) of $f(x)$ with values in $\mathbb{R}^k$ and if g is differentiable at $f(x)$, then we know that $g \circ f$ is differentiable at x and in terms of Jacobian matrices we have

$$J(g \circ f)(x) = Jg(f(x)) \times Jf(x).$$

For instance, in the special case where $n = p = 2$ and $k = 1$, applying the previous matrix expression, we have

$$\partial_1(g \circ f)(x_1, x_2) = \partial_1 g(f(x_1, x_2))\partial_1 f_1(x_1, x_2)$$
$$+ \partial_2 g(f(x_1, x_2))\partial_1 f_2(x_1, x_2),$$
$$\partial_2(g \circ f)(x_1, x_2) = \partial_1 g(f(x_1, x_2))\partial_2 f_1(x_1, x_2)$$
$$+ \partial_2 g(f(x_1, x_2))\partial_2 f_2(x_1, x_2).$$

If f is an homeomorphism from a neighborhood of x to a neighborhood of $f(x)$ and if $Jf(x)$ is invertible (which implies $n = p$) then f^{-1} is differentiable at $f(x)$ and

$$Jf^{-1}(f(x)) = [Jf(x)]^{-1}.$$

In the case of real-valued functions, i.e., when $F = \mathbb{R}$, $Jf(x)$ is the row-vector:

$$Jf(x) := \big(\partial_1 f(x), \ldots, \partial_n f(x)\big),$$

hence

$$f'(x)(h) = \langle \nabla f(x), h \rangle = \sum_{j=1}^{n} \partial_j f(x) h_j, \ \forall h \in \mathbb{R}^n,$$

so that the gradient of f at x has the form:

$$\nabla f(x) = \begin{pmatrix} \partial_1 f(x) \\ \vdots \\ \partial_n f(x) \end{pmatrix} = Jf(x)^T.$$

In the polar case where $E = \mathbb{R}$ and $F = \mathbb{R}^p$ (so that $\{f(t), \ t \in \Omega\}$ describes a curve in $\mathbb{R}^p$), denoting $f = (f_1, \ldots, f_p)$, if f is differentiable at t, then $f'(t) \in \mathcal{L}(\mathbb{R}, \mathbb{R}^p)$:

$$f'(t)(h) = (f_1'(t), \ldots, f_p'(t))h, \text{ for every } h \in \mathbb{R}$$

and one simply identifies $f'(t)$ to the (velocity) vector of $\mathbb{R}^p$, $(f_1'(t), \ldots, f_p'(t))$.

Exercise 2.4. *Show that $M \in \mathcal{M}_n(\mathbb{R}) \mapsto \det(M)$ is of class C^1 and compute its derivative. (Hint: show that the partial derivative of $\det$ at M with respect to the entry M_{ij} is the corresponding cofactor of M.)*

2.2. Second-order differential calculus

2.2.1. *Definitions*

Let E and F be two normed vector spaces, Ω be an open subset of E and f be a map defined on Ω with values in F. Then, the second-order differentiability is defined as follows.

Definition 2.10. Let $x \in \Omega$. f is called twice (Fréchet) differentiable at x if there exists an open set $U \subset \Omega$ with $x \in \Omega$ such that

- f is differentiable on U;
- the map $y \in U \mapsto f'(y) \in \mathcal{L}_c(E, F)$ is differentiable at x.

In this case, the second derivative of f at x, denoted $f''(x)$, is defined by

$$f''(x) := (f')'(x) \in \mathcal{L}_c(E, \mathcal{L}_c(E, F)).$$

The previous definition can be expressed as

$$f'(x + h) - f'(x) = f''(x)(h) + o(h) \text{ in } \mathcal{L}_c(E, F),$$

where again the Landau notation $o(h)$ means that

$$\frac{\|o(h)\|_{\mathcal{L}_c(E,F)}}{\|h\|} \to 0 \text{ as } h \to 0, \, h \neq 0$$

or, equivalently, as $o(h) = \|h\|\varepsilon(h)$ with $\varepsilon(h) \in \mathcal{L}_c(E, F)$ is such that

$$\|\varepsilon(h)\|_{\mathcal{L}_c(E,F)} \to 0 \text{ as } h \to 0.$$

Thanks to Theorem 1.14 from Section 1.2.2, we can identify $\mathcal{L}_c(E, \mathcal{L}_c(E, F))$ with $\mathcal{L}_{2,c}(E \times E, F)$, this enables us to view $f''(x)$ as a continuous bilinear map (also denoted $f''(x)$):

$$f''(x)(h, k) = (f''(x)(h))(k) \text{ for every } (h, k) \in E^2.$$

We will always do this identification in the sequel.

If we fix $k \in E$ and assume that f is twice differentiable at x, then

$$(f'(x + h) - f'(x))(k) = (f''(x)(h))(k) + o(h)(k) = f''(x)(h, k) + o(h).$$

The map $f'(.)(k) : y \mapsto f'(y)(k)$ is therefore differentiable at x with:

$$(f'(.)(k))'(x)(h) = f''(x)(h, k).$$

This means that $f''(x)(h,k)$ is the derivative at x of $f'(\cdot)(k)$ in the direction h. Also remark that if f is twice differentiable at x then f' is continuous at x.

Definition 2.11. The map f is said to be of class C^2 on Ω (which is denoted $f \in C^2(\Omega, F)$) if f is twice differentiable on Ω (i.e., twice differentiable at each point of Ω) and $f'' \in C(\Omega, \mathcal{L}_{2,c}(E \times E, F))$.

We can of course define higher derivatives, but in this (mainly optimization-oriented) book, we will mainly restrict ourselves to second derivatives and refer the interested reader to Cartan's book [32] for higher derivatives, higher Taylor formula, etc.

Exercise 2.5. *Let E be a Hilbert space and let*

$$f(x) := \frac{1}{2}\|x\|^2 = \frac{1}{2}x \cdot x.$$

Show that f is of class C^2 and compute its first and second derivatives. Let $A \in \mathcal{L}_c(E)$, $b \in E$ and

$$g(x) := \frac{1}{2}\|Ax - b\|^2.$$

Show that g is of class C^2 and compute its first and second derivatives.

2.2.2. Schwarz's symmetry theorem

An important property of second derivatives due to Schwarz is their symmetry. Before proving Schwarz's theorem, we will need the following preliminary result.

Proposition 2.6. *If f is twice differentiable at x, then the quantity:*

$$\frac{1}{(\|h\| + \|k\|)^2} \left(f(x+h+k) - f(x+k) - f(x+h) + f(x) - f''(x)(h,k) \right)$$

tends to 0 when (h,k) tends to $(0,0)$ with $(h,k) \in E \times E \setminus \{(0,0)\}$.

Proof. By definition, for each $\varepsilon > 0$, there exists $\delta_\varepsilon > 0$ such that $B(x, \delta_\varepsilon) \subset \Omega$ and for every $v \in B(0, \delta_\varepsilon)$

$$\|f'(x+v) - f'(x) - f''(x)(v)\| \leq \frac{\varepsilon}{2}\|v\|. \tag{2.19}$$

Let $r > 0$ be such that $B(x, r) \subset \Omega$ and f is differentiable on $B(x, r)$. Define then for every $(h, k) \in E \times E$ such that $\|h\| + \|k\| \leq r$:

$$\Phi(h, k) := f(x + h + k) - f(x + k) - f(x + h) + f(x) - f''(x)(h, k).$$

On the one hand, Φ is differentiable, $\Phi(h, 0) = 0$ so that thanks to the mean value inequality theorem, we have

$$\|\Phi(h, k)\| = \|\Phi(h, k) - \Phi(h, 0)\| \leq \sup_{u \in [0, k]} \|\partial_2 \Phi(h, u)\| \|k\|. \qquad (2.20)$$

On the other hand, we have

$$\partial_2 \Phi(h, u) = f'(x + h + u) - f'(x + u) - f''(x)(h)$$

and by linearity of $f''(x)$, $f''(x)(h) = f''(x)(h + u) - f''(x)(u)$ hence

$$\partial_2 \Phi(h, u) = (f'(x + h + u) - f'(x) - f''(x)(h + u)) - (f'(x + u) - f'(x) - f''(x)(u)).$$

If $\|h\| + \|k\| \leq \delta_\varepsilon$, then for every $u \in [0, k]$, $\|u\| \leq \|h\| + \|u\| \leq \|h\| + \|k\| \leq \delta_\varepsilon$, using (2.19) we thus get

$$\|\partial_2 \Phi(h, u)\| \leq \frac{\varepsilon}{2} (\|h + u\| + \|u\|) \leq \varepsilon(\|h\| + \|k\|).$$

Thanks to (2.20), we deduce that whenever $\|h\| + \|k\| \leq \delta_\varepsilon$, there holds

$$\|\Phi(h, k)\| \leq \varepsilon(\|h\| + \|k\|)\|k\| \leq \varepsilon(\|h\| + \|k\|)^2,$$

which gives the desired result. $\qquad \square$

The symmetry theorem of Schwarz then reads as follows.

Theorem 2.10 (Symmetry of second derivatives, Schwarz's theorem). *If f is twice differentiable at x, then $f''(x)$ is symmetric, i.e.,*

$$f''(x)(h, k) = f''(x)(k, h) \text{ for every } (h, k) \in E \times E.$$

Proof. For $(h, k) \in E^2$ small enough, define

$$S(h, k) = (f(x + h + k) - f(x + k) - f(x + h) + f(x)).$$

S is symmetric (in the sense that $S(h, k) = S(k, h)$) and by Proposition 2.6, for every $\varepsilon > 0$ there is a $\delta_\varepsilon > 0$ such that whenever $\|h\| + \|k\| \leq \delta_\varepsilon$ one has

$$\|S(h, k) - f''(x)(h, k)\| \leq \frac{\varepsilon}{2}(\|h\| + \|k\|)^2.$$

If $\|h\| + \|k\| \leq \delta_\varepsilon$, since $S(h, k) = S(k, h)$ we thus have

$$\|f''(x)(h, k) - f''(x)(k, h)\| \leq \|S(h, k) - f''(x)(h, k)\|$$

$$+ \|S(k, h) - f''(x)(k, h)\| \leq \varepsilon(\|h\| + \|k\|)^2.$$

Let $(u, v) \in E \times E$ and $t > 0$ be such that $t(\|u\| + \|v\|) \leq \delta_\varepsilon$, since $f''(x)$ is bilinear we have $f''(x)(tu, tv) = t^2 f''(x)(u, v)$, $f''(x)(tv, tu) = t^2 f''(x)(v, u)$ and then

$$\|f''(x)(tv, tu) - f''(x)(tu, tv)\| = t^2 \|f''(x)(v, u) - f''(x)(u, v)\|$$

$$\leq \varepsilon t^2 (\|u\| + \|v\|)^2$$

so that $\|f''(x)(v, u) - f''(x)(u, v)\| \leq \varepsilon(\|u\| + \|v\|)^2$, $\varepsilon > 0$ being arbitrary, we finally conclude that $f''(x)(v, u) = f''(x)(u, v)$. $\qquad\square$

2.2.3. *Second-order partial derivatives*

Consider now the case where $E = E_1 \times \cdots \times E_p$ is the product of the normed vector spaces $E_1, \ldots, E_p$. We know that if f is differentiable at $x = (x_1, \ldots, x_p) \in \Omega$, then for each $k \in \{1, \ldots, p\}$, f admits a k-th derivative at x, $\partial_k f(x) \in \mathcal{L}_c(E_k, F)$, and for $h = (h_1, \ldots, h_p) \in E$, one has

$$f'(x)(h) = \sum_{k=1}^{p} \partial_k f(x)(h_k) \text{ and } \partial_k f(x)(h_k)$$

$$= f'(x)((\underbrace{0, \ldots, 0, h_k, 0, \ldots, 0}_{k})).$$

For i and j in $\{1, \ldots, p\}$, one can consider the partial derivative of $\partial_j f$ with respect to its ith variable.

Definition 2.12. Let $x \in \Omega$, $(i, j) \in \{1, \ldots, p\}^2$, f is said to admit a second partial derivative of indices (i, j) at x if:

- there is an open neighborhood U of x in Ω on which f admits a partial derivative with respect to its jth variable;
- the map $y \in U \mapsto \partial_j f(y)$ admits a partial derivative with respect to its ith variable at x.

In this case, the second partial derivative of indices (i, j) at x, denoted $\partial^2_{ij} f(x)$, is defined by

$$\partial^2_{ij} f(x) := \partial_i(\partial_j f)(x).$$

Since $\partial_j f(x) \in \mathcal{L}_c(E_j, F)$, $\partial_{ij}^2 f(x) \in \mathcal{L}_c(E_i, \mathcal{L}_c(E_j, F))$. Using again Theorem 1.14, one identifies $\mathcal{L}_c(E_i, \mathcal{L}_c(E_j, F))$ with $\mathcal{L}_{2,c}(E_i \times E_j, F)$ and views $\partial_{ij}^2 f(x)$ as a bilinear map (again denoted $\partial_{ij}^2 f(x)$):

$$\partial_{ij}^2 f(x)(h_i, k_j) = (\partial_{ij}^2 f(x)(h_i))(k_j) \text{ for every } (h_i, k_j) \in E_i \times E_j.$$

The relations between second derivatives and partial second derivatives are summarized by the following proposition.

Proposition 2.7. *If f is twice differentiable at x, then for every $(i,j) \in \{1, \ldots, p\}^2$, f admits a second partial derivative of indices (i,j) at x and for every $(h_i, k_j) \in E_i \times E_j$ one has*

$$\partial_{ij}^2 f(x)(h_i, k_j) = f''(x)(\underbrace{(0, \ldots, 0, h_i, 0, \ldots, 0)}_{i}, \underbrace{(0, \ldots, 0, k_j, 0, \ldots, 0)}_{j}),$$

$$(2.21)$$

as well as the symmetry relations:

$$\partial_{ij}^2 f(x)(h_i, k_j) = \partial_{ji}^2 f(x)(k_j, h_i). \tag{2.22}$$

Moreover, for every $h = (h_1, \ldots, h_p)$ and $k = (k_1, \ldots, k_p)$ in E one has

$$f''(x)(h, k) = \sum_{1 \leq i,j \leq p} \partial_{ij}^2 f(x)(h_i, k_j). \tag{2.23}$$

Proof. If f is twice differentiable at x, it is differentiable on an open neighborhood U of x and therefore admits partial derivatives on U, let then $y \in U$ and $k_j \in E_j$, since:

$$\partial_j f(y)(k_j) = f'(y)(\underbrace{(0, \ldots, 0, k_j, 0, \ldots, 0)}_{j})$$

and the right-hand side is differentiable at x:

$$(\partial_j f(.)(k_j))'(x)(h) = f''(x)(h)(0, \ldots, 0, k_j, 0, \ldots, 0) \text{ for every } h \in E,$$

hence admits a partial derivative with respect to its ith variable; f therefore admits a second partial derivative of indices (i,j) at x, and taking $h = (0, \ldots, 0, h_i, 0, \ldots, 0)$ in the previous identity gives:

$$\partial_{ij}^2 f(x)(h_i, k_j) = \partial_i(\partial_j f(.)(k_j))(x)(h_i)$$
$$= f''(x)(\underbrace{(0, \ldots, 0, h_i, 0, \ldots, 0)}_{i}, \underbrace{(0, \ldots, 0, k_j, 0, \ldots, 0)}_{j}).$$

The symmetry relations then follow from (2.21) and Schwarz's theorem. Finally, (2.23) follows from (2.21) and the bilinearity of $f''(x)$. $\square$

In the case where $E = \mathbb{R}^n$ and $F = \mathbb{R}$, denoted by $(e_1, \ldots, e_n)$ the canonical basis of $\mathbb{R}^n$, one has

$$\partial^2_{ij} f(x)(h_i, k_j) = f''(x)(h_i e_i, k_j e_j)$$
$$= h_i \cdot k_j f''(x)(e_i, e_j) \text{ for every } (h_i, k_j) \in \mathbb{R}^2.$$

One therefore identifies $\partial^2_{ij} f(x)$ to the real $f''(x)(e_i, e_j)$. The symmetry theorem of Schwarz then takes the form $\partial^2_{ij} f(x) = \partial^2_{ji} f(x)$, $f''(x)$ therefore is a symmetric bilinear form which, thanks to formula (2.23), reads as follows:

$$f''(x)(h, k) = \sum_{1 \leq i,j \leq n} \partial^2_{ij} f(x) h_i \cdot k_j.$$

The Hessian matrix of f at x, denoted $D^2 f(x)$, is then by definition the matrix of the bilinear form $f''(x)$ in the canonical basis, its entries are given by the second partial derivatives $f''(x)(e_i, e_j) = \partial^2_{ij} f(x)$. $D^2 f(x)$ therefore is a symmetric matrix and its expression is

$$D^2 f(x) := \begin{pmatrix} \partial^2_{11} f(x) & \cdots & \partial^2_{1n} f(x) \\ \vdots & \ddots & \vdots \\ \partial^2_{n1} f(x) & \cdots & \partial^2_{nn} f(x) \end{pmatrix}.$$

Note also that for $(h, k) \in \mathbb{R}^n \times \mathbb{R}^n$, one has

$$f''(x)(h, k) = \langle D^2 f(x)(h), k \rangle = \sum_{1 \leq i,j \leq n} \partial^2_{ij} f(x) h_i \cdot k_j.$$

2.2.4. *Taylor formula*

Theorem 2.11. *If f is twice differentiable at x, then*

$$f(x + h) = f(x) + f'(x)(h) + \frac{1}{2} f''(x)(h, h) + o(\|h\|^2) \tag{2.24}$$

with

$$\frac{o(\|h\|^2)}{\|h\|^2} \to 0 \text{ as } h \to 0, \ h \neq 0.$$

Proof. We have to show that for every $\varepsilon > 0$, there exists a $\delta_\varepsilon > 0$ such that, if $\|h\| \leq \delta_\varepsilon$ then

$$\left\| f(x + h) - f(x) - f'(x)(h) - \frac{1}{2} f''(x)(h, h) \right\| \leq \varepsilon \|h\|^2. \tag{2.25}$$

Define

$$R(h) := f(x+h) - f(x) - f'(x)(h) - \frac{1}{2}f''(x)(h,h).$$

R is a well defined and C^1 map in a neighborhood of 0 in E, $R(0) = 0$ and

$$R'(h) = f'(x+h) - f'(x) - \frac{1}{2}(f''(x)(h,.) + f''(x)(\cdot,h))$$

and since $f''(x)$ is symmetric, this rewrites as

$$R'(h) = f'(x+h) - f'(x) - f''(x)(h).$$

Hence, there exists $\delta_\varepsilon > 0$ such that whenever $\|u\| \le \delta_\varepsilon$, one has

$$\|R'(u)\| \le \varepsilon\|u\|.$$

So if $\|h\| \le \delta_\varepsilon$, the mean value inequality gives

$$\|R(h)\| = \|R(h) - R(0)\| \le \sup_{u \in [0,h]} \|R'(u)\|\|h\|$$

$$\le \varepsilon\|h\|^2,$$

which proves (2.24). □

If f is twice differentiable on Ω (or simply in a neighborhood of x), one has a slightly more precise result.

Theorem 2.12. *Let $h \in E$ such that $[x, x+h] \subset \Omega$ and f is twice differentiable on Ω. then*

$$\left\| f(x+h) - f(x) - f'(x)(h) - \frac{1}{2}f''(x)(h,h) \right\|$$

$$\le \sup_{z \in [x,x+h]} \|f''(z) - f''(x)\|\|h\|^2. \tag{2.26}$$

Proof. Define $R(h)$ as in the proof of Theorem 2.11 we have to show that

$$\|R(h)\| \le \sup_{z \in [x,x+h]} \|f''(z) - f''(x)\|\|h\|^2. \tag{2.27}$$

Since $R(0) = 0$, the mean value inequality yields:

$$\|R(h)\| \leq \sup_{u \in [0,h]} \|R'(u)\| \|h\|. \qquad (2.28)$$

But since

$$R'(u) = f'(x+u) - f'(x) - f''(x)(u)$$

if $u \in [0, h]$, applying the mean value inequality of Corollary 2.4 to f' between x and $x + u$, we get

$$\|R'(u)\| \leq \sup_{z \in [x,x+u]} \|f''(z) - f''(x)\| \|u\|$$

$$\leq \sup_{z \in [x,x+h]} \|f''(z) - f''(x)\| \|h\|.$$

Plugging this inequality in (2.28) gives (2.27). $\qquad\qquad\qquad\square$

Let us finish this section with an exact formula, namely the second-order Taylor formula with an integral remainder, in the case where $F = \mathbb{R}^p$.

Theorem 2.13. *Let* $f : \Omega \to \mathbb{R}^p$ *and* $(x, h) \in \Omega \times E$ *such that* $[x, x + h] \subset \Omega$. *If* f *is twice differentiable at each point of* $[x, x+h]$ *and* $t \mapsto f''(x+th)$ *is continuous on* $[0, 1]$, *then we have*

$$f(x+h) = f(x) + f'(x)(h) + \int_0^1 (1-t)f''(x+th)(h,h)dt, \qquad (2.29)$$

which, denoting by $f_1, \ldots, f_p$ *the components of* f, *means*

$$f_i(x+h) = f_i(x) + f_i'(x)(h) + \int_0^1 (1-t)f_i''(x+th)$$

$$\times (h,h)dt \text{ for } i = 1, \ldots, p. \qquad (2.30)$$

Proof. Since (2.29) is a componentwise identity, it is enough to prove it when $p = 1$. For $t \in [0, 1]$, let $g(t) := f(x + th)$; g is twice differentiable with:

$$g'(t) = f(x+th)(h), \quad g''(t) = f''(x+th)(h,h)$$

an integration by parts thus gives:

$$f(x+h) - f(x) = g(1) - g(0) = \int_0^1 g'(t)dt$$

$$= \int_0^1 (1-t)g''(t)dt - [(1-t)g'(t)]_0^1$$

$$= \int_0^1 (1-t)f''(x+th)(h,h)dt + g'(0)$$

$$= f'(x)(h) + \int_0^1 (1-t)f''(x+th)(h,h)dt.$$

$\square$

2.3. The inverse function and implicit function theorems

2.3.1. *The inverse function theorem*

The inverse function theorem, stated below, expresses that, under suitable assumptions, if (the linear map) $f'(a)$ is invertible then (the nonlinear map) f is locally invertible, i.e., is invertible in a neighborhood of a.

Theorem 2.14 (Inverse function theorem). *Let E and F be two Banach spaces, Ω be an open subset of E, $f \in C^1(\Omega, F)$ and $a \in \Omega$. If $f'(a)$ is invertible there exist open neighborhoods U and V, respectively, of a and $f(a)$ such that the restriction $f : U \to V$ is a diffeomorphism of class C^1.*

Proof. Step 1: reduction

For $x \in \Omega - a$, define

$$g(x) := [f'(a)]^{-1}(f(x+a) - f(a)).$$

g is of class C^1 on the open set $\Omega - a \subset E$ with values in E. Moreover, $g(0) = 0$ and $g'(0) = \mathrm{id}$. Since g is obtained by composing f with affine continuous and invertible operations, it is enough to show that g is locally a diffeomorphism between two open neighborhoods of 0.

For $x \in \Omega - a$, define

$$\Phi(x) := x - g(x).$$

Since $\Phi'(0) = 0$, there exists $r > 0$ such that for all $x \in \overline{B}(0, r)$ one has[a]:

$$\|\Phi'(x)\| \leq 1/2. \tag{2.31}$$

---

[a]Here $\|\Phi'(x)\|$ of course denotes $\|\Phi'(x)\|_{\mathcal{L}_c(E)}$.

Step 2: g has a pseudo inverse on $\overline{B}(0, r/2)$

Let $y \in \overline{B}(0, r/2)$, for $x \in \overline{B}(0, r)$ set:

$$\Phi_y(x) := \Phi(x) + y = x - g(x) + y.$$

Observe then that

$$g(x) = y \Leftrightarrow \Phi_y(x) = x. \tag{2.32}$$

For x_1 and x_2 in $\overline{B}(0, r)$ and $y \in \overline{B}(0, r/2)$, using (2.31) and Corollary 2.2, we have

$$\|\Phi_y(x_1) - \Phi_y(x_2)\| = \|\Phi(x_1) - \Phi(x_2)\|$$

$$\leq \sup_{z \in [x_1, x_2]} \|\Phi'(z)\| \|x_1 - x_2\| \leq \frac{1}{2} \|x_1 - x_2\|.$$

In particular since $\Phi_y(0) = y$, for every $x \in \overline{B}(0, r)$, there holds

$$\|\Phi_y(x)\| \leq \|y\| + \frac{1}{2} \|x\| \leq r/2 + r/2 = r.$$

This shows that, for every $y \in \overline{B}(0, r/2)$, Φ_y is a contraction of $\overline{B}(0, r)$. It then follows from the fixed-point theorem (Theorem 1.4) that Φ_y admits a unique fixed-point in $\overline{B}(0, r)$. Together with (Theorem (2.32)), this enables us to deduce that for every $y \in \overline{B}(0, r/2)$, there is a unique $x \in \overline{B}(0, r)$ such that $g(x) = y$. For $y \in \overline{B}(0, r/2)$, we then denote (with a slight abuse of notations) by $x = g^{-1}(y)$ the unique root in $\overline{B}(0, r)$ of the equation $g(x) = y$, g^{-1} is then a well-defined map from $\overline{B}(0, r/2) \to \overline{B}(0, r)$, it is a *pseudo*-inverse of g in the sense that $g^{-1}(g(x)) = x$ for every $x \in \overline{B}(0, r)$.

Step 3: g^{-1} is 2-Lipschitz on $\overline{B}(0, r/2)$

Let $(y_1, y_2) \in \overline{B}(0, r/2)^2$, $x_1 := g^{-1}(y_1)$ and $x_2 := g^{-1}(y_2)$. Using the same notations as in Step 2, we have $x_1 := \Phi_{y_1}(x_1)$ and $x_2 := \Phi_{y_2}(x_2)$. It follows that

$$\|g^{-1}(y_1) - g^{-1}(y_2)\| = \|x_1 - x_2\| = \|\Phi_{y_1}(x_1) - \Phi_{y_2}(x_2)\|$$

$$= \|y_1 - y_2 + \Phi(x_1) - \Phi(x_2)\|$$

$$\leq \|y_1 - y_2\| + \frac{1}{2} \|x_1 - x_2\|$$

$$= \|y_1 - y_2\| + \frac{1}{2} \|g^{-1}(y_1) - g^{-1}(y_2)\|.$$

This gives the desired Lipschitz estimate

$$\|g^{-1}(y_1) - g^{-1}(y_2)\| \leq 2\|y_1 - y_2\|.$$

In particular g^{-1} is continuous on $B(0, r/2)$ and thus induces an homeomorphism between $B(0, r/2)$ and $g^{-1}(B(0, r/2))$, which is open since g is continuous.

Step 4: g^{-1} is differentiable on $B(0, r/2)$

Let $y \in B(0, r/2)$ and $x := g^{-1}(y) \in \overline{B}(0, r)$, first recall that, thanks to (2.31), there holds

$$\|\Phi'(x)\| = \|\text{id} - g'(x)\| \leq \frac{1}{2} < 1,$$

which in particular implies that $g'(x)$ is invertible (use Exercise 1.11). It then follows from Theorem 2.3 that g^{-1} is differentiable at y and more precisely $(g^{-1})'(y) = [g'(x)]^{-1} = [g'(g^{-1}(y))]^{-1}$. Finally, since $[g^{-1}]'$ is continuous,[b] g is a C^1 diffeomorphism between $g^{-1}(B(0, r/2))$ and $B(0, r/2)$, which completes the proof. $\qquad\square$

One can immediately deduce from the inverse function theorem the following result which is of global nature in the case where $f'(a)$ is invertible for every $a \in \Omega$.

Theorem 2.15. *Let E and F be two Banach spaces, Ω be an open set of E and $f \in C^1(\Omega, F)$. If $f'(a)$ is invertible for every $a \in \Omega$, then $f(\Omega)$ is open in F. If, in addition, f is injective then f is a C^1-diffeomorphism from Ω to $f(\Omega)$.*

For more on the fascinating topic of inverse functions theorem(s), we warmly recommend the book [65].

2.3.2. *The implicit function theorem*

The implicit function theorem enables one to pass from an implicit condition between variables x and y of the form $f(x, y) = c$ to an explicit relation of the form $y = g(x)$.

Theorem 2.16 (Implicit function theorem). *Let E, F and G be three Banach spaces, A and B be open subsets of E and F respectively,*

---

[b]Use the classic fact that if E is a Banach space on the (open) set of invertible elements Ψ of $\mathcal{L}_c(E)$, the map $\Psi \mapsto \Psi^{-1}$ is continuous, see Exercise 1.11.

$(a, b) \in A \times B$, $f \in C^1(A \times B, G)$ and $c := f(a, b)$. *If $\partial_2 f(a, b)$ is invertible (in $\mathcal{L}_c(F, G)$) then there exists an open neighborhood U of a in E, an open neighborhood V of b in F and $g \in C^1(U, V)$ such that*

$$\{(x, y) \in U \times V : f(x, y) = c\} = \{(x, g(x)) : x \in U\}.$$

This implies in particular that $g(a) = b$ and $f(x, g(x)) = c$ for every $x \in U$.

Proof. Let $\Phi(x, y) := (x, f(x, y))$, $\forall (x, y) \in A \times B$. Then $\Phi \in C^1(A \times B, E \times G)$. For $(h, k) \in E \times F$, $\Phi'(a, b)(h, k) = (h, \partial_1 f(a, b)h + \partial_2 f(a, b)k)$. For $(u, v) \in E \times G$, since $\partial_2 f(a, b)$ is invertible, one has

$$\Phi'(a, b)(h, k) = (u, v) \Leftrightarrow (h, k) = (u, [\partial_2 f(a, b)]^{-1}(v - \partial_1 f(a, b)u)).$$

Therefore, $\Phi'(a, b)$ is invertible, thanks to the inverse function theorem (Theorem 2.14), we deduce that there is an open neighborhood M of (a, b) in $A \times B$ and an open neighborhood N of $\Phi(a, b) = (a, c)$ such that Φ is a C^1-diffeomorphism between M and N. Without loss of generality, we may assume that $M = M_a \times M_b$, with M_a (respectively, M_b) an open neighborhood of a in A (respectively, of b in B). Let us then write $\Phi^{-1} : N \to M_a \times M_b$ in the form $\Phi^{-1}(x, z) =: (u(x, z), v(x, z)) = (x, v(x, z))$. Let us define

$$U := \{x \in M_a : (x, c) \in N \text{ and } v(x, c) \in M_b\}, \quad V := M_b.$$

By construction, U is an open neighborhood of a, V is an open neighborhood of b. For every $x \in U$, define $g(x) := v(x, c)$, then $g \in C^1(U, V)$ and if $(x, y) \in U \times V$, then $(x, c) \in N$, $g(x) = v(x, c) \in V$, hence:

$$f(x, y) = c \Leftrightarrow \Phi(x, y) = (x, c) \Leftrightarrow (x, y) = \Phi^{-1}(x, c)$$

$$\Leftrightarrow (x, y) = (x, v(x, c)) \Leftrightarrow (x, y) = (x, g(x)). \qquad \square$$

Remark 2.6. Differentiating the identity $f(x, g(x)) = c$, one gets

$$\partial_1 f(x, g(x)) + \partial_2 f(x, g(x)) \circ g'(x) = 0$$

and since in a neighborhood of a, $\partial_2 f(x, g(x))$ is invertible, this yields:

$$g'(x) = -[\partial_2 f(x, g(x))]^{-1} \circ (\partial_1 f(x, g(x))). \qquad (2.33)$$

Remark 2.7. We have used the inverse function theorem to prove the implicit function theorem. We let the reader check, as an exercise, that the two statements are in fact equivalent.

Remark 2.8. The assumption that $\partial_2 f(a, b)$ is invertible cannot be weakened. Indeed consider the trigonometric circle $S^1 := \{(x, y) \in \mathbb{R}^2 : f(x, y) := x^2 + y^2 = 1\}$, $\partial_2 f(1, 0) = 0$ and there is no neighborhood of $(1, 0)$ on which S^1 can be realized as a graph over the first variable, $x \mapsto g(x)$.

Remark 2.9. If in Theorem 2.16, f is further assumed to be of class C^2 then g is of class C^2 as well (use (2.33) and the differentiability of the inverse, see Exercise 2.11).

2.3.3. A local surjection theorem via Ekeland's variational principle

A very interesting consequence of Ekeland's variational theorem (Theorem 1.6) is the following local surjection theorem, whose elegant proof (which is neither reliant on Banach's fixed-point theorem nor on Newton's method) is taken from Ekeland [45]:

Theorem 2.17. *Let E and F be Banach spaces, $f : E \to F$ be continuous and Gateaux-differentiable, let $x_0 \in E$, $y_0 := f(x_0)$. Assume that there exist $r > 0$ and $M > 0$ such that*

- *for every $x \in \overline{B}(x_0, r)$, $D_G f(x)$ has a continuous right-inverse, i.e., there exists $L(x) \in \mathcal{L}_c(F, E)$ such that $D_G f(x) \circ L(x) = \mathrm{id}_F$;*
- *$\|L(x)\|_{\mathcal{L}_c(F,E)} < M$ for every $x \in \overline{B}(x_0, r)$.*

Then, for every $y \in \overline{B}(y_0, \frac{r}{M})$, there exists $x \in E$ such that

$$f(x) = y, \ \|x - x_0\| \le M\|y - y_0\|. \tag{2.34}$$

Proof. There is no loss of generality in assuming $x_0 = 0$ and $y_0 = f(0) = 0$. To simplify things, we will also assume that $(F, \langle \cdot, \cdot \rangle)$ is a Hilbert space (and postpone the general case, which requires some convex analysis from Chapter 3, to Exercise 3.25). Of course if $y = 0$, $x = 0$ satisfies (2.34), let us then take $y \in \overline{B}(y_0, \frac{r}{M}) \setminus \{0\}$ and define $\varphi(z) := \|f(z) - y\|$ for every $z \in E$. Since φ is continuous and non-negative, it follows from Ekeland's theorem (Theorem 1.6)[c] that there exists $x \in E$ such that $\|x\| \le M\|y\| \le r$ and

$$\varphi(z) \ge \varphi(x) - \frac{\varphi(0)}{M\|y\|}\|z - x\| = \varphi(x) - \frac{1}{M}\|z - x\|, \ \forall z \in E. \tag{2.35}$$

---

[c]A applied to φ with $\varepsilon = \varphi(0) = \|y\|$ and $k = \frac{1}{M\|y\|}$.

Given $h \in E$, $t > 0$, choosing $z = x + th$ above and using the triangle inequality, we first get

$$\|f(x) - y + tD_G f(x)(h)\| + o(t) \geq \|f(x + th) - y\|$$

$$\geq \|f(x) - y\| - t\frac{\|h\|}{M}.$$

Our aim is to prove that $f(x) = y$. If, on the contrary, $f(x) \neq y$, dividing the previous inequality by t, using (2.6) and letting $t \to 0^+$ would yield

$$\left\langle \frac{f(x) - y}{\|f(x) - y\|}, D_G f(x)(h) \right\rangle \geq -\frac{\|h\|}{M}$$

which, choosing $h = -L(x)(f(x) - y)$, would give

$$\|f(x) - y\| \leq \frac{1}{M}\|L(x)(f(x) - y)\| \leq \frac{\|L(x)\|}{M}\|f(x) - y\| < \|f(x) - y\|,$$

which is the desired contradiction.   □

2.4.   Smooth functions on $\mathbb{R}^d$, regularization, integration by parts

Throughout this section, we will work in $\mathbb{R}^d$ (and denote by $|\cdot|$ the Euclidean norm on it and $x \cdot y = \sum_{i=1}^{d} x_i y_i$ the usual scalar product of two vectors x and y in $\mathbb{R}^d$) and Ω will be a non-empty open subset of $\mathbb{R}^d$. Given $u \in C(\Omega) = C(\Omega, \mathbb{R})$, the support of u, denoted $\mathrm{supp}(u)$, is the complement of the largest open set on which $u = 0$ it is therefore also the closure of $\{x \in \Omega : u(x) \neq 0\}$, then u is said to have compact support (in Ω) if its support is a compact subset of Ω. The space of continuous functions with compact support in Ω is denoted $C_c(\Omega)$. Of course if $u \in C_c(\Omega)$ the extension of u by 0 outside Ω belongs to $C_c(\mathbb{R}^d)$. If A is a non-empty subset of $\mathbb{R}^d$, the distance to A will be denoted $\mathrm{dist}(\cdot, A)$, i.e., $\mathrm{dist}(x, A) := \inf_{a \in A} |x - a|$. We also denote by $\mathbf{1}_A$ the characteristic function of the set A:

$$\mathbf{1}_A(x) := \begin{cases} 1 & \text{if } x \in A, \\ 0 & \text{if } x \notin A. \end{cases}$$

Given U a non-empty open subset of $\mathbb{R}^d$ and K a non-empty compact subset of U, there exists an open set O such that $K \subset O \subset U$ with $\bar{O}$

compact (for instance $O := \{x \in \mathbb{R}^d, \, \text{dist}(x, K) < \delta\}$ with δ small enough), so that the function

$$\eta(x) := \frac{\text{dist}(x, \mathbb{R}^d \setminus O)}{\text{dist}(x, \mathbb{R}^d \setminus O) + \text{dist}(x, K)}$$

has the following properties (this is the classi Urysohn's lemma) :

$$\eta \in C_c(\mathbb{R}^d), \, \mathbf{1}_K \leq \eta \leq \mathbf{1}_U.$$

An exhaustive sequence of compact subsets of Ω is a sequence of compact subsets K_j included in Ω such that $K_j \subset \text{int}(K_{j+1})$ and $\Omega = \bigcup_j K_j$ (for instance $K_j := \{x \in \mathbb{R}^d \, : \, |x| \leq j, \, \text{dist}(x, \mathbb{R}^d \setminus \Omega) \geq 1/j\}$).

A generalization of Urysohn's lemma is as follows.

Proposition 2.8 (Partition of unity). *Let K be a non-empty compact subset of $\mathbb{R}^d$ and $V_1, \ldots, V_n$ be an open covering of K. Then, there exist $g_1, \ldots, g_n \in C_c(\mathbb{R}^d)^n$ such that $0 \leq g_i \leq 1$, $\text{supp}(g_i) \subset V_i$ and*

$$\sum_{i=1}^n g_i(x) = 1, \, \forall x \in K.$$

The family $g_1, \ldots, g_n$ is called a partition of unity subordinate to the open cover $V_1, \ldots, V_n$ of K.

Proof. For every $x \in K$, there exists an open neighborhood W_x of x such that $\overline{W}_x$ is compact and included in one of the V_i's. Since K is compact, it can be covered by a finite subfamily W_{x_j}, $j = 1, \ldots, p$. We then define for $i = 1, \ldots, n$

$$K_i := \bigcup_{j \, : \, \overline{W}_{x_j} \subset V_i} \overline{W}_{x_j}.$$

It follows from Urysohn's lemma that there is some $f_i \in C_c(\mathbb{R}^d)$ such that $\mathbf{1}_{K_i} \leq f_i \leq \mathbf{1}_{V_i}$. Setting

$$g_1 := f_1, \, g_2 := f_2(1 - f_1), \ldots, \, g_n := f_n(1 - f_{n-1}) \ldots (1 - f_1)$$

we then have

$$\sum_{i=1}^n g_i(x) = 1 - (1 - f_1(x)) \ldots (1 - f_n(x))$$

and if $x \in K$, $x \in K_i$ for a certain i and then $\sum_{i=1}^n g_i(x) = 1$. $\qquad\square$

2.4.1. *Test-functions, mollification*

So far, we have only dealt with $C^1(\Omega)$ or $C^2(\Omega)$ functions but we can define inductively functions of class $C^k(\Omega)$ for $k \geq 3$ as well and say that $u \in C^{k+1}(\Omega)$ if $u \in C^k(\Omega)$ and $\partial_i u \in C^k(\Omega)$ for every $i = 1, \ldots, m$. Partial derivatives of order $k + 1$ are then obtained as follows. If $u \in C^{k+1}(\Omega)$, $(\alpha_1, \ldots, \alpha_d) \in (\mathbb{N}^*)^d$ is such that $\sum_{l=1}^{d} \alpha_l = k$ and $i \in \{1, \ldots, d\}$,

$$\frac{\partial^{k+1} u}{\partial x_1^{\alpha_1} \ldots \partial x_i^{\alpha_i + 1} \ldots \partial x_d^{\alpha_d}} = \left(\frac{\partial^k \partial_i u}{\partial x_1^{\alpha_1} \ldots \partial x_i^{\alpha_i} \ldots \partial x_d^{\alpha_d}} \right).$$

Note that by Schwarz's symmetry theorem (Theorem 2.10), we also have

$$\frac{\partial^{k+1} u}{\partial x_1^{\alpha_1} \ldots \partial x_i^{\alpha_i + 1} \ldots \partial x_d^{\alpha_d}} = \partial_i \left(\frac{\partial^k u}{\partial x_1^{\alpha_1} \ldots \partial x_i^{\alpha_i} \ldots \partial x_d^{\alpha_d}} \right).$$

Likewise we define $C^\infty(\Omega) := \bigcap_{k \in \mathbb{N}^*} C^k(\Omega)$ which is the space of smooth functions, i.e., functions which admit continuous partial derivatives up to any order. If $u \in C^\infty(\Omega) \cap C_c(\Omega)$ then since u vanishes outside of a compact subset of Ω all its derivatives as well. The space of test-functions is by definition $C_c^\infty(\Omega) := C^\infty(\Omega) \cap C_c(\Omega)$, i.e., the space of smooth functions with compact support in Ω.

A multi-index α is an element $(\alpha_1, \ldots, \alpha_d)$ of $\mathbb{N}^d$, its length denoted $|\alpha|$ is by definition $|\alpha| = \sum_{i=1}^{d} \alpha_i$. For $\alpha = (\alpha_1, \ldots, \alpha_d)$ and $\beta = (\beta_1, \ldots, \beta_d)$ two multi-indices, we shall write

- $\alpha \leq \beta$ if $\alpha_i \leq \beta_i$, for $i = 1, \ldots, d$,
- $\alpha \pm \beta := (\alpha_1 \pm \beta_1, \ldots, \alpha_d \pm \beta_d)$,
- $\alpha! := \alpha_1! \ldots \alpha_d!$, and for $\beta \leq \alpha$

$$\binom{\alpha}{\beta} := \frac{\alpha!}{\beta!(\alpha - \beta)!},$$

- for $u \in C^k(\mathbb{R}^d)$, and $\alpha \neq (0, \ldots, 0)$ with $|\alpha| \leq k$, define

$$\partial^\alpha u := \frac{\partial^{|\alpha|} u}{\partial x_1^{\alpha_1} \ldots \partial x_d^{\alpha_d}},$$

and $\partial^\alpha u = u$ if $\alpha = (0, \ldots, 0)$.

We let the reader check (by induction on k) the well-known Leibniz formula: whenever $u, v \in C^k(\mathbb{R}^d)^2$ then $uv \in C^k(\mathbb{R}^d)$ and for $|\alpha| \leq k$, one has

$$\partial^\alpha(uv) = \sum_{\beta \leq \alpha} \binom{\alpha}{\beta} \partial^{\alpha-\beta} u \, \partial^\beta v. \tag{2.36}$$

The usual way to approximate functions by smoother ones is by convolution with a so-called family of mollifiers. Given[d] f and g in $L^1(\mathbb{R}^d)$ the convolution of f and g is the function denoted by $f \star g$ and defined by

$$(f \star g)(x) := \int_{\mathbb{R}^d} f(x) g(x-y) dy, \ \forall x \in \mathbb{R}^d. \tag{2.37}$$

It easily follows from Fubini and Tonelli's theorems that $f \star g$ is well-defined and that $f \star g = g \star f \in L^1(\mathbb{R}^d)$ with

$$\|f \star g\|_{L^1} \leq \|f\|_{L^1} \|g\|_{L^1}.$$

Define the spaces $L^1_{\text{loc}}(\Omega)$ and $L^1_c(\Omega)$ by

$$L^1_{\text{loc}}(\Omega) := \{ f : \Omega \to \mathbb{R} : f \mathbf{1}_K \in L^1(\Omega), \ \forall K \text{ compact included in } \Omega \} \tag{2.38}$$

(for instance $x \mapsto \frac{1}{x} + \frac{1}{(1-x)^2}$ belongs to $L^1_{\text{loc}}((0,1))$ but is not integrable) and

$$L^1_c(\Omega) := \{ f \in L^1(\Omega), \exists K \text{ compact included in } \Omega : f = f\mathbf{1}_K \}. \tag{2.39}$$

For $(f, g) \in L^1_{\text{loc}}(\mathbb{R}^d) \times L^1_c(\mathbb{R}^d)$, one can also define $f \star g \in L^1_{\text{loc}}(\mathbb{R}^d)$ by (2.37). Note that $L^1_c(\Omega)$ is dense in $L^1(\Omega)$.[e] We also recall that $C_c(\Omega)$ is dense in $L^1(\Omega)$ (see [88]). A standard way to approximate rough (L^1 say) functions by smooth ones is by convolution with a mollifying family, we leave as an exercise the following (which is classic but very important in practice).

Exercise 2.6 (Mollification). *For $x \in \mathbb{R}^d$, set*

$$\eta(x) := \begin{cases} e^{\frac{1}{|x|^2-1}} & \text{if } |x| < 1, \\ 0 & \text{otherwise.} \end{cases}$$

---

[d]Here we assume the reader is familiar with integration theory and Lebesgue spaces, see [63] or [88], for $f \in L^1(\mathbb{R}^d)$, the L^1-norm of f is by definition $\|f\|_{L^1} := \int_{\mathbb{R}^d} |f|$.

[e]Indeed, take an exhaustive sequence of compact subsets K_n of Ω, use Urysohn's lemma to find $g_n \in C_c(\Omega)$ such that $\mathbf{1}_{K_n} \leq g_n \leq \mathbf{1}_{\text{int}(K_{n+1})}$, then for $f \in L^1(\Omega)$, Lebesgue's dominated convergence theorem implies that $f_n := g_n f \in L^1_c(\Omega)$ converges to f in $L^1(\Omega)$.

Show that $\eta \in C_c^\infty(\mathbb{R}^d)$ with $\mathrm{supp}(\eta) \subset \overline{B}_1$ (we therefore have constructed a nonnegative and nontrivial test function).

1) *Show that if $f \in L^1(\mathbb{R}^d)$ and $g \in C_c(\mathbb{R}^d)$, then $f \star g \in C(\mathbb{R}^d)$.*
2) *Show that if $f \in L_c^1(\mathbb{R}^d)$ and $g \in C_c(\mathbb{R}^d)$, then $f \star g \in C_c(\mathbb{R}^d)$.*
3) *Show that if $f \in L^1(\mathbb{R}^d)$ and $g \in C_c^1(\mathbb{R}^d)$, then $f \star g \in C^1(\mathbb{R}^d)$ and $\partial_i(f \star g) = f \star (\partial_i g)$ for $i = 1, \ldots, m$.*
4) *Let $k \in \mathbb{N}^* \cup \{+\infty\}$. Show that if $f \in L^1(\mathbb{R}^d)$ and $g \in C_c^k(\mathbb{R}^d)$, then $f \star g \in C^k(\mathbb{R}^d)$ and $\partial^\alpha(f \star g) = f \star (\partial^\alpha g)$ for every multi-index α of length less than k.*
5) *Let $\rho \in C_c^1(\mathbb{R}^d)$ with $\rho \geq 0$, $\mathrm{supp}(\rho) \subset \bar{B}(0,1)$ and $\int_{\mathbb{R}^d} \rho = 1$, for each $\varepsilon > 0$ and $x \in \mathbb{R}^d$, set $\rho_\varepsilon(x) := \varepsilon^{-d}\rho(\frac{x}{\varepsilon})$ (such a family $(\rho_\varepsilon)_\varepsilon$ is called a regularizing or mollifying family). Let $f \in C(\mathbb{R}^d)$ be uniformly continuous. Show that $f_\varepsilon := \rho_\varepsilon \star f$ converges uniformly to f.*
6) *Let $f \in L^1(\mathbb{R}^d)$. Show that $f_\varepsilon := \rho_\varepsilon \star f$ converges to f in L^1.*
7) *Show that if Ω is a nonempty open subset of $\mathbb{R}^d$. $C_c^\infty(\Omega)$ is dense in $L^1(\Omega)$.*

Let us list some useful consequences of the regularization by convolution method from Exercise 2.6:

- Given U, a non-empty open subset of $\mathbb{R}^d$, and K a compact non-empty subset of U, there exists $\eta \in C_c^\infty(\mathbb{R}^d)$ such that $\mathbf{1}_K \leq \eta \leq \mathbf{1}_U$ (we first chose $\delta > 0$ small enough so that the compact set $K_\delta := \{x \in \mathbb{R}^d : \mathrm{dist}(x, K) \leq \delta\}$ and the open set $U_\delta := \{x \in \mathbb{R}^d : \mathrm{dist}(x, \mathbb{R}^d \setminus U) > \delta\}$ satisfy $K_\delta \subset U_\delta$, we then use Urysohn's lemma to find $\eta \in C_c(\mathbb{R}^d)$ such that $\mathbf{1}_{K_\delta} \leq \eta \leq \mathbf{1}_{U_\delta}$. We then take a mollifying family $(\rho_\varepsilon)_\varepsilon$ and set $\eta_\varepsilon := \rho_\varepsilon \star \eta$ so that $\eta_\varepsilon \in C_c^\infty(\mathbb{R}^d)$ and by construction for $\varepsilon \leq \delta$ one has $\mathbf{1}_K \leq \eta_\varepsilon \leq \mathbf{1}_U$).
- It therefore follows from the proof of Proposition 2.8, that if $V_1, \ldots, V_n$ is an open cover of the compact K, then there is a partition of unity subordinate to this cover consisting of $C_c^\infty(\mathbb{R}^d)$ functions.
- If K_n is an exhaustive sequence of compact subsets of the open set Ω, one can find $\eta_n \in C_c^\infty(\mathbb{R}^d)$ (smooth truncations) such that $\mathbf{1}_{K_n} \leq \eta_n \leq \mathbf{1}_{\mathrm{int}(K_{n+1})}$.

An easy consequence is the following proposition.

Proposition 2.9. *Let* Ω *be a non-empty bounded subset of* $\mathbb{R}^d$ *and* $f \in L^1_{\text{loc}}(\Omega)$ *such that*

$$\int_\Omega f\varphi = 0, \quad \forall \varphi \in C^\infty_c(\Omega).$$

Then

$$f = 0 \text{ a.e. in } \Omega.$$

Proof. Let K be a compact subset of Ω, extend $f\mathbf{1}_K$ by 0 outside K, let $(\rho_\delta)_\delta$ be a sequence of mollifiers. For $\varepsilon > 0$ and $\delta > 0$ define

$$\varphi_{\varepsilon,\delta} := \rho_\delta \star \varphi_\varepsilon \text{ with } \varphi_\varepsilon := \frac{\mathbf{1}_K f}{\varepsilon + |f|}$$

so that $\varphi_{\varepsilon,\delta} \in C^\infty_c(\Omega)$ and then

$$\int_\Omega f\varphi_{\varepsilon,\delta} = 0, \; \forall \varepsilon > 0, \; \delta > 0.$$

For fixed $\varepsilon > 0$, $\varphi_{\varepsilon,\delta}$ converges in $L^1(\mathbb{R}^d)$ to φ_ε as $\delta \to 0^+$. Hence (see [23] Theorem IV.9]) there is a sequence $\delta_n \to 0$ such that $\varphi_{\varepsilon,\delta_n}$ converges a.e. to φ_ε and since $|f\varphi_{\varepsilon,\delta_n}| \leq \mathbf{1}_{K+\sigma\overline{B}_1}|f|$ for $\sigma > 0$ an upper bound on δ_n, Lebesgue's dominated convergence first gives letting $n \to +\infty$

$$\int_K \frac{f^2}{\varepsilon + |f|} = 0$$

so letting $\varepsilon \to 0$ and using again the dominated convergence theorem, we find that $f = 0$ a.e. in K. Finally, applying the previous reasoning to an exhaustive sequence of compact subsets of Ω, we deduce that $f = 0$ a.e. in Ω. $\qquad\square$

2.4.2. *The divergence theorem and other integration by parts formulas*

Let Ω be a non-empty subset of $\mathbb{R}^d$, one says that Ω is of class C^1 if there exists $\Phi \in C^1(\mathbb{R}^d, \mathbb{R})$ such that

$$\Omega = \{x \in \mathbb{R}^d \; : \; \Phi(x) < 0\}, \; \partial\Omega = \{x \in \mathbb{R}^d \; : \; \Phi(x) = 0\} \qquad (2.40)$$

and

$$\nabla\Phi(x) \neq 0, \ \forall x \in \partial\Omega. \tag{2.41}$$

For every $x \in \partial\Omega$, the unit outward normal to Ω at x is denoted $\nu(x)$:

$$\nu(x) := \frac{\nabla\Phi(x)}{|\nabla\Phi(x)|}.$$

The surface measure σ on $\partial\Omega$ is then constructed as follows. Let $x_0 \in \partial\Omega$ and $e_d := \nu(x_0)$, identifying the (tangent) hyperplane $e_d^{\perp}$ with $\mathbb{R}^{d-1}$ let us write $x \in \mathbb{R}^d$ in the form $x = (x', x_d)$ with $x_d = x \cdot e_d$ and x' the vector of coordinates of the orthogonal projection of x in an orthonormal basis of $e_d^{\perp}$. We then have $\partial_d\Phi(x_0) = \nabla\Phi(x_0) \cdot e_d = 1$ and we deduce from the inverse function theorem that there exists an open subset U of $\mathbb{R}^d$ containing x_0, Q' an open subset of $\mathbb{R}^{d-1}$ containing x_0' and $\varepsilon > 0$ such that $x \mapsto (x', \Phi(x))$ is a C^1-diffeomorphism between U and $Q' \times (-\varepsilon, \varepsilon)$. Let us denote the inverse of this diffeomorphism in the form $(x', t) \in Q' \times (-\varepsilon, \varepsilon) \mapsto (x', g(x', t))$ and set $g_0(x') := g(x', 0)$ for every $x' \in Q'$, so that

$$\{\Phi = t\} \cap U = \{(x', g(x', t)), \ x' \in Q'\}, \ \forall t \in (-\varepsilon, \varepsilon), \tag{2.42}$$

hence

$$\Omega \cap U = \{(x', g(x', t)), \ x' \in Q', \ t \in (-\varepsilon, 0)\}, \tag{2.43}$$

and

$$\partial\Omega \cap U = \{(x', g_0(x')), \ x' \in Q'\}. \tag{2.44}$$

For $f \in C_c(U)$, we then define

$$\int_{\partial\Omega} f(x)d\sigma(x) := \int_{Q'} f(x', g_0(x'))\sqrt{1 + |\nabla g_0(x')|^2}dx'. \tag{2.45}$$

For $f \in C_c(\mathbb{R}^d)$, let us cover $\partial\Omega \cap \text{supp}(f)$ by finitely many open sets U_j on each of which $\partial\Omega$ can be represented in the form (2.42), let finally θ_j be a partition of unity of $\partial\Omega \cap \text{supp}(f)$ subordinate to the cover by the open sets U_j. Since $\theta_j f$ has support in U_j, one defines $\int_{\partial\Omega} \theta_j f d\sigma$ as in (2.45), and we define

$$\int_{\partial\Omega} f d\sigma = \sum_j \int_{\partial\Omega} \theta_j f d\sigma.$$

At this point, it is unclear whether this construction is independent of the cover U_j, of the local parameterization of $\partial\Omega$ and of the partition of unity θ_j.

The next results clarifies this point and confirms the geometric intuition that surface integrals are limits of solid integrals on small enlargements of the boundary divided by their width:

Lemma 2.2. *Let Ω be a non-empty open subset of class C^1 with Φ as in (2.40)–(2.41), let the integral with respect to the surface measure σ be defined as above and let $f \in C_c(\mathbb{R}^d)$. Then*

$$\int_{\partial\Omega} f(x)d\sigma(x) = \lim_{\delta\to 0^+} \frac{1}{\delta} \int_{\Omega_\delta} |\nabla\Phi(x)| f(x)dx,$$

where

$$\Omega_\delta := \{x \in \Omega \ : \ \Phi(x) > -\delta\}.$$

Proof. Let us assume, without loss of generality, that $\mathrm{supp}(f) \subset U$ where U is an open set for which $U \cap \{\Phi = t\}$ has the form (2.42) for $t \in (-\varepsilon, \varepsilon)$ for small enough $\varepsilon > 0$. For $\delta < \varepsilon$, we thus have $\Omega_\delta \cap U = \{(x', g(x', t)), \ x' \in Q', \ t \in (-\delta, 0)\}$. Observing that the Jacobian determinant of the change of coordinates $(x', t) \mapsto (x', g(x', t))$ is $\partial_t g(x', t)$, it follows from the change of variables formula (see [88]) that

$$\int_{\Omega_\delta} |\nabla\Phi(x)| f(x)dx = \int_{-\delta}^{0} \int_{Q'} |\nabla\Phi(x', g(x', t))|$$

$$\times |f(x', g(x', t))| |\partial_t g(x', t)| dx' dt. \qquad (2.46)$$

By construction $\Phi(x', g(x', t)) = t$, for every $(x', t) \in Q' \times (-\varepsilon, \varepsilon)$, differentiating with respect to t and x' yields

$$\partial_d \Phi(x', g(x', t)) \partial_t g(x', t) = 1, \ \nabla_{x'} \Phi(x', g(x', t))$$

$$= -\partial_d \Phi(x', g(x', t)) \nabla_{x'} g(x', t).$$

We thus deduce

$$|\nabla\Phi(x', g(x', t))|^2 = |\partial_d \Phi(x', g(x', t))|^2 + |\nabla_{x'} \Phi(x', g(x', t))|^2$$

$$= |\partial_d \Phi(x', g(x', t))|^2 \left(1 + |\nabla_{x'} g(x', t)|^2\right)$$

$$= \frac{1 + |\nabla_{x'} g(x', t)|^2}{|\partial_t g(x', t)|^2}.$$

Substituting the previous relation in (2.46), we get

$$\frac{1}{\delta} \int_{\Omega_\delta} |\nabla \Phi(x)| f(x) dx = \frac{1}{\delta} \int_{-\delta}^{0} \int_{Q'} f(x', g(x', t)) \sqrt{1 + |\nabla_{x'} g(x', t)|^2} dx' dt.$$

The desired claim easily follows by invoking Fubini and Lebesgue's dominated convergence theorems. $\square$

Arguing as in the proof above and using Lebesgue's dominated convergence theorem, for $f \in C_c(\mathbb{R}^d)$ and for $\varepsilon \geq 0$ small enough, one has

$$0 = \lim_{\delta \to 0^+} \int_{\Phi^{-1}(-\varepsilon-\delta, -\varepsilon+\delta)} f = \int_{\Phi^{-1}(-\varepsilon)} f, \qquad (2.47)$$

so that the level set $\Phi^{-1}(-\varepsilon)$ (and in particular $\partial\Omega$) is Lebesgue negligible.

Observe that if $u \in C_c^1(\Omega, \mathbb{R})$ (extended by 0 outside Ω) and for $i = 1, \ldots, d$ one has

$$\int_{\Omega} \partial_i u = \int_{\mathbb{R}^d} \partial_i u = 0.$$

Hence if $u \in C_c^1(\Omega, \mathbb{R})$ and $v \in C^1(\Omega, \mathbb{R})$ since $uv \in C_c^1(\Omega, \mathbb{R})$ and $\partial_i(uv) = \partial_i uv + u\partial_i v$ we have the integration by parts formula

$$\int_{\Omega} \partial_i uv = -\int_{\Omega} u\partial_i v. \qquad (2.48)$$

The divergence formula from Theorem 2.18 below extends the previous integration by parts formula to functions which do not necessarily vanish close to $\partial\Omega$. The divergence of a vector field $\varphi = (\varphi_1, \ldots, \varphi_d) \in C^1(\Omega, \mathbb{R}^d)$ is by definition

$$\operatorname{div}(\varphi(x)) := \sum_{i=1}^{d} \frac{\partial \varphi_i}{\partial x_i}(x) = \operatorname{tr}(J\varphi(x)), \ \forall x \in \Omega.$$

Thanks to (2.48), if $(u, \varphi) \in C_c^1(\Omega, \mathbb{R}^d) \times C^1(\mathbb{R}^d, \mathbb{R}^d)$ of if $(u, \varphi) \in C^1(\Omega, \mathbb{R}^d) \times C_c^1(\mathbb{R}^d, \mathbb{R}^d)$ then

$$\int_{\Omega} \operatorname{div}(\varphi) u = -\int_{\Omega} \varphi \cdot \nabla u.$$

In particular if $\varphi \in C_c^1(\Omega, \mathbb{R}^d)$ then

$$\int_{\Omega} \operatorname{div}(\varphi) = 0. \qquad (2.49)$$

Theorem 2.18 (Divergence formula). *Let Ω be a non-empty open subset of $\mathbb{R}^d$ of class C^1 and $\varphi \in C_c^1(\mathbb{R}^d, \mathbb{R}^d)$. Then*

$$\int_\Omega \operatorname{div}(\varphi) = \int_{\partial\Omega} \varphi(x) \cdot \nu(x) d\sigma(x).$$

Proof. Note first that if $\eta \in C^1(\mathbb{R}^d, \mathbb{R})$ and $\eta \equiv 1$ in a neighborhood of $\partial\Omega$ then using (2.49) we have

$$\int_\Omega \operatorname{div}(\varphi) = \int_\Omega \operatorname{div}(\eta\varphi) = \int_\Omega \eta \operatorname{div}(\varphi) + \int_\Omega \nabla\eta \cdot \varphi,$$

since $(\eta - 1)\varphi|_\Omega$ has compact support in Ω.

For $\delta \in (0, 1/2)$ and $t \in \mathbb{R}$ let $\rho_\delta(t) := \frac{1}{\delta}\rho(\frac{t}{\delta})$ where $\rho \in C_c^\infty(\mathbb{R})$, $\rho \geq 0$, $\int_\mathbb{R} \rho = 1$. Let $g_\delta : \mathbb{R} \to [0, 1]$ be the continuous even function which equals 1 on $[-\delta, \delta]$, 0 on $\mathbb{R} \setminus [-1, 1]$ and is affine between δ and 1. Then, for $\varepsilon > 0$, set $\eta_{\varepsilon,\delta} = f_\delta(\varepsilon^{-1}\Phi)$ with $f_\delta = \rho_{\delta/2} \star g_\delta$. Since $\eta_{\varepsilon,\delta} = 1$ on $\Phi^{-1}(-\varepsilon\delta/2, \varepsilon\delta/2)$, we first have

$$\int_\Omega \operatorname{div}(\varphi) = \int_\Omega \eta_{\varepsilon,\delta} \operatorname{div}(\varphi) + \int_\Omega \nabla\eta_{\varepsilon,\delta} \cdot \varphi. \tag{2.50}$$

Since by construction $\eta_{\varepsilon,\delta}$ vanishes when $\Phi < -(1 + \delta/2)\varepsilon$ and $\delta < 1/2$ we have (again denoting $\Omega_\theta := \Phi^{-1}((-\theta, 0))$)

$$\left| \int_\Omega \eta_{\varepsilon,\delta} \operatorname{div}(\varphi) \right| \leq \int_{\Omega_{5\varepsilon/4}} |\operatorname{div}(\varphi)|, \tag{2.51}$$

and since $\partial\Omega$ is Lebesgue negligible, this term tends to 0 as $\varepsilon \to 0^+$, uniformly in $\delta \in (0, 1/2)$. Let us now rewrite the second term in the right-hand side of (2.50) as

$$\frac{1}{\varepsilon} \int_{\Omega_{\varepsilon(1+\delta/2)}} f_\delta'(\varepsilon^{-1}\Phi)\nabla\Phi \cdot \varphi. \tag{2.52}$$

It now follows from (2.47) that for $\varepsilon > 0$ small enough, $\Phi^{-1}(-\varepsilon) \cap \operatorname{supp}(\varphi)$ has zero Lebesgue measure. For such an ε, applying again Lebesgue's dominated convergence theorem, it is easy to deduce from the construction of f_δ that the quantity in (2.52) converges as $\delta \to 0^+$ to

$$\frac{1}{\varepsilon} \int_{\Omega_\varepsilon} \nabla\Phi \cdot \varphi.$$

Using Lemma 2.2, we obtain

$$\lim_{\varepsilon \to 0^+} \frac{1}{\varepsilon} \int_{\Omega_\varepsilon} \nabla \Phi \cdot \varphi = \int_{\partial\Omega} \frac{\nabla \Phi}{|\nabla \Phi|} \cdot \varphi d\sigma = \int_{\partial\Omega} \varphi \cdot \nu d\sigma$$

which, together with (2.50) and (2.51), ends the proof of the divergence formula. □

Let us now mention some integration by parts formulas which directly follow from the divergence theorem. For u and v in $C_c^1(\mathbb{R}^d, \mathbb{R})$ and $i = 1, \ldots, d$, we have the integration by parts formula

$$\int_\Omega u \, \partial_i v = -\int_\Omega \partial_i u \, v + \int_{\partial\Omega} uv \, \nu_i d\sigma. \tag{2.53}$$

For $\varphi \in C_c^1(\mathbb{R}^d, \mathbb{R}^d)$ and $u \in C_c^1(\mathbb{R}^d, \mathbb{R})$, using the identity $\operatorname{div}(u\varphi) = u \operatorname{div}(\varphi) + \nabla u \cdot \varphi$, the divergence formula yields

$$\int_\Omega u \operatorname{div}(\varphi) = -\int_\Omega \nabla u \cdot \varphi + \int_{\partial\Omega} u \, \varphi \cdot \nu d\sigma. \tag{2.54}$$

In particular for $\varphi = \nabla v$ with $v \in C_c^2(\mathbb{R}^d, \mathbb{R})$, defining the laplacian of v by $\Delta v := \operatorname{div}(\nabla v)$ and setting $\frac{\partial v}{\partial \nu} := \nabla v \cdot \nu$ (the normal derivative of v on $\partial\Omega$) on $\partial\Omega$, we obtain the Green formulas:

$$\int_\Omega \nabla v \cdot \nabla u = -\int_\Omega \Delta v \, u + \int_{\partial\Omega} u \frac{\partial v}{\partial \nu} d\sigma, \tag{2.55}$$

and

$$\int_\Omega \Delta v \, u = \int_\Omega \Delta u \, v + \int_{\partial\Omega} \left(u \frac{\partial v}{\partial \nu} - v \frac{\partial u}{\partial \nu} \right) d\sigma. \tag{2.56}$$

2.5. Exercises

Exercise 2.7. *Let u and v both in $C^1(\mathbb{R}, \mathbb{R})$, $L \in C^1(\mathbb{R}^2, \mathbb{R})$, study the continuity and the differentiability of the map f defined by*

$$f(t) = \int_{u(t)}^{v(t)} L(t, s) ds, \quad \forall t \in \mathbb{R}.$$

Exercise 2.8 (Gronwall's lemma). *Let $x \in C([0, T], \mathbb{R}^d)$ satisfy for some constants a and b*

$$|x(t)| \leq a + b \int_0^t |x(s)| ds, \forall t \in [0, T]. \tag{2.57}$$

Then

$$|x(t)| \leq ae^{bt}, \forall t \in [0, T].$$ (2.58)

Exercise 2.9. *Find all the functions $f \in C^1(\mathbb{R}^2, \mathbb{R})$ such that*

$$\partial_x f(x, y) = \partial_y f(x, y), \; \forall (x, y) \in \mathbb{R}^2.$$

(Hint: consider the behavior of f on lines $x + y = t$.)

Exercise 2.10. *Let $E := C([0, 1], \mathbb{R})$ equipped with the uniform norm, $f \in C^1(\mathbb{R}^2, \mathbb{R})$ for $x \in E$ define the function $N_f(x)$ via:*

$$N_f(x)(t) := f(t, x(t)) \text{ for every } t \in [0, 1].$$

Study the continuity and differentiability of N_f on E.

Exercise 2.11. *Let E be a Banach space and $\mathrm{GL}_c(E)$ be the set of invertible elements of $\mathcal{L}_c(E)$, we know from Exercise 1.11 that $\mathrm{GL}_c(E)$ is open and that the map $u \mapsto u^{-1}$ is an homeomorphism of $\mathrm{GL}_c(E)$. Show that it is differentiable and compute its derivative.*

Exercise 2.12. *Show that for every $(x, y) \in \mathbb{R}^2$ one has*

$$|\sin(y) - \sin(x) - \cos(x)(y - x)| \leq |y - x|^2.$$

Exercise 2.13. *Let $a < b$ be two reals, $f \in C^2((a, b), \mathbb{R}^n)$, assume that there exist positive α and β such that*

$$\|f(t)\| \leq \alpha \text{ and } \|f''(t)\| \leq \beta \text{ for every } t \in (a, b).$$

Let $t \in (a, b)$ and $r > 0$ such that $(t - r, t + r) \subset (a, b)$. Show that

$$\|f'(t)\| \leq \frac{\alpha}{r} + \frac{\beta r}{2}.$$

Exercise 2.14. *Let $f : \mathcal{M}_n(\mathbb{R}) \to \mathcal{M}_n(\mathbb{R})$ be defined by $f(A) = A^3 + 2A^2 - A$ for every $A \in \mathcal{M}_n(\mathbb{R})$. Prove that f is of class C^2 and compute its first and second derivatives.*

Exercise 2.15. *Let $(E, (\cdot, \cdot))$ be a Hilbert Space, $f \in C^1(E, E)$ such that there exists $\alpha > 0$ such that:*

$$(f'(x)(h), h) \geq \alpha(h, h) \; \forall (x, h) \in E^2.$$

1) *Show that for every* $(a, b) \in E^2$, *one has*

$$(f(b) - f(a), b - a) \geq \alpha(b - a, b - a).$$

2) *Show that* f *is injective.*

Exercise 2.16. *Let* $f \in C^2(\mathbb{R}^d, \mathbb{R})$, f *is called quasiconvex if for every* $\alpha \in \mathbb{R}$ *the sub-level set* $\{x \in \mathbb{R}^d : f(x) \leq \alpha\}$ *is convex.*

1) *Prove that if* f *is quasiconvex then for every* $x \in \mathbb{R}^d$ *and every* h *orthogonal to* $\nabla f(x)$ *one has* $f''(x)(h, h) \geq 0$. *(Hint: argue by contradiction and find points* $x \pm th$ *such that* $f(x \pm th) < f(x)$.*)*
2) *Prove that if for every* $x \in \mathbb{R}^d$ *and every* h *orthogonal to* $\nabla f(x)$ *and such that* $|h| = 1$ *one has* $f''(x)(h, h) > 0$ *then* f *is quasi-convex (Hint: consider a maximum point of* $t \in [0, 1] \mapsto f(x_0 + t(x_1 - x_0))$.*)*
3) *Find examples that show that the condition of question 1) is not sufficient and that the condition of question 2) is not necessary.*

Exercise 2.17. *Let* $f \in C^1(\mathbb{R}^n, \mathbb{R}^n)$ *with* $f(0) = 0$ *show that for* $M > 0$ *large enough there are open neighborhoods of* 0, U *and* V *such that* $f + M \mathrm{id}$ *is a* C^1-*diffeomorphism between* U *and* V.

Exercise 2.18. *Let* Ω *be an open subset of* $\mathbb{R}^n$, $f \in C^1(\Omega, \mathbb{R}^n)$, $a \in \Omega$ *and* $L := f'(a)$. *Assume that that there are sequences* $(x_k)_k$ *and* $(y_k)_k$ *in* Ω *such that*

$$x_k \neq y_k, \ f(x_k) = f(y_k), \ \forall k \ \text{and} \ \lim_k x_k = \lim_k y_k = a.$$

Show that the rank of L *is less than* $n - 1$.

Exercise 2.19. *Let* E *and* F *be Banach spaces,* U *be an open subset of* E *containing* 0 *and* $A \in C^1(U, \mathcal{L}_c(E, F))$. *For* $x \in U$, *define* $B(x) := A(x)(x) \in F$. *Show that if* $A(0)$ *est invertible, there are open neighborhoods of* 0 *in* E *and* F, *respectively,* V *and* W, *such that (the restriction of)* B *is a* C^1-*diffeomorphism between* V *and* W.

Exercise 2.20. *For* $(x, y, z) \in \mathbb{R}^3$, *define*

$$F(x, y, z) := (y + x^4 z + y z^5 + z, x + x^2 y^2 - y z^5 - z)$$

and define $M := F^{-1}(\{(0, 0)\})$.

1) *Show that there are open intervals containing* 0, U_1, U_2 U_3, *functions* $f \in C^1(U_3, U_1)$, $g \in C^1(U_3, U_2)$ *such that*

$$M \cap (U_1 \times U_2 \times U_3) = \{(f(z), g(z), z), z \in U_3\}.$$

2) *Compute* $f(0)$, $g(0)$, $f'(0)$ *and* $g'(0)$.

Exercise 2.21. *Let* $(n, p) \in (\mathbb{N}^*)^2$, Ω *an open subset of* $\mathbb{R}^n$ *containing* 0, $f \in C^1(\Omega, \mathbb{R}^{n+p})$ *such that* $f(0) = 0$ *and* $f'(0)$ *is injective. Show that there exist* U *and* V, *open neighborhoods of* 0 *in* $\mathbb{R}^{n+p}$, Ψ *a* C^1-*diffeomorphism between* U *and* V *and* Ω_0 *an open neighborhood of* 0 *in* Ω *such that:*

$$\Psi(f(x_1, \ldots, x_n)) = (x_1, \ldots, x_n, 0, \ldots, 0), \ \forall x \in \Omega_0.$$

Exercise 2.22. *Let* E *be a Banach space, for* $u \in \mathcal{L}_c(E)$ *define*

$$\exp(u) := \sum_{k=0}^{\infty} \frac{u^k}{k!} \ (u^k = u \circ \cdots \circ u \ k \ times).$$

1) *Show that* $\exp(u)$ *is a well-defined element of* $\mathcal{L}_c(E)$.
2) *Show that for every integer* n, *the map* $u \mapsto u^n$ *is of class* C^1 *on* $\mathcal{L}_c(E)$ *and compute its derivative.*
3) *Show that the map* $u \mapsto \exp(u)$ *is of class* C^1 *on* $\mathcal{L}_c(E)$ *and compute its derivative.*
4) *Show that* $u \mapsto \exp(u)$ *is a* C^1-*diffeomorphism between an open neighborhood of* 0 *and an open neighborhood of* id.

Exercise 2.23. *Let* Ω *be an open bounded subset of* $\mathbb{R}^2$ *of class* C^1. *Let* $\varphi = (\varphi_1, \varphi_2)$ *and* $\psi = (\psi_1, \psi_2)$ *be in* $C^2(\mathbb{R}^2, \mathbb{R}^2)$ *and denote by* $D\varphi$, $D\psi$ *their respective Jacobian matrices.*

1) *Show that* $\det(D\varphi) = \operatorname{div}(\sigma_\varphi)$ *where* $\sigma_\varphi = (-\varphi_2 \partial_2 \varphi_1, \varphi_2 \partial_1 \varphi_1)$ *and deduce that*

$$\int_\Omega \det(D\varphi) = \int_{\partial\Omega} \varphi_2 \nabla \varphi_1 \cdot \nu^\perp d\sigma,$$

where $\nu(x) = (\nu_1(x), \nu_2(x))$ *is the outward unit normal to* Ω *at* $x \in \partial\Omega$ *and* $\nu^\perp = (\nu_2, -\nu_1)$.

2) *Show that if* $\varphi(x) = \psi(x)$ *for every* $x \in \partial\Omega$ *then* $\nabla(\varphi_1 - \psi_1)(x)$ *is proportional to* $\nu(x)$ *and deduce that*

$$\int_\Omega \det(D\varphi) = \int_\Omega \det(D\psi). \tag{2.59}$$

3) *Show that (2.59) holds for every* $\varphi = (\varphi_1, \varphi_2)$ *and* $\psi = (\psi_1, \psi_2)$ *in* $C^1(\mathbb{R}^2, \mathbb{R}^2)$ *such that* $\varphi(x) = \psi(x)$ *for every* $x \in \partial\Omega$.

Exercise 2.24 (Brouwer's fixed point theorem in dimension 2). *Let us denote by* B_2 *the open unit ball of* $\mathbb{R}^2$, *by* $\overline{B}_2$ *the closed unit ball of* $\mathbb{R}^2$ *and by* $S_1 := \partial B_2$ *its boundary. The aim of this exercise is to prove the following results:*

- *If* $\varphi \in C(\overline{B}_2)$, *and* $\varphi(\overline{B}_2) \subset \overline{B}_2$, *then* φ *admits a fixed point, i.e., there exists* $x \in \overline{B}_2$ *such that* $\varphi(x) = x$ *(Brouwer's theorem).*
- *There does not exist any* $\psi \in C^1(B_2) \cap C(\overline{B}_2)$ *such that* $\psi(\overline{B}_2) \subset S_1$ *and* $\psi(x) = x$ *for every* $x \in S_1$ *(no retraction theorem).*

We shall first prove the no-retraction theorem (by contradiction and using Exercise 2.23) and deduce from it (again by contradiction) Brouwer's theorem:

1) *Assume that* $\psi \in C^1(B_2) \cap C(\overline{B}_2)$ *such that* $\psi(B_2) \subset S_1$ *and* $\psi(x) = x$ *for every* $x \in S_1$. *For* $t \in [0,1]$ *define* $\psi_t := t\,\mathrm{id} + (1-t)\psi$ *and*

$$P(t) := \int_{B_2} \det(D\psi_t)$$

compute $P(1)$, *and show that* $P(t) = P(1)$ *for every* $t \in [0,1]$.

2) *Show that* $\det(D\psi(x)) = 0$ *for every* $x \in B_2$ *(Hint: argue by contradiction and use the inverse function theorem) and then prove the no retraction theorem.*

3) *Assume now that* $\varphi \in C^1(B_2) \cap C(\overline{B}_2)$ *with* $\varphi(\overline{B}_2) \subset \overline{B}_2$ *and* $\varphi(x) \neq x$ *for every* $x \in \overline{B}_2$. *For every* $x \in \overline{B}_2$, *let* $\psi(x)$ *be the point where the half line* $\{x + \lambda(\varphi(x) - x), \ \lambda \geq 0\}$ *intersects* S^1, *show that* $\psi \in C^1(B_2) \cap C(\overline{B}_2)$ *and* $\psi(x) = x$ *for every* $x \in S_1$. *Derive the desired contradiction.*

4) *Prove Brouwer's theorem (to remove the assumption that* φ *is* C^1 *in the previous question, use a regularization argument, see Exercise 2.6).*

Chapter 3

Convexity

3.1. Hahn–Banach theorems

3.1.1. *The analytic form of Hahn–Banach theorem*

Let us start with the analytic form of the Hahn–Banach theorem. We omit the proof (which, with the degree of generality below, requires the use of Zorn's lemma) and refer the reader to the book of Brezis [23].

Theorem 3.1. *Let E be a real vector space and $p : E \to \mathbb{R}$ satisfy*

$$p(\lambda x) = \lambda p(x), \ \forall x \in E, \ \forall \lambda > 0, \ p(x + y) \le p(x) + p(y), \forall (x, y) \in E^2.$$
(3.1)

Let G be a linear subspace of E and g be a linear form on G such that

$$g(x) \le p(x), \quad \forall x \in G.$$

Then there exists a linear form f on E which extends g (i.e., $f(x) = g(x)$, $\forall x \in G$) and such that

$$f(x) \le p(x), \quad \forall x \in E.$$

A function $p : E \to \mathbb{R}$ which satisfies (3.1) is called *positively homogeneous and sublinear*. Note that such functions are convex and that if p_1 and p_2 are positively homogeneous and sublinear then so is $\max(p_1, p_2)$. There are many relevant examples of such functions: norms and linear forms in the first place, but also absolute values of linear forms, max of finitely many linear forms, etc. In the following section, we will see that gauges of convex sets having 0 in their interior also belong to this class, which will be the keypoint to deduce separation theorems from Theorem 3.1.

A first useful consequence of Theorem 3.1 is the fact that continuous linear forms on subspaces admit linear continuous extensions.

Corollary 3.1. *Let $(E, \|\cdot\|)$ be a normed space, G be a linear subspace of E (equipped with the norm $\|\cdot\|$) and $g \in G^*$. Then, there exists $f \in E^*$, which extends g (i.e., $f(x) = g(x)$, $\forall x \in G$) and such that*

$$\|f\|_{E^*} = \|g\|_{G^*}.$$

Proof. Apply Theorem 3.1 to $p(x) := \|g\|_{G^*} \|x\|$. $\qquad\square$

Another consequence is that the norm of E can be represented as a supremum over the unit ball of E^* and that this supremum is in fact attained.

Corollary 3.2. *Let $(E, \|\cdot\|)$ be a normed space and $x_0 \in E$. There exists $f \in E^*$ such that $\|f\|_{E^*} = 1$ and $f(x_0) = \|x_0\|$. In particular, this implies that, for every $x \in E$,*

$$\|x\| = \max\{f(x), \ f \in E^*, \ \|f\|_{E^*} \leq 1\}.$$

Proof. Apply Corollary 3.1 to $G = \mathbb{R}x_0$, and g defined by $g(tx_0) = t\|x_0\|$ for every $t \in \mathbb{R}$. By construction, $\|g\|_{G^*} = 1$ and $g(x_0) = \|x_0\|$. The second assertion easily follows. $\qquad\square$

3.1.2. *Separation of convex sets*

The gauge (or Minkowski functional) of an open convex set containing 0 is defined as follows.

Definition 3.1. Let $(E, \|\cdot\|)$ be a normed space and C be a non-empty open convex set of E such that $0 \in C$. Then, the gauge of C is defined by

$$p_C(x) := \inf\left\{\lambda > 0 : \frac{x}{\lambda} \in C\right\}, \quad \forall x \in E. \tag{3.2}$$

In the special case where C is the open unit ball of E, it is obvious that the gauge of C is the norm $\|\cdot\|$ itself. It is a general fact that Minkowski functionals are positively homogeneous and sublinear.

Lemma 3.1. *Let C be a non-empty open convex set of the normed space E such that $0 \in C$. Its gauge p_C defined by (3.2) fulfills the following properties:*

1) *there exists $\lambda > 0$ such that $p_C(x) \leq \lambda\|x\|$ for every $x \in E$;*
2) *p_C is positively homogeneous and sublinear, i.e., satisfies (3.1);*
3) *C can be recovered from its gauge through:*

$$C = \{x \in E : p_C(x) < 1\}. \tag{3.3}$$

Proof. There is an $r > 0$ such that $B(0, r) \subset C$ which clearly implies the first assertion with $\lambda = 1/r$. The positive homogeneity of p_C is obvious. Next, we observe that if x and y are in E, and $\varepsilon > 0$, both $(p_C(x) + \varepsilon)^{-1}x$ and $(p_C(y) + \varepsilon)^{-1}y$ belong to C and

$$\frac{x + y}{p_C(x) + p_C(y) + 2\varepsilon} = \left(\frac{p_C(x) + \varepsilon}{p_C(x) + p_C(y) + 2\varepsilon} \right) \frac{x}{p_C(x) + \varepsilon}$$

$$+ \left(\frac{p_C(y) + \varepsilon}{p_C(x) + p_C(y) + 2\varepsilon} \right) \frac{y}{p_C(y) + \varepsilon}.$$

Since C is convex the right-hand side belongs to C, hence $p_C(x + y) \leq p_C(x) + p_C(y) + 2\varepsilon$, letting ε tend to 0, we get the desired sublinearity inequality:

$$p_C(x + y) \leq p_C(x) + p_C(y). \tag{3.4}$$

Let us finally prove the third assertion. The fact that $p_C(x) < 1$ implies that $x \in C$ follows from the very definition of the gauge p_C. If $x \in C$, since C is open for $\varepsilon > 0$ small enough $(1 + \varepsilon)x \in C$ hence $p_C(x) \leq \frac{1}{1+\varepsilon} < 1$. $\square$

We are now in the position to prove a first separation theorem between disjoint convex sets, one of which being open.

Theorem 3.2 (Hahn–Banach separation theorem). *Let E be a real normed vector space, let A and B be two convex and disjoint subsets of E with A open. Then there exists $f \in E^*$ such that*

$$f(a) < f(b), \quad \forall(a, b) \in A \times B.$$

Proof. Let $x_0 \in B - A := \{b - a, (a, b) \in A \times B\}$ and set $C := A - B + x_0$. Then C is convex and open (indeed $C = \bigcup_{b \in B}(A - b + x_0)$, it is therefore open, as a union of open sets) and by construction $0 \in C$. Note also that, since $A \cap B = \emptyset$, $x_0 \notin C$. Let $G = \mathbb{R}x_0$, $g \in G'$ be defined by $g(tx_0) = t$ for all $t \in \mathbb{R}$. Since $x_0 \notin C$ we have $g(x_0) = 1 \leq p_C(x_0)$ so by homogeneity, $g(tx_0) \leq p_C(tx_0)$ for every $t \geq 0$. Now, for $t \leq 0$, $g(tx_0) = t \leq 0 \leq p_C(tx_0)$; we deduce that $g \leq p_C$ on G. It therefore follows from Theorem 3.1 and Lemma 3.1 that there exists $f \in E'$ such that $f = g$ on G and $f \leq p_C$ on E. It then follows from the first assertion of Lemma 3.1 that there is a $\lambda > 0$ such that $f \leq \lambda\|x\|$, hence $f \in E^*$. Now let $(a, b) \in A \times B$, since $a - b + x_0 \in C$ we have $p_C(a - b + x_0) < 1$ and then

$$f(a - b + x_0) = f(a) - f(b) + 1 \leq p_C(a - b + x_0) < 1,$$

i.e., $f(a) < f(b)$ which is the desired conclusion. $\square$

In geometric terms, Theorem 3.2 means that A and B can be separated by a closed affine hyperplane of the form $f^{-1}(\alpha)$ in the sense that A lies in the open half-space $f^{-1}((-\infty, \alpha))$ whereas B lies in $f^{-1}([\alpha, +\infty))$. Of course, non-convex disjoint sets cannot in general be separated by an affine hyperplane (even in the plane, think of A being the open unit ball and B the crown $\{x \in \mathbb{R}^2 : 2 < |x| < 3\}$).

Remark 3.1. A singleton being a convex set, a consequence of Theorem 3.2, is that if A is an open convex set and $x_0 \notin A$, there is an $f \in E^*$ such that $f(x) < f(x_0)$ for every $x \in A$. Since this inequality is strict, f is non-trivial, i.e., $f \neq 0$. One can wonder whether, if one drops the assumption that A is open, one can find $f \in E^* \setminus \{0\}$ such that $f(x) \leq f(x_0)$ for every $x \in E$. We shall see in Proposition 3.1 that this is the case in finite dimensions. This fails however, in general, in infinite dimensions. Indeed, assume that A is a dense but strict vector subspace of E (there are plenty of examples in infinite dimensions) and $x_0 \notin A$. If $f \in E^*$ is such that $f(x) \leq f(x_0)$ for every $x \in A$, f has to vanish identically on A hence on E by density. In other words, it is impossible to separate x_0 from A.

Let us now state a strict separation result (note the differences in the assumptions compared to Theorem 3.2).

Theorem 3.3 (Hahn–Banach strict separation theorem). *Let E be a real normed vector space, let A and B be two convex and disjoint subsets of E with A compact and B closed. Then there exists $f \in E^*$ and $\varepsilon > 0$ such that*

$$f(a) \leq f(b) - \varepsilon, \quad \forall (a, b) \in A \times B.$$

Proof. First of all, we claim that there exists $\delta > 0$ such that the (open and convex) set $A_\delta := A + B(0, \delta) = \{a + \delta h, \, a \in A, h \in E, \|h\| < 1\}$ is disjoint from B. Indeed, if it was not the case, for every $n \in \mathbb{N}^*$, we could find $b_n \in B \cap A_{1/n}$, i.e., there would exist $a_n \in A$ with $\|a_n - b_n\| \leq 1/n$. But since A is compact, up to a subsequence, we can assume that a_n converges to some $a \in A$, we would then also have the convergence of b_n to a, but since B is closed this would imply $a \in A \cap B$, which is the desired contradiction. Since $A_\delta \cap B = \emptyset$ and A_δ is convex and open, it follows from Theorem 3.2 that there exists $f \in E^*$ such that

$$f(a + \delta h) = f(a) + \delta f(h) < f(b),$$

$$\forall (a, b, h) \in A \times E \text{ with } \|h\| < 1. \tag{3.5}$$

Maximizing the left-hand side with respect to h in the unit ball gives

$$f(a) \le f(b) - \delta \|f\|_{E^*},$$

which gives the desired conclusion with $\varepsilon = \delta \|f\|_{E^*}$ (which is positive since, because of (3.5), f cannot be identically 0). □

As mentioned in Remark 3.1, in finite dimensions, it is always possible to separate disjoint convex sets (no topological assumption!).

Proposition 3.1. *Let A and B be two disjoint convex sets of $\mathbb{R}^d$. Then, there exists $p \in \mathbb{R}^d$. Then, with $p \ne 0$ such that*

$$p \cdot a \le p \cdot b, \quad \forall (a,b) \in A \times B.$$

Proof. Define the (convex) set $C := A - B$, our aim is to show the existence of a non-zero vector p such that $p \cdot x \le 0$ for every $x \in C$. Let S be the unit sphere of $\mathbb{R}^d$. For $x \in C$ define

$$F_x := \{p \in S \ : \ p \cdot x \le 0\}.$$

We have to show that $\bigcap_{x \in C} F_x \ne \emptyset$. First, observe that for each $x \in C$, F_x is a non-empty closed subset of the compact S. We now claim that the family of non-empty closed subsets $\{F_x\}_{x \in C}$ of S has the finite intersection property, i.e., if $N \in \mathbb{N}^*$ and $(x_1, \ldots, x_N) \in C^N$ then $\bigcap_{i=1}^N F_{x_i} \ne \emptyset$. Indeed, define the convex hull of the points $(x_1, \ldots, x_N)$, i.e., the set

$$C_{x_1,\ldots,x_N} := \left\{ \sum_{i=1}^N \lambda_i x_i, \ \lambda_i \ge 0, \ \sum_{i=1}^N \lambda_i = 1 \right\}.$$

Then $C_{x_1,\ldots,x_N}$ is convex and compact and, by convexity of C, it is included in C so that $0 \notin C_{x_1,\ldots,x_N}$. Thanks to the strict separation theorem (Theorem 3.3) there is an $f \in \mathbb{R}^d$ and an $\varepsilon > 0$ such that $f \cdot x \le -\varepsilon$ for every $x \in C_{x_1,\ldots,x_N}$. In particular $f \ne 0$ and setting $p := f/|f|$ we have in particular $p \cdot x_i \le 0$ for $i = 1, \ldots, N$, i.e., $p \in \bigcap_{i=1}^N F_{x_i} \ne \emptyset$. Since S is compact, using Corollary 1.1, we deduce that $\bigcap_{x \in C} F_x \ne \emptyset$. □

3.2. Convex sets

3.2.1. *Basic properties*

Let E be a real normed space and C be a non-empty convex subset of E.

Lemma 3.2. *The closure $\overline{C}$ of the non-empty convex set C is convex.*

Proof. This directly follows from the fact that $\overline{C} = \bigcap_{n \in \mathbb{N}^*}(C + B(0, 1/n))$. $\qquad\square$

Regarding the interior (when nonempty) of a convex set, we first have the following lemma.

Lemma 3.3. *Let C be a non-empty convex subset of E and let $x \in \mathrm{int}(C)$, $y \in C$. Then*

$$[x, y) := \{(1 - \lambda)x + \lambda y, \ \lambda \in [0, 1)\} \subset \mathrm{int}(C).$$

Proof. Let $r > 0$ be such that $B(x, r) \subset C$, let $\lambda \in [0, 1)$ and $z := (1 - \lambda)x + \lambda y$. For $u \in B(0, 1)$ we have

$$z + r(1 - \lambda)u = (1 - \lambda)(x + ru) + \lambda y$$

so $z + r(1 - \lambda)u \in C$ by convexity. This shows that $B(z, r(1 - \lambda)) \subset C$ hence $z \in \mathrm{int}(C)$. $\qquad\square$

Proposition 3.2. *If the non-empty convex set C has non-empty interior, i.e., $\mathrm{int}(C) \neq \emptyset$, then $\mathrm{int}(C)$ is convex and dense in C.*

Proof. If both y and x are in $\mathrm{int}(C)$, it follows from Lemma 3.3 that both $[x, y)$ and $[y, x)$ are included in $\mathrm{int}(C)$, then so is their union, i.e., $[x, y] \subset \mathrm{int}(C)$. Now let $x \in C$ and $y \in \mathrm{int}(C)$, for any $n \geq 1$, by Lemma 3.3 $x_n := (1 - 1/n)x + y/n$ belongs to $\mathrm{int}(C)$. Since $\lim_n x_n = x$, the density claim follows. $\qquad\square$

3.2.2. *Linear inequalities*

Let C be a non-empty closed convex subset of the normed vector space E. The support function of C is the function $\sigma_C \colon E^* \mapsto \mathbb{R} \cup \{+\infty\}$ defined by

$$\sigma_C(f) := \sup_{x \in E} f(x), \quad \forall f \in E^*.$$

Note that σ_C is positively homogeneous and, as a supremum of linear forms, it is convex.

Theorem 3.4. *Let C be a non-empty closed convex subset of the normed vector space E and σ_C be its support function. Then C can be represented as*

$$C = \bigcap_{f \in E^*} \{x \in E \ : \ f(x) \leq \sigma_C(f)\}. \tag{3.6}$$

Proof. Let

$$A := \bigcap_{f \in E^*} \{x \in E \ : \ f(x) \leq \sigma_C(f)\}.$$

By construction, we obviously have $C \subset A$ and A is convex and closed (as an intersection of closed half-spaces). To prove that $A \subset C$, we argue by contradiction and assume that, on the contrary, there is an $x_0 \in A$ with $x_0 \notin C$. Then, thanks to Theorem 3.3 there exits an $f \in E^*$ and an $\varepsilon > 0$ such that

$$f(x) \leq f(x_0) - \varepsilon, \forall x \in C$$

so that $f(x_0) \geq \sigma_C(f) + \varepsilon$, a contradiction to the fact that $x_0 \in A$. $\qquad\square$

The representation formula (3.6) expresses the geometric fact that if C is closed and convex it is the intersection of the closed half-spaces which contain it; in particular it can be described by (infinitely many) linear inequalities, a useful consequence is the following corollary.

Corollary 3.3. *In a normed vector space, closed (for the norm) convex sets are closed and sequentially closed for the weak topology.*

Proof. The proof directly follows from (3.6) since each set $\{x \in E \ : \ f(x) \leq \sigma_C(f)\}$ is weakly closed and weakly sequentially closed. $\qquad\square$

A situation we will frequently encounter in constrained optimization is the following: given a real vector space E, an integer N and $(f, f_1, \ldots, f_N) \in E'^{N+1}$, the N linear inequalities

$$f_i(x) \geq 0, i = 1, \ldots, N,$$

imply the $(N + 1)$th one ($x \in E$ arbitrary)

$$f(x) \geq 0.$$

An obvious case where this implication holds is when there are non-negative reals $\lambda_1, \ldots, \lambda_N$ such that $f = \sum_{i=1}^{N} \lambda_i f_i$. The Minkowski–Farkas theorem below shows that this is actually a necessary condition (note that the trick in the proof is to apply a separation argument in finite dimensions which explains why we have no topological assumption here!).

Theorem 3.5 (Minkowski–Farkas). *Let $(f, f_1, \ldots, f_N) \in E'^{N+1}$. Then the following assertions are equivalent:*

1) $\{x \in E \; : \; f_i(x) \geq 0, i = 1, \ldots, N\} \subset \{x \in E \; : \; f(x) \geq 0\}$.
2) *there exists* $(\lambda_1, \ldots, \lambda_N) \in \mathbb{R}_+^N$ *such that* $f = \sum_{i=1}^{N} \lambda_i f_i$.

Proof. The second assertion straightforwardly implies the first one. To prove the converse implication, we shall proceed by induction on N, starting with the case $N = 1$, i.e., assuming that $\{x \in E \; : \; f_1(x) \geq 0\} \subset \{x \in E \; : \; f(x) \geq 0\}$. This means that $A \cap B = \emptyset$ where A and B are the subsets of $\mathbb{R}^2$:

$$A := \{(f_1(x), f(x)), x \in E\}, \; B := \mathbb{R}_+ \times (-\infty, 0).$$

Both sets A and B are convex hence it follows from Proposition 3.1 that there exists $(p_1, q) \in \mathbb{R}^2 \setminus \{(0,0)\}$ such that

$$\sup_{x \in E}\{p_1 f_1(x) + q f(x)\} \leq \inf_{\alpha \geq 0, \beta > 0} \{p_1 \alpha - q \beta\}. \tag{3.7}$$

This implies in particular that the linear form $p_1 f_1 + q f$ is bounded from above hence identically 0: $p_1 f_1 + q f = 0$. The fact that the infimum in the right-hand side of (3.7) is 0 then implies that $p_1 \geq 0$ and $q \leq 0$. If $q \neq 0$ then $f = \frac{p_1}{-q} f_1$ which yields the desired conclusion. Now if $q = 0$ then $p_1 > 0$ and then $f_1 = 0$ so that $f(x) \geq 0$ for every $x \in E$ hence $f = 0$ as well. This proves the desired implication when $N = 1$. Let us now assume that $N \geq 2$, and that for every vector space G, any collection $(g_1, \ldots, g_{N-1}, g) \in G'^N$ the implication

$$\{x \in G \; : \; g_i(x) \geq 0, i = 1, \ldots, N-1\} \subset \{x \in G \; : \; g(x) \geq 0\}$$

$$\Rightarrow \exists (\mu_1, \ldots, \mu_{N-1}) \in \mathbb{R}_+^{N-1} \; : \; g = \sum_{i=1}^{N-1} \mu_i g_i$$

holds true. Take now $(f_1, \ldots, f_N, f) \in E'^{N+1}$ such that $\{x \in E \; : \; f_i(x) \geq 0, i = 1, \ldots, N\} \subset \{x \in E \; : \; f(x) \geq 0\}$, arguing as above, by separating, thanks to Proposition 3.1, the subspace $\{(f_1(x), \ldots, f_N(x), f(x)), x \in E\}$ from the convex set $\mathbb{R}_+^N \times (-\infty, 0)$ in $\mathbb{R}^{N+1}$ gives the existence of $p = (p_1, \ldots, p_N, q) \in \mathbb{R}^{N+1} \setminus \{0\}$ such that

$$\sup_{x \in E}\left\{\sum_{i=1}^{N} p_i f_i(x) + q f(x)\right\} \leq \inf_{\alpha_i \geq 0, \alpha > 0} \left\{\sum_{i=1}^{N} p_i \alpha_i - q \alpha\right\} \tag{3.8}$$

from which we deduce as above that

$$-qf = \sum_{i=1}^{N} p_i f_i, \ p_i \geq 0, \ q \leq 0. \tag{3.9}$$

If $q < 0$, we get the desired conclusion by taking $\lambda_i = -\frac{p_i}{q}$. Let us then assume that $q = 0$ then $\sum_{i=1}^{N} p_i f_i = 0$ and one of the coefficients p_i is strictly positive. Up to a permutation of indices, let us assume that $p_N > 0$, then f_N is a linear combination of $f_1, \ldots, f_N$ with non-positive coefficients:

$$f_N = -\sum_{i=1}^{N-1} \frac{p_i}{p_N} f_i. \tag{3.10}$$

This implies that, defining $G := \ker(f_N)$, one has the equivalence:

$$f_i(x) \geq 0, i = 1, \ldots, N \Leftrightarrow f_i(x) \geq 0, i = 1, \ldots, N-1 \text{ and } x \in G.$$

Let us then denote by $(g_1, \ldots, g_{N-1}, g)$ the restrictions of $(f_1, \ldots, f_{N-1}, f)$ to G, our induction hypothesis gives the existence of $(\mu_1, \ldots, \mu_{N-1}) \in \mathbb{R}_+^{N-1}$ such that $g = \sum_{i=1}^{N-1} \mu_i g_i$, which means that the linear form $f - \sum_{i=1}^{N-1} \mu_i f_i$ vanishes on $G := \ker(f_N)$ hence is a multiple of f_N so that there is a $\mu \in \mathbb{R}$ such that

$$f = \sum_{i=1}^{N-1} \mu_i f_i + \mu f_N.$$

If $\mu \geq 0$, we are done, if $\mu < 0$, recalling (3.10), we get

$$f = \sum_{i=1}^{N-1} \lambda_i f_i, \text{ with } \lambda_i = \mu_i - \frac{\mu p_i}{p_N} \geq 0,$$

which ends the induction step. $\square$

3.2.3. *Extreme points*

In the previous section, we have described convex sets by linear inequalities. An alternative description is in terms of convex combinations (or barycenters) of a smaller set of points. For instance, in the plane, a triangle is the set of convex combinations of its vertices, the ball is the convex hull of the sphere, etc. Throughout this section, for the sake of simplicity we shall work in finite dimensions, i.e., in $\mathbb{R}^d$. Given A a non-empty subset of $\mathbb{R}^d$ the set of convex combinations of points of A is called the convex hull of A and denoted $\operatorname{co}(A)$:

$$\operatorname{co}(A) := \left\{ \sum_{i=1}^{N} \lambda_i a_i, \ N \in \mathbb{N}^*, \ \lambda_i \geq 0, \ \sum_{i=1}^{N} \lambda_i = 1, \ a_i \in A \right\}. \quad (3.11)$$

It is easy to check that $\operatorname{co}(A)$ is convex and, in fact, it is the smallest convex set containing A. A less straightforward fact is the Carathéodory theorem which allows one to express elements of $\operatorname{co}(A)$ as convex combinations of at most $d + 1$ points.

Theorem 3.6 (Carathéodory's theorem). *Let A be a non-empty subset of $\mathbb{R}^d$. Then*

$$\operatorname{co}(A) := \left\{ \sum_{i=1}^{d+1} \lambda_i a_i, \ \lambda_i \geq 0, \ \sum_{i=1}^{d+1} \lambda_i = 1, \ a_i \in A \right\}.$$

Proof. It is enough to show that whenever $N > d + 1$ and x is a convex combination of N points of A, it can in fact be expressed as a convex combination of $N - 1$ of these points. Assume then that $N > d+1$ and $x = \sum_{i=1}^{N} \lambda_i a_i$ with $a_i \in A$, $\lambda_i > 0$ and $\sum_{i=1}^{N} \lambda_i = 1$. Since we are in dimension d, the vectors $(a_1 - a_N, \ldots, a_{N-1} - a_N)$ are not linearly independent, so that there is $(\mu_1, \ldots, \mu_{N-1}) \in \mathbb{R}^{N-1} \backslash \{0\}$ such that $\sum_{k=1}^{N-1} \mu_k (a_k - a_N) = 0$, hence for every $t \in \mathbb{R}$ we have

$$x = \sum_{k=1}^{N} \alpha_k(t) a_k$$

with

$$\alpha_k(t) = \lambda_k + t\mu_k \text{ for } k = 1, \ldots, N-1 \text{ and } \alpha_N(t) = \lambda_N - t \sum_{k=1}^{N-1} \mu_k.$$

Let us also define the continuous function $\alpha(t) := \min_{k=1,\ldots,N} \alpha_k(t)$ and remark that $\alpha(0) > 0$ and that α tends to $-\infty$ as $t \to \infty$ (because either $\sum_{k=1}^{N-1} \mu_k > 0$ or $\mu_k < 0$ for some $k \leq N - 1$). Now, choosing $t_0 := \inf\{t > 0 : \alpha(t) < 0\}$, we see that $x = \sum_{k=1}^{N} \alpha_k(t_0) a_k$ is a convex combinations of $N - 1$ points among $a_1, \ldots, a_N$ (by construction, the coefficients α_k always sum to 1 and t_0 has been chosen such that they are also all non-negative and at least one of them vanishes). $\qquad \square$

A direct consequence of Carathéodory's theorem is the following corollary.

Corollary 3.4. *Let A be a non-empty subset of $\mathbb{R}^d$. if A is compact so is* $\text{co}(A)$.

Proof. Defining the compact set $\Delta := \{\lambda \in \mathbb{R}^d_+, \sum_{i=1}^{d+1} \lambda_i = 1\}$, it follows from Carathéodory's theorem that $\text{co}(A)$ is the image of the compact set $\Delta \times A^{d+1}$ by the continuous map $(\lambda, a_1, \ldots, a_{d+1}) \in \Delta \times A^{d+1} \mapsto \sum_{i=1}^{d+1} \lambda_i a_i$. It is therefore compact. $\qquad\square$

Definition 3.2. Let C be a non-empty convex subset of $\mathbb{R}^d$. A point $x \in C$ is called an extreme point of C if whenever $(y, z, \lambda) \in C \times C \times (0,1)$ are such that $x = (1 - \lambda)y + \lambda z$ then $y = z$. The set of extreme points of C is denoted by $\text{Ext}(C)$.

Note that $x \in \text{Ext}(C)$ if and only if $x \in C$ and $C \setminus \{x\}$ is convex. We now wish to prove that if C is a non-empty convex compact subset of $\mathbb{R}^d$, then $\text{Ext}(C) \neq \emptyset$. To do so, the following observation will be useful.

Lemma 3.4. *Let C be a non-empty convex compact subset of $\mathbb{R}^d$ and $p \in \mathbb{R}^d \setminus \{0\}$. Defining the (non-empty convex compact) subset of C*

$$C_p := \{x \in C \,:\, p \cdot x \leq p \cdot y, \,\forall y \in C\},$$

one has $\text{Ext}(C_p) \subset \text{Ext}(C)$.

Proof. Let $x \in \text{Ext}(C_p)$, if $(y, z, \lambda) \in C \times C \times (0,1)$ are such that $x = (1-\lambda)y + \lambda z$ so $p \cdot x = (1-\lambda)p \cdot y + \lambda p \cdot z$, then since $p \cdot y \geq p \cdot x$ and $p \cdot z \geq p \cdot x$, these two inequalities should in fact be equalities, i.e., $(y, z) \in C_p \times C_p$ but since $x \in \text{Ext}(C_p)$ this implies $y = z$ so that $x \in \text{Ext}(C)$. $\qquad\square$

Proposition 3.3. *If C is a non-empty convex compact subset of $\mathbb{R}^d$, then* $\text{Ext}(C) \neq \emptyset$.

Proof. Let $C_0 := C$ and $x_0 \in C_0$. If $x_0 \in \text{Ext}(C_0)$ we have nothing to prove. If $x_0 \notin \text{Ext}(C_0)$ we can write $x_0 = (1 - \lambda)y + \lambda z$ for some $(y, z, \lambda) \in C_0 \times C_0 \times (0,1)$ with $y \neq z$. In particular, there is a non-zero vector $p_0 \in \mathbb{R}^d$ for which $p_0 \cdot y < p_0 \cdot z$. Let us then define the (nonempty convex and compact) set

$$C_1 := \{x \in C_0 \,:\, p_0 \cdot x \leq p_0 \cdot y, \,\forall y \in C_0\}$$

and let $x_1 \in C_1$. If $x_1 \in \text{Ext}(C_1)$ then by Lemma 3.4, x_1 is also an extreme point of C_0 and we are done, if not, we can repeat the argument above and find another non-zero vector p_1 such that $x_1 \notin C_2$ where C_2 is the subset of C_1 where the linear form $x \mapsto p_1 \cdot x$ achieves its minimum on C_1. Now we observe that each time we can iterate this process, the set C_k obtained at the kth step lies in an affine subspace of dimension at most $d - k$, so after at most d steps, the set C_k has to be reduced to a singleton, $C_k = \{x_k\}$ in particular $x_k \in \text{Ext}(C_k)$ and by a repeated use of Lemma 3.4, $x_k \in \text{Ext}(C_{k-1}) \subset \text{Ext}(C)$. $\qquad\square$

Let us mention that the conclusion of Proposition 3.3 also holds true in infinite dimensions but the proof (similar in spirit to the one above) requires the use of Zorn's lemma (see [91] for details). Combining Proposition 3.3 and Lemma 3.4, we deduce the following proposition.

Proposition 3.4. *If C is a nonempty convex compact subset of $\mathbb{R}^d$ and $p \in \mathbb{R}^d \setminus \{0\}$ the linear form $x \mapsto p \cdot x$ achieves its minimum on C at least one extreme point of C.*

In other words, when minimizing a linear form on a convex compact subset C of $\mathbb{R}^d$, it is enough to minimize it on the (much smaller!) set $\text{Ext}(C)$. This is particularly interesting when C is a convex polytope (i.e., when C has finitely many extreme points). This observation is crucial for linear programming (the simplex algorithm in the first place).

We are now in position to prove the following theorem due to Minkowski.

Theorem 3.7 (Minkowski's theorem). *If C is a non-empty convex compact subset of $\mathbb{R}^d$, then $C = \text{co}(\text{Ext}(C))$.*

Proof. Since C is convex, it is obvious that $\text{co}(\text{Ext}(C)) \subset C$. To show that $C \subset \text{co}(\text{Ext}(C))$, we shall use an induction argument on the dimension of C that is the dimension k of the affine subspace spanned by C. If $k = 0$, then C is reduced to a singleton and the result is obvious. Now assume that every convex compact set of dimension at most k is the convex hull of its extreme points. Let C be a convex compact set of dimension $k + 1$. Assume that $x \in C$ and $x \notin \text{co}(\text{Ext}(C))$. By performing a translation if necessary, we may assume that $x = 0$ and that the affine space spanned by C is $\mathbb{R}^{k+1}$. We thus consider now $\mathbb{R}^{k+1}$ as the ambient space. Since $\text{co}(\text{Ext}(C))$ is convex, using Proposition 3.1, we can find $p_0 \in \mathbb{R}^{k+1} \setminus \{0\}$ such that $0 \leq p_0 \cdot y$ for every $y \in \text{Ext}(C)$. With Proposition 3.4, we thus have

$$0 \in C_0 := \left\{ y \in C \; : \; p_0 \cdot y = 0 = \min_{z \in C} p_0 \cdot z \right\}.$$

Since C_0 has dimension at most k, our induction hypothesis ensures that $0 \in$ co(Ext(C_0)) but since Ext(C_0) $\subset$ Ext(C) by Lemma 3.4 we have completed the induction step. $\qquad\square$

The previous result extends to infinite dimensions where one has to take the closed convex hull of the set of extreme points, this is the Krein–Milman theorem, see [66].

3.3. Convex functions

Let E be a real vector space, C be a non-empty convex subset of E and f : $C \to \mathbb{R} \cup \{+\infty\}$. Then f is said to be convex on C if for every $(x, y, t) \in C \times C \times [0, 1]$

$$f(tx + (1 - t)y) \leq tf(x) + (1 - t)f(y) \tag{3.12}$$

(with the convention $0(+\infty) = 0$). If the inequality in (3.12) is strict for every $t \in (0, 1)$ and every $x \neq y$ in C such that the right-hand side is finite, f is said to be strictly convex.

By a direct induction argument, f is convex on C if whenever $(x_1, \ldots, x_k) \in C^k$, $(\lambda_1, \ldots, \lambda_k) \in \mathbb{R}_+^k$ are such that $\sum_{i=1}^{k} \lambda_i = 1$ then

$$f\left(\sum_{i=1}^{k} \lambda_i x_i\right) \leq \sum_{i=1}^{k} \lambda_i f(x_i). \tag{3.13}$$

Note that both sides of either (3.12) or (3.13) may take the value $+\infty$ but they imply that the set where f is finite is a convex subset of E. Given A a subset of E, the characteristic function[a] of the set A, denoted by χ_A is defined for every $x \in E$ by:

$$\chi_A(x) := \begin{cases} 0 & \text{if } x \in A, \\ +\infty & \text{otherwise.} \end{cases}$$

We let the reader check the following as an easy exercise.

Exercise 3.1. *Let E be a normed space and A be a subset of E. show that χ_A is lsc (respectively, convex) if and only if A is closed (respectively, convex). Prove that linear combinations with non-negative coefficients of lsc (respectively, convex) functions remain lsc (respectively, convex).*

---

[a]Not to be confused with the characteristic function used in measure theory which is 1 on the set of interest and 0 outside.

If f is convex on the convex subset C of E, up to extending f by $+\infty$ outside C, we may actually assume that f is defined on the whole of E. A trivial but useful observation is then that $f \colon E \to \mathbb{R} \cup \{+\infty\}$ is convex (as a function) if and only if its epigraph, $\operatorname{epi}(f)$ (see (1.4)), is a convex subset of $\mathbb{R} \times E$. Since $\operatorname{epi}(\sup_{i \in I} f_i) = \bigcap_{i \in I} \operatorname{epi}(f_i)$ and intersections of convex sets are convex, we deduce that the pointwise supremum of a family of convex functions is itself convex (possibly identically $+\infty$), in particular since affine functions are convex, suprema of affine functions are convex.

Lemma 3.5. *Let E be a normed space and $f : E \to \mathbb{R} \cup \{+\infty\}$ be convex. If there is a point $x_0 \in E$ at which f is finite and continuous, then $\operatorname{epi}(f)$ has non-empty interior and $\operatorname{int}(\operatorname{epi}(f))$ is convex and dense in $\operatorname{epi}(f)$.*

Proof. Since f is continuous at x_0 the point $(f(x_0) + 1, x_0)$ belongs to the interior of $\operatorname{epi}(f)$, the claim then directly follows from Proposition 3.2. $\square$

A useful consequence of Corollary 3.3 is the following theorem.

Theorem 3.8. *Let E be a normed space and $f : E \mapsto \mathbb{R} \cup \{+\infty\}$ be convex. If f is strongly lsc, it is also lsc for the weak topology and weakly sequentially lsc.*

Proof. This is a direct consequence of Corollary 3.3 applied to the convex set $\operatorname{epi}(f)$ for which, being strongly and weakly closed are equivalent. $\square$

Although we shall avoid as much as possible dealing with convex functions achieving the value $-\infty$, it is sometimes unavoidable *a priori* because some interesting convex functions (infimal convolutions, value functions, etc. as we shall see later) are defined as infima. One therefore has to define convexity for such functions and find easy to check criteria which *a posteriori* rule out the value $-\infty$.

Definition 3.3. Let E be a vector space and $f : E \to \overline{\mathbb{R}} := \mathbb{R} \cup \{-\infty, +\infty\}$. Then, f is said to be convex if for every x and y in E and every $t \in [0, 1]$ one has

$$f(tx + (1 - t)y) \leq tf(x) + (1 - t)f(y)$$

provided the right-hand side is well-defined in $\overline{\mathbb{R}}$ (equivalently, except when $t \in (0, 1)$ and $\{f(x), f(y)\} = \{-\infty, +\infty\}$).

Of course, convex functions with values in $\overline{\mathbb{R}}$ as defined above may be weird and not very amenable to any analysis, for instance $f \equiv +\infty$, $f \equiv -\infty$

or a function which is everywhere $+\infty$ except on a convex set where it takes the value $-\infty$ are examples of functions that fit the previous definition. An easy way to rule out such pathological behaviors is the following lemma.

Lemma 3.6. *Let E be a normed space and $f: E \to \overline{\mathbb{R}}$ be a convex function. If $x_0 \in E$ is such that $f(x_0) \in \mathbb{R}$ and f is bounded from above in a neighborhood of x_0, then f never takes the value $-\infty$.*

Proof. Performing a translation if necessary we may assume that $x_0 = 0$ and $f \le M$ on $\overline{B}(0, r)$. Assume then that there is a $y \in E$ such that $f(y) = -\infty$ so that $\|y\| > 0$. We now write 0 as a convex combination of y and $-\frac{r}{\|y\|} y \in \overline{B}(0, r)$ and use the convexity of f together with the fact that $f\left(-\frac{r}{\|y\|} y\right) \le M$ to deduce that $f(0) = -\infty$ which is the desired contradiction. $\qquad\square$

A more subtle way to eliminate the value $-\infty$ by a separation argument as follows.

Proposition 3.5. *Let E be a normed space and $f : E \to \overline{\mathbb{R}}$ be an lsc convex function. If f takes the value $-\infty$, it nowhere takes a finite value.*

Proof. Assume on the contrary that $f(x_0) \in \mathbb{R}$ then separating strictly (in $E \times \mathbb{R}$) the point $(f(x_0) - 1, x_0)$ from the closed convex set $\text{epi}(f)$, we find, thanks to theorem 3.3, $p \in E^*$, $\alpha \in \mathbb{R}$ and $\varepsilon > 0$ such that

$$p(x_0) + \alpha(f(x_0) - 1) \le p(x) + \alpha\lambda - \varepsilon, \ \forall (\lambda, x) \in \text{epi}(f)$$

taking $x = x_0$ and $\lambda = f(x_0)$ first gives $\alpha > 0$. But taking then $x \in E$ such that $f(x) = -\infty$ we have $(\lambda, x) \in \text{epi}(f)$ for any $\lambda \in \mathbb{R}$, letting λ tend to $-\infty$ in the previous inequality therefore provides the desired contradiction. $\qquad\square$

3.3.1. *Continuity properties*

Proposition 3.6. *Let E be a normed space and $f: E \to \overline{\mathbb{R}}$ be convex. If $x_0 \in E$ is such that $f(x_0) \in \mathbb{R}$ and f is bounded from above in a neighborhood of x_0, then f is continuous at x_0.*

Proof. It follows from Lemma 3.6 that f does not take the value $-\infty$. We may assume without loss of generality that $x_0 = 0$, $f(0) = 0$ and $f \le M$ on

$\overline{B}(0,r)$. Let $u \in \overline{B}(0,1)$ and $\varepsilon \in [0,r]$. The convexity of f first gives writing εu as a convex combination of ru and 0:

$$f(\varepsilon u) = f\left(\frac{\varepsilon}{r}ru + \left(1 - \frac{\varepsilon}{r}\right)0\right).$$
$$\leq \frac{\varepsilon}{r}f(ru) \leq \frac{\varepsilon M}{r}$$

so that

$$f(y) = f(y) - f(0) \leq \frac{M\|y\|}{r}, \ \forall y \in \overline{B}(0,r). \tag{3.14}$$

For a lower bound, we write 0 as a convex combination of $-ru$ and εu and use the convexity of f to get

$$f(0) = 0 = f\left(\frac{r}{\varepsilon + r}\varepsilon u - \frac{\varepsilon}{\varepsilon + r}ru\right)$$
$$\leq \frac{1}{\varepsilon + r}(rf(\varepsilon u) + \varepsilon M)$$

so that $f(\varepsilon u) \geq -\frac{\varepsilon M}{r}$. Together with (3.14) this gives

$$|f(y) - f(0)| \leq \frac{M\|y\|}{r}, \ \forall y \in \overline{B}(0,r)$$

so that f is continuous at 0. $\qquad\square$

Note that the proof above gives a more precise local Lipschitz bound on the convex function f, namely

$$f - f(x_0) \leq M \text{ on } \overline{B}(x_0,r) \Rightarrow |f(x_0) - f(x)|$$
$$\leq \frac{M}{r}\|x - x_0\|, \ \forall x \in \overline{B}(x_0,r). \tag{3.15}$$

Corollary 3.5. *Let E be a normed space and $f : E \to \mathbb{R} \cup \{+\infty\}$ be convex. If f is bounded from above on a non-empty open set, then $\mathrm{dom}(f)$ has non-empty interior and f is continuous on $\mathrm{int}(\mathrm{dom}(f))$.*

Proof. Assume that $f \leq M$ on $\overline{B}(0,r)$ (so that 0 is an interior point of $\mathrm{dom}(f)$) and let $x_0 \in \mathrm{int}(\mathrm{dom}(f))$. To prove that f is continuous at x_0, it is enough, by virtue of Proposition 3.6, to show that f is bounded from above in a neighborhood of x_0. Since $x_0 \in \mathrm{int}(\mathrm{dom}(f))$, there is a $t > 0$ such that

$(1+t)x_0 \in \text{dom}(f)$. Now, let $u \in \overline{B}(0,1)$. The convexity of f and the fact that $(1+t)x_0 \in \text{dom}(f)$ give

$$f\left(x_0 + \frac{tr}{1+t}u\right) = f\left(\frac{1}{1+t}(1+t)x_0 + \frac{t}{1+t}ru\right)$$

$$\leq \frac{1}{1+t}f((1+t)x_0) + \frac{t}{1+t}f(ru)$$

$$\leq \frac{1}{1+t}f((1+t)x_0) + \frac{Mt}{1+t} := M_t \in \mathbb{R}$$

so $f \leq M_t$ on $\overline{B}(x_0, \frac{tr}{1+t})$, allowing us to conclude that f is continuous at x_0. $\square$

Note also that under the assumptions of Corollary 3.5, $\text{int}(\text{epi}(f))$ is non-empty, convex and dense in $\text{epi}(f)$ (Proposition 3.2). This observation will be particularly useful in the situation where x_0 is a continuity point of f so that one can separate $(f(x_0), x_0)$ from $\text{int}(\text{epi}(f))$.

In the finite-dimensional case, we have:

Proposition 3.7. *Let* $f : \mathbb{R}^d \to \mathbb{R} \cup \{+\infty\}$ *be convex. If* $\text{int}(\text{dom}(f))$ *is non-empty, then* f *is continuous on* $\text{int}(\text{dom}(f))$.

Proof. Since $\text{int}(\text{dom}(f))$ is non-empty, $\text{dom}(f)$ contains $d + 1$-affinely independent points $x_1, \ldots, x_{d+1}$ so that $\text{co}(\{x_1, \ldots, x_{d+1}\})$ has non-empty interior but if $x \in \text{co}(\{x_1, \ldots, x_{d+1}\})$ one deduces from the convexity of f that $f(x) \leq \max\{f(x_1), \ldots, f(x_{d+1})\} < +\infty$ so that f is bounded from above on a non-empty open set. It therefore follows from Corollary 3.5 that f is continuous on $\text{int}(\text{dom}(f))$. $\square$

Note that anything can happen on the boundary of $\text{dom}(f)$ (consider the case where f is 0 on $B(0,1)$, 1 on $\partial B(0,1)$ and $+\infty$ outside $\overline{B}(0,1)\ldots$), but if f is everywhere finite, then we have the following corollary.

Corollary 3.6. *If* $f : \mathbb{R}^d \to \mathbb{R}$ *is convex, then* $f \in C(\mathbb{R}^d)$ *and* f *is locally Lipschitz, i.e., Lipschitz in the neighborhood of every point (hence Lipschitz on every compact set).*

3.3.2. Differentiable characterizations

Proposition 3.8. *Let* E *be normed space,* $f : E \to \mathbb{R} \cup \{+\infty\}$ *be convex and let* $x_0 \in E$ *be a point where* f *is finite and continuous. Then, for every*

$h \in E$, f *admits a right-derivative at* x_0 *in the direction* h, $D^+f(x_0; h)$. *Moreover, for every* $y \in E$ *one has*

$$f(y) \geq f(x_0) + D^+f(x_0; y - x_0). \tag{3.16}$$

Proof. To simplify the notations, let us assume $x_0 = 0$, $f(0) = 0$ and fix $r > 0$ and M such that $f \leq M$ on $\overline{B}(0, r)$. Let $h \in \overline{B}(0, 1)$, our aim is to show that the difference quotient $\theta_h(t) := \frac{1}{t} f(th)$ admits a limit as $t \to 0^+$. Let $0 < s \leq t \leq r$, writing sh as a convex combination of 0 and th as $sh = \frac{s}{t} th + (1 - \frac{s}{t}) 0$ and using the convexity of f immediately gives $f(sh) \leq \frac{s}{t} f(th)$ so that θ_h is non-decreasing on $(0, r]$. Let $t \in (0, r]$ and writing 0 as a convex combination of th and $-rh$, we get $0 \leq \frac{r}{t+h} f(th) + \frac{t}{t+h} f(-rh)$ so that $\theta_h(t) \geq -\frac{M}{r}$. The non-decreasing function θ_h is bounded from below near 0^+ hence admits a limit as $t \mapsto 0^+$.

If $f(y) = +\infty$, (3.16) is obvious. We may therefore assume that $f(y) \in \mathbb{R}$ so that f is continuous on the segment $[x_0, y]$. The same argument as above implies that $t \in (0, 1] \mapsto \frac{1}{t} [f(x_0 + t(y - x_0)) - f(x_0)]$ is non-decreasing so that

$$f(y) - f(x_0) \geq \lim_{t \to 0^+} \frac{1}{t} [f(x_0 + t(y - x_0)) - f(x_0)] = D^+f(x_0; y - x_0).$$

$\square$

In case f is convex and Gateaux-differentiable at x_0, inequality (3.16) expresses that the graph of f lies above the tangent plane to the graph of f at x_0. In particular, if $D_G f(x_0) = 0$ then $f(y) \geq f(x_0)$ for every $y \in E$: any critical point of a convex function f is a global minimizer of f! If f is convex and Gateaux-differentiable, then summing the inequalities $f(y) - f(x) \geq D_G f(x)(y - x)$ and $f(x) - f(y) \geq D_G f(y)(x - y)$ we get

$$(D_G f(x) - D_G f(y))(x - y) \geq 0, \tag{3.17}$$

which expresses the monotonicity of the Gateaux derivative of a convex function.

Proposition 3.9. *Let* E *be a normed space,* C *be a non-empty open convex subset of* E *and* $f : C \to \mathbb{R}$ *be a continuous and Gateaux-differentiable function. Then, the following conditions are equivalent:*

1) f *is convex on* C;
2) *(above the tangent property) for every* $(x, y) \in C \times C$, $f(y) - f(x) \geq D_G f(x)(y - x)$;

3) *(monotonicity of the derivative)* for every $(x, y) \in C \times C$, $(D_G f(x) - D_G f(y))(x - y) \geq 0$.

In particular, if $x \in C$ such that $D_G f(x) = 0$, then $f(y) \geq f(x)$ for every $y \in C$, i.e., x is a minimizer of f on C.

Proof. We already know from Proposition 3.8 that convexity implies the inequality in 2) (i.e., the fact that the graph of f lies above its tangent hyperplane), and have observed above that this implies the monotonicity of the derivative. It thus remains to prove that if f has a monotone derivative it is convex, let us fix $(x, y) \in C^2$ and define $g(t) := f(x + t(y - x))$ for every $t \in [0, 1]$. We then have to show that for every $t \in (0, 1)$ one has $g(t) \leq (1 - t)g(0) + tg(1)$ which we rewrite as

$$\frac{g(t) - g(0)}{t} \leq \frac{g(1) - g(t)}{1 - t}. \tag{3.18}$$

Observe that g is differentiable on $[0, 1]$ and that $g'(t) = D_G f(x + t(y - x))(y - x)$ so that if $1 \geq s \geq t \geq 0$, $g'(s) - g'(t) = (D_G f(x + s(y - x)) - D_G f(x + t(y - x)))(y - x)$; hence, by the monotonicity of the derivative, $g'(s) \geq g'(t)$, i.e., g' is non-decreasing on $[0, 1]$. It follows from the mean value theorem that there exist t_1 and t_2 such that $0 < t_1 < t < t_2 < 1$ and

$$\frac{1}{t}(g(t) - g(0)) = g'(t_1) \quad \text{and} \quad \frac{g(1) - g(t)}{1 - t} = g'(t_2)$$

but since $g'(t_2) \geq g'(t_1)$ this shows inequality (3.18). □

The interest of strict convexity for optimization is that it gives free uniqueness results. Indeed, if C is convex and $f : C \to \mathbb{R} \cup \{+\infty\}$ is strictly convex and proper, then it admits at most one minimizer; if both x_1 and x_2 minimize f over C and $x_1 \neq x_2$, by strict convexity $f(\frac{x_1}{2} + \frac{x_2}{2}) < \frac{1}{2}f(x_1) + \frac{1}{2}f(x_2) = f(x_1)$, a contradiction with the optimality of x_1.

We let the reader check as an exercise that strict convexity is equivalent to a *strictly above the tangent* property.

Exercise 3.2. *Let E be a normed space, C be a non-empty open convex subset of E and $f : C \to \mathbb{R}$ be a continuous and Gateaux-differentiable function. Then the following conditions are equivalent:*

1) *f is strictly convex on C;*
2) *(strictly above the tangent property) for every $(x, y) \in C \times C$ with $x \neq y$ one has $f(y) - f(x) > D_G f(x)(y - x)$;*

3) *(strict monotonicity of the derivative) for every* $(x, y) \in C \times C$, *with* $x \neq y$ *one has* $(D_G f(x) - D_G f(y))(x - y) > 0$.

In the event the function of interest is twice differentiable, we then have the following proposition.

Proposition 3.10. *Let E be a normed space, C be a non-empty open convex subset of E and $f : C \to \mathbb{R}$ be a twice differentiable function. Then the following conditions are equivalent:*

1) *f is convex on C;*
2) *(non-negativity of the second derivative) for every $(x, h) \in C \times E$, $f''(x)(h, h) \geq 0$.*

Proof. If f is convex, then it has a monotone derivative so that for every $(x, h) \in C \times E$ and $t > 0$ small enough (so that $x + th$ remains in C) we have $(f'(x + th) - f'(x))) h \geq 0$, dividing by t and letting t tend to 0 thus yields $f''(x)(h, h) \geq 0$. Conversely, assume that f has a non-negative second derivative. To show that it is convex, using Proposition 3.9, it is enough to show that $(f'(y) - f'(x))(y - x) \geq 0$, but

$$(f'(y) - f'(x))(y - x) = \int_0^1 \frac{d}{dt} f'(x + t(y - x))(y - x) \mathrm{d}t$$

$$= \int_0^1 f''(x + t(y - x))(y - x)(y - x) \mathrm{d}t$$

which is non-negative by assumption.   □

Linear-quadratic functions appear in various applied settings (least squares problems, etc.). In finite dimensions, these are functions of the form

$$f(x) := \frac{1}{2} Ax \cdot x + p \cdot x, \ x \in \mathbb{R}^d,$$

where $p \in \mathbb{R}^d$ and A is a symmetric $d \times d$ matrix.[b] The Hessian of f is A so that f is convex if and only if $Ah \cdot h \geq 0$ for every $h \in \mathbb{R}^d$, i.e., A is positive semidefinite.

Exercise 3.3. *Let $f \in C^2(\mathbb{R}^d, \mathbb{R})$ be such that $D^2 f(x)$ is positive definite for every $x \in \mathbb{R}^d$. Show that f is strictly convex. Let $f(x) := |x|^4$. Show that*

---

[b]This is without loss of generality since, denoting by A^T the transpose of A one has $Ax \cdot x = \frac{1}{2}(A + A^T)x \cdot x$ and $\frac{1}{2}(A + A^T)$ is symmetric.

f is strictly convex, compute $D^2 f(0)$ and conclude that having everywhere a positive definite Hessian is not necessary for a C^2 function to be strictly convex.

3.4. The Legendre transform

The Legendre transform (also often called Legendre-Fenchel or convex conjugate) plays a key role in various fields and, in the first place, in the theory of convex duality for convex minimization, as we shall see in Chapter 6. It was originally introduced by Adrien–Marie Legendre, motivated by classic mechanics issues (derivation of the Hamiltonian formalism from the Lagrangian one), as a device to invert a gradient map. Indeed, imagine that $f: \mathbb{R}^d \to \mathbb{R}$ is a smooth function and that given $p \in \mathbb{R}^d$, we wish to solve the equation $\nabla f(x) = p$, which amounts to finding a critical point x of $y \mapsto f(y) - p \cdot y$, if f is a convex function such critical points are global minimizers, which naturally leads to

$$\inf_{y \in \mathbb{R}^d} \{f(y) - p \cdot y\} \text{ or equivalently } f^*(p) := \sup_{y \in \mathbb{R}^d} \{p \cdot y - f(y)\}. \quad (3.19)$$

The value $f^*(p)$ of the maximization problem above is by definition the value of the Legendre transform of f at p. Another slightly different motivation is in terms of affine minorants for f (again take $f: \mathbb{R}^d \to \mathbb{R}$), let us fix a vector $p \in \mathbb{R}^d$ and ask ourselves whether f admits an affine minorant with *slope p*, i.e., whether there is a finite real γ for which $f(x) \geq p \cdot x - \gamma$ for every $x \in \mathbb{R}^d$. The smallest such γ is of course given by

$$\gamma = \sup_{x \in \mathbb{R}^d} \{p \cdot x - f(x)\} = f^*(p)$$

so that f admits an affine minorant with slope p precisely when $f^*(p) < +\infty$. A key discovery is that when f is convex and smooth and ∇f is invertible, its inverse map is a gradient: the gradient of the Legendre transform! We shall prove this assertion more precisely (even for non-smooth convex functions) thanks to the Fenchel reciprocity formula (see (3.32)).

3.4.1. *Basic properties*

Definition 3.4. Let E be a normed space and $f : E \to \overline{\mathbb{R}}$. The Legendre transform of f is the function $f^*: E^* \to \overline{\mathbb{R}}$ defined by

$$f^*(p) := \sup_{x \in E} \{p(x) - f(x)\}, \ \forall p \in E^*. \quad (3.20)$$

If f attains the value $-\infty$ then $f^* \equiv +\infty$ and if $f \equiv +\infty$ then $f^* \equiv -\infty$, since these two extreme cases are pathological, we will mainly restrict ourselves to consider f^* for f in some restricted classes:

Definition 3.5. Let E be a normed space and $f : E \mapsto \mathbb{R} \cup \{+\infty\}$. Then f is called proper if $\mathrm{dom}(f) \neq \emptyset$. We will denote by $\Gamma_0(E)$ the set of proper, convex and lsc functions $f : E \mapsto \mathbb{R} \cup \{+\infty\}$.

Here are some obvious properties of the Legendre transform:

- if $f : E \mapsto \mathbb{R} \cup \{+\infty\}$ is proper, then $f^* : E^* \to \mathbb{R} \cup \{+\infty\}$, f^* is convex and lsc (as a supremum of affine continuous functions[c]);
- $f \leq g$ implies $f^* \geq g^*$, if λ is a constant $(f - \lambda)^* = f^* - \lambda$;
- $f^*(0) = \sup_E -f = -\inf_E f$;
- if $f : E \mapsto \mathbb{R} \cup \{+\infty\}$ is proper, then, for every $(x, p) \in E \times E^*$, one has Young–Fenchel inequality:

$$f(x) + f^*(p) \geq p(x). \tag{3.21}$$

Example 3.1. Let C be a non-empty subset of E. Then

$$\chi_C^*(p) = \sup_{x \in C} p(x) = \sigma_C(p).$$

The Legendre transform of the characteristic function of C is the support function of C. In particular, the Legendre transform of the unit ball of E is the dual norm $\|\cdot\|_{E^*}$. Now, consider the polar case of the Legendre transform of the norm

$$(\|\cdot\|)^*(p) = \sup_{x \in E} \{p(x) - \|x\|\}.$$

If $\|p\|_{E^*} \leq 1$ then $p(x) \leq \|x\|$ so that $\|\cdot\|^*(p) = 0$. Now if $\|p\|_{E^*} > 1$ by the very definition of $\|p\|_{E^*}$ there exists $x_0 \in \overline{B}_E$ such that $p(x_0) > 1 \geq \|x_0\|$ taking $x = tx_0$ and letting $t \to \infty$ we find that $(\|.\|)^*(p) = +\infty$. This shows that

$$(\|.\|)^*(p) = \chi_{\overline{B}_{E^*}}(p) = \begin{cases} 0 & \text{if } \|p\|_{E^*} \leq 1, \\ +\infty & \text{otherwise.} \end{cases}$$

---

[c]For the same reason, f^* is lsc for the weak-$*$ topology of E^*, i.e., the coarsest topology making $f \mapsto f(x)$ continuous for every $x \in E$ and $\liminf f^*(p_n) \geq f^*(p)$ whenever $p_n \overset{*}{\rightharpoonup} p$ (recall that $p_n \overset{*}{\rightharpoonup} p$ if $p_n(x) \to p(x)$ for every $x \in E$).

Example 3.2. Let E be a Hilbert space (identified with its dual so that the Legendre transform of a function defined on E can also be seen as a function defined on E) and let

$$f(x) := \frac{1}{2}\|x\|^2 = \frac{1}{2}x \cdot x.$$

Firstly, then,

$$f^*(y) = \sup_{x \in E}\{x \cdot y - f(x)\} \geq \|y\|^2 - f(y) = \frac{1}{2}\|y\|^2.$$

Secondly

$$x \cdot y - \frac{1}{2}\|x\|^2 = \frac{1}{2}\|y\|^2 - \frac{1}{2}\|x - y\|^2 \leq \frac{1}{2}\|y\|^2,$$

taking the supremum in x for fixed y thus gives $f^*(y) \leq \frac{1}{2}\|y\|^2$. This shows that

$$\left(\frac{1}{2}\|\cdot\|^2\right)^* = \frac{1}{2}\|\cdot\|^2.$$

Exercise 3.4. *Let Ω be an open subset of $\mathbb{R}^d$, equipped with the Lebesgue measure. Let $p \in (1, +\infty)$, for $u \in L^p(\Omega)$ define*

$$f_p(u) := \frac{1}{p}\int_\Omega |u|^p.$$

Identifying $(L^p(\Omega))^*$ with $L^q(\Omega)$ (with q the conjugate exponent of p, i.e., $q = \frac{p}{p-1}$) show that

$$f_p^*(v) = \frac{1}{q}\int_\Omega |v|^q, \ \forall v \in L^q(\Omega).$$

Exercise 3.5. *Let $f : E \mapsto \overline{\mathbb{R}}$. Show that*

1) *for $\lambda > 0$, one has*

$$(\lambda f)^*(p) = \lambda f^*\left(\frac{p}{\lambda}\right), \ \forall p \in E^*;$$

2) *for $x_0 \in E$ (defining the corresponding translation $f(x_0 + \cdot)$ of f by $f(x_0 + \cdot)(x) := f(x_0 + x)$), one has*

$$f(x_0 + \cdot)^*(p) = f^*(p) - p(x_0), \quad \forall p \in E^*;$$

3) *for* $(p_0, \alpha_0) \in E^* \times \mathbb{R}$, *let* γ_0 *be the affine continuous function* $x \mapsto \gamma_0(x) := p_0(x) + \alpha_0$, *one has*

$$(f - \gamma)^*(p) = f^*(p + p_0) + \alpha_0.$$

We already know that if f is proper then f^* is convex and lsc. Regarding the properness of f^*, we have the following proposition.

Proposition 3.11. *Let* $f \in \Gamma_0(E)$. *Then* $\mathrm{dom}(f^*)$ *is non-empty, thus* f *admits a continuous affine minorant.*

Proof. Let $x_0 \in \mathrm{dom}(f)$ and $\lambda_0 \in (-\infty, f(x_0))$, since $(\lambda_0, x_0) \notin \mathrm{epi}(f)$ and $\mathrm{epi}(f)$ is closed and convex, we can strictly separate (λ_0, x_0) from $\mathrm{epi}(f)$: there exist $p \in E^*$, $\alpha \in \mathbb{R}$ and $\varepsilon > 0$ such that

$$p(x_0) + \alpha \lambda_0 \leq p(x) + \alpha \lambda - \varepsilon, \quad \forall (\lambda, x) \in \mathrm{epi}(f). \tag{3.22}$$

Taking $x = x_0$ and $\lambda = f(x_0)$ in the right-hand side immediately gives that $\alpha > 0$. Defining then $q := -\frac{p}{\alpha}$, equation (3.22) implies

$$q(x) - f(x) \leq q(x_0) - \lambda_0 - \varepsilon, \quad \forall x \in E,$$

so that $f^*(q) < +\infty$. $\qquad \square$

The Legendre transform is a highly nonlinear operation but it behaves nicely, with respect to a certain operation on functions, called infimal convolution. Given f_1 and f_2: $E \to \mathbb{R} \cup \{+\infty\}$, the infimal convolution of f_1 and f_2 denoted $f_1 \square f_2$ is the function defined by

$$f_1 \square f_2(x) := \inf_{y \in E} \{f_1(x - y) + f_2(y)\}, \quad \forall x \in E. \tag{3.23}$$

The infimal convolution $f_1 \square f_2$ is a function from E to $\overline{\mathbb{R}}$ and it can a *priori* take the value $-\infty$ (for instance, if $f_i(x_i) = p_i(y)$ with $p_i \in E^*$ and $p_1 \neq p_2$, then $f_1 \square f_2 \equiv -\infty$). One way to rule this out is to assume that

$$\mathrm{dom}(f_1^*) \cap \mathrm{dom}(f_2^*) \neq \emptyset. \tag{3.24}$$

Indeed, in this case, taking $p \in \mathrm{dom}(f_1^*) \cap \mathrm{dom}(f_2^*)$ we have thanks to Young–Fenchel inequality

$$f_1(x - y) + f_2(y) \geq p(x - y) - f_1^*(p) + p(y) - f_2^*(p)$$
$$= p(x) - f_1^*(p) - f_2^*(p)$$

so that

$$(f_1 \square f_2)(x) \geq p(x) - f_1^*(p) - f_2^*(p) \text{ hence } p \in \mathrm{dom}(f_1 \square f_2).$$

Obviously, if (3.24) holds and both f_1 and f_2 are proper then so is $f_1 \square f_2$ because if $x_i \in \mathrm{dom}(f_i)$ then $f_1 \square f_2(x_1 + x_2) \leq f_1(x_1) + f_2(x_2) < +\infty$. The link between infimal convolution and Legendre transforms is as follows:

Lemma 3.7. *Let f_1 and $f_2 : E \to \mathbb{R} \cup \{+\infty\}$ be two proper functions. Then*

$$(f_1 \square f_2)^* = f_1^* + f_2^*. \tag{3.25}$$

Proof. Let $p \in E^*$. Then

$$(f_1 \square f_2)^*(p) = \sup_{x \in E} \left\{ p(x) - \inf_{y \in E} \{ f_1(x - y) + f_2(y) \} \right\}$$

$$= \sup_{x \in E} \left\{ p(x) - \inf_{(x_1, x_2) \in E^2 \,:\, x_1 + x_2 = x} \{ f_1(x_1) + f_2(x_2) \} \right\}$$

$$= \sup_{x \in E} \left\{ p(x) + \sup_{(x_1, x_2) \in E^2 \,:\, x_1 + x_2 = x} \{ -f_1(x_1) - f_2(x_2) \} \right\}$$

$$= \sup_{(x_1, x_2) \in E^2} \{ p(x_1 + x_2) - f_1(x_1) - f_2(x_2) \}$$

$$= \sup_{x_1 \in E} \{ p(x_1) - f_1(x_1) \} + \sup_{x_2 \in E} \{ p(x_2) - f_2(x_2) \}$$

$$= f_1^*(p) + f_2^*(p). \qquad \square$$

Proposition 3.12. *Let f_1 and f_2 be in $\Gamma_0(E)$ and satisfy (3.24). Then $f_1 \square f_2$ is convex. Moreover, if $x_1 \in \mathrm{dom}(f_1)$ and f_2 is finite and continuous at x_2, then $f_1 \square f_2$ is finite and continuous at $x = x_1 + x_2$.*

Proof. Let $(y, z, t) \in E^2 \times [0, 1]$. If $(y_1, y_2, z_1, z_2) \in E^4$ are such that $y_1 + y_2 = y$ and $z_1 + z_2 = z$, then $((1 - t)y_1 + tz_1) + ((1 - t)y_2 + tz_2) = (1 - t)y + tz$ so that

$$(f_1 \square f_2)((1 - t)y + tz) \leq f_1((1 - t)y_1 + tz_1) + f_2((1 - t)y_2 + tz_2)$$

$$\leq (1 - t)(f_1(y_1) + f_2(y_2)) + t(f_1(z_1) + f_2(z_2)),$$

where we have used the convexity of f_1 and f_2 in the second line. Taking the infimum with respect to $(y_1, y_2, z_1, z_2) \in E^4$ such that $y_1 + y_2 = y$ and $z_1 + z_2 = z$ gives the desired convexity inequality

$$(f_1 \square f_2)((1 - t)y + tz) \leq (1 - t)(f_1 \square f_2)(y) + t(f_1 \square f_2)(z).$$

If $x_1 \in \text{dom}(f_1)$ and f_2 is finite and continuous at x_2 then $(f_1 \square f_2)(x_1 + x_2 + h) \leq f_1(x_1) + f_2(x_2 + h)$ but since f_2 is finite and continuous at x_2, $f_1 \square f_2$ is bounded from above in a neighborhood of $x_1 + x_2$. Since $f_1 \square f_2$ is convex and nowhere $-\infty$ thanks to (3.24), this enables us to conclude that it is continuous at $x_1 + x_2$ thanks to Proposition 3.6. $\square$

3.4.2.   *The biconjugate*

Definition 3.6. Given $f : E \to \mathbb{R} \cup \{+\infty\}$, the biconjugate of f, denoted f^{**}, is the function defined on E with values in $\overline{\mathbb{R}}$ given by

$$f^{**}(x) := \sup_{p \in E^*} \{p(x) - f^*(p)\}, \ \forall x \in E.$$

Some basic properties of f^{**} are as follows:

- if $f \geq g$, then $f^{**} \geq g^{**}$;
- if γ_0 is a continuous affine function: $\gamma_0(x) = p_0(x) + \alpha_0$ with $(p_0, \alpha_0) \in E^* \times \mathbb{R}$ since (see Exercise 3.5)

$$(f - \gamma_0)^*(p) = f^*(p + p_0) + \alpha_0, \ \forall p \in E^*,$$

we have

$$(f - \gamma_0)^{**}(x) = \sup_{p \in E^*} \{p(x) - f^*(p + p_0)\} - \alpha_0$$

$$= f^{**}(p) - \gamma_0(x); \tag{3.26}$$

- Young–Fenchel inequality gives that $f(x) \geq p(x) - f^*(p)$ for every $p \in E^*$ hence taking the supremum in p gives:

$$f \geq f^{**};$$

in particular, if f is proper so is f^{**} (and if $f \equiv +\infty$ then $f^* \equiv -\infty$ so $f^{**} = +\infty$);
- f^{**} is convex and lsc, hence also weakly lsc.

Note that since the family of affine functions $x \mapsto p(x) - f^*(p)$ are the affine minorants of f, f^{**} is by construction the upper envelope of such functions. If $f \in \Gamma_0(E)$, then f^* is proper (Proposition 3.11) so f^{**} does not take the value $-\infty$ hence $f^{**} \in \Gamma_0(E)$. In fact, we have the following theorem.

Theorem 3.9 (Fenchel–Moreau theorem). *If* $f \in \Gamma_0(E)$, *then* $f = f^{**}$.

Proof. Let $p_0 \in \text{dom}(f^*)$ ($\text{dom}(f^*) \neq \emptyset$ by Proposition 3.11). Then $f \geq \gamma_0$ where $\gamma_0(x) := p_0(x) - f^*(p_0)$ so that $g := f - \gamma_0 \geq 0$. Of course $g \in \Gamma_0(E)$ and $g^{**} = f^{**} - \gamma_0$ by (3.26), so that we have to show $g^{**} = g$. Since $g^{**} \leq g$, it is enough to show that this inequality cannot be strict. Assume on the contrary that there is $x_0 \in E$ such that $g^{**}(x_0) < g(x_0)$ (so that $g^{**}(x_0) < +\infty$ whereas we cannot exclude that $g(x_0) = +\infty$), one can therefore strictly separate the point $(g^{**}(x_0), x_0)$ from the non-empty convex and closed set $\text{epi}(g)$, i.e., there exist $(p, \alpha) \in E^* \times \mathbb{R}$ and $\varepsilon > 0$ such that

$$p(x_0) + \alpha g^{**}(x_0) \leq p(x) + \alpha\lambda - \varepsilon, \quad \forall(\lambda, x) \in \text{epi}(g). \tag{3.27}$$

Clearly, equation (3.27) implies that $\alpha \geq 0$ but the case where $\alpha = 0$ cannot be excluded (and this is where we will use the fact that $g \geq 0$). It follows from (3.27) that $p(x_0) + \alpha g^{**}(x_0) \leq p(x) + \alpha g(x) - \varepsilon$ for every $x \in E$, hence for every $\delta > 0$ we also have

$$p(x_0) + \alpha g^{**}(x_0) \leq p(x) + (\alpha + \delta)g(x) - \varepsilon.$$

Minimizing the right-hand side with respect to x yields

$$p(x_0) + \alpha g^{**}(x_0) \leq -(\alpha + \delta)g^*\left(\frac{-p}{\alpha + \delta}\right) - \varepsilon. \tag{3.28}$$

But by Young–Fenchel inequality, we also have

$$-(\alpha + \delta)g^*\left(\frac{-p}{\alpha + \delta}\right) \leq p(x_0) + (\alpha + \delta)g^{**}(x_0). \tag{3.29}$$

Combining (3.28) and (3.29) thus gives $\delta g^{**}(x_0) \geq \varepsilon > 0$ for every $\delta > 0$; the desired contradiction is therefore reached by letting $\delta \to 0^+$. $\qquad\square$

An obvious consequence of the Fenchel–Moreau theorem is the following corollary.

Corollary 3.7. *Let* f *and* g *be in* $\Gamma_0(E)$. *Then* $f = g$ *if and only if* $f^* = g^*$.

Let us close this section by observing that it is useless to iterate the Legendre transform more than twice.

Lemma 3.8. *If $f: E \mapsto \mathbb{R} \cup \{+\infty\}$, then $(f^{**})^* = f^*$.*

Proof. Since $f^{**} \leq f$ we first have $(f^{**})^* \geq f^*$. Now by Young–Fenchel inequality $p(x) - f^{**}(x) \leq f^*(p)$, taking the supremum in p gives the converse inequality. $\qquad\square$

3.4.3.   *Subdifferentiability*

We have seen that if $f : E \to \mathbb{R}$ is convex and differentiable at $x \in E$ then $p = f'(x) \in E^*$ such that $f(y) \geq f(x) + p(y-x)$ for every $y \in E$. On the one hand, this inequality is really strong since it gives an affine lower estimate on f which holds everywhere; on the other hand, it is only one-sided. It naturally leads to the notion of subgradient.

Definition 3.7. Let E be a normed space, $f \in \Gamma_0(E)$, $x \in E$, the subdifferential of f at x, denoted $\partial f(x)$, is the set of $p \in E^*$ such that

$$f(y) \geq f(x) + p(y - x), \quad \forall y \in E.$$

If $\partial f(x) \neq \emptyset$, then f is said to be subdifferentiable at x and elements of $\partial f(x)$ are called subgradients of f at x.

Note that since f is proper if it is subdifferentiable at x then $x \in \text{dom}(f)$. The obvious observation that

$$0 \in \partial f(x) \iff f(y) \geq f(x), \ \forall x \in E \tag{3.30}$$

gives a first motivation of the notion of subdifferential for minimization problems. A second observation is that $p \in \partial f(x)$ can be equivalently expressed in terms of the Legendre transform of f as

$$p \in \partial f(x) \iff f(x) - p(x) = \inf_{y \in E}\{f(y) - p(y)\} = -f^*(p)$$

so that

$$p \in \partial f(x) \iff f(x) + f^*(p) = p(x). \tag{3.31}$$

So, if we define in a similar fashion, for $p \in E^*$ the set

$$\partial f^*(p) := \{x \in E \ : \ f^*(q) \geq f^*(p) + (q - p)(x), \ \forall q \in E^*\}$$
$$= \{x \in E \ : \ f^{**}(x) + f^*(p) = p(x)\},$$

using the fact that $f = f^{**}$ by virtue of the Fenchel–Moreau theorem, we obtain thus the Fenchel reciprocity formula:

$$p \in \partial f(x) \iff x \in \partial f^*(p). \tag{3.32}$$

Now let $(x_1, x_2) \in E^2$ be points where f is subdifferentiable and let $(p_1, p_2) \in \partial f(x_1) \times \partial f(x_2)$. Summing the inequalities $f(x_2) \geq f(x_1) + p_1(x_2 - x_1)$ and $f(x_1) \geq f(x_2) + p_2(x_1 - x_2)$ shows that subgradients satisfy the monotonicity property:

$$(p_1, p_2) \in \partial f(x_1) \times \partial f(x_2) \Rightarrow (p_2 - p_1)(x_2 - x_1) \geq 0. \tag{3.33}$$

Example 3.3. Let us consider the case where $E = \mathbb{R}^d$ and f is the l^1-norm

$$f(x) = |x|_1 := \sum_{i=1}^{d} |x_i|.$$

Since f is a norm, as seen in Example 3.1 its Legendre transform is the characteristic function of the dual unit ball which is simply the hypercube $[-1, 1]^d$. We then have $f^* = \chi_{[-1,1]^d}$. So $p \in \partial f(x)$ if and only if $p \in [-1, 1]^d$ and $|x|_1 = \sum_{i=1}^{d} p_i x_i$ which gives $p_i x_i = |x_i|$ for every $i = 1, \ldots, d$ and since $p_i \in [-1, 1]$, we get

$$p_i \begin{cases} = 1 & \text{if } x_i > 0, \\ \in [-1, 1] & \text{if } x_i = 0, \\ = -1 & \text{if } x_i < 0. \end{cases}$$

A simple criterion for subdifferentiability is as follows.

Theorem 3.10. *Let $f : E \mapsto \mathbb{R} \cup \{+\infty\}$ be convex, $x \in E$ be such that f is finite and continuous at x. Then f is subdifferentiable at x. Moreover, $\partial f(x)$ is convex and closed.*

Proof. Note that $\mathrm{epi}(f)$ has non-empty interior, and $(f(x), x)$ can be separated from the convex open set $\mathrm{int}(\mathrm{epi}(f))$, i.e., there exists $(p, \alpha) \in E \times \mathbb{R}$ with $(p, \alpha) \neq (0, 0)$ such that

$$p(x) + \alpha f(x) \leq p(y) + \alpha \lambda, \quad \forall (\lambda, y) \in \mathrm{int}(\mathrm{epi}(f)). \tag{3.34}$$

By density of $\mathrm{int}(\mathrm{epi}(f))$ in $\mathrm{epi}(f)$, equation (3.34) actually holds for every $(\lambda, y) \in \mathrm{epi}(f)$. Once again (taking $y = x$, $\lambda = f(x) + M$ and letting $M \to +\infty$), it is clear that $\alpha \geq 0$. Let us prove in fact that $\alpha > 0$. If $\alpha = 0$ then $p \neq 0$ and $p(x) \leq p(y)$ for every $y \in \mathrm{dom}(f)$, but since f is continuous at x, there exists $r > 0$ such that $\overline{B}(x, r) \subset \mathrm{dom}(f)$ so $p(x) \leq p(y)$ for every $y \in \overline{B}(x, r)$ yielding $p = 0$ which is the desired contradiction. This shows that $\alpha > 0$ and (3.34) gives

$$\frac{p(x)}{\alpha} + f(x) \leq \frac{p(y)}{\alpha} + f(y), \quad \forall y \in E,$$

i.e., $\frac{-p}{\alpha} \in \partial f(x)$. The fact that $\partial f(x)$ is convex and closed follows from the fact that it is defined by continuous linear inequalities. $\qquad \square$

In particular, if $f \colon \mathbb{R}^d \to \mathbb{R}$, is a convex function: it is continuous everywhere and thus subdifferentiable everywhere as well. The link between subdifferential and directional derivatives is as follows.

Proposition 3.13. *Let $f \in \Gamma_0(E)$ and $x \in E$ such that f is finite and continuous at x (so that it is subdifferentiable and admits a right derivative in every direction at x). Then, for every $h \in E$, one has*

$$D^+ f(x; h) = \sigma_{\partial f(x)}(h) := \sup\{p(h),\ p \in \partial f(x)\}. \tag{3.35}$$

Proof. If $p \in \partial f(x)$ then $f(x+th) - f(x) \geq tp(h)$ for every $t > 0$. Dividing by t and letting $t \to 0^+$ gives $D^+ f(x; h) \geq p(h)$, this shows $D^+ f(x; h) \geq \sup\{p(h),\ p \in \partial f(x)\}$. To prove the converse inequality, recall that by virtue of Proposition 3.8, one has

$$f(x + th) \geq f(x) + D^+ f(x; th) = f(x) + tD^+ f(x; h),\ \forall t \geq 0$$

so that the half line $\{((f(x) + tD^+ f(x; h), x + th),\ t \geq 0\}$ is disjoint from $\mathrm{int}(\mathrm{epi}(f))$ in $\mathbb{R} \times E$. A half line being convex and $\mathrm{int}(\mathrm{epi}(f))$ being open and convex (and dense in $\mathrm{epi}(f)$). We can use Theorem 3.2 to separate these to sets. Therefore, there exists $(\alpha, p) \in \mathbb{R} \times E^*$, $(\alpha, p) \neq (0,0)$ such that

$$p(x + th) + \alpha(f(x) + tD^+ f(x; h)) \leq p(y) + \alpha\lambda,\ \forall(\lambda, y, t) \in \mathrm{epi}(f) \times \mathbb{R}_+. \tag{3.36}$$

One easily checks that $\alpha > 0$. Dividing (3.36) by α and setting $q := \frac{p}{\alpha}$, one obtains

$$q(x) + f(x) + t(D^+ f(x; h) + q(h)) \leq q(y) + f(y),\ \forall t \in \mathbb{R}_+,\ \forall y \in E.$$

Taking $t = 0$ immediately gives $-q \in \partial f(x)$ and the fact that $\sup_{t \in \mathbb{R}_+}\{t(D^+ f(x; h) + q(h))\}$ is finite implies that $D^+ f(x; h) \leq -q(h)$ so $D^+ f(x; h) \leq \sup\{p(h),\ p \in \partial f(x)\}$ which concludes the proof. $\qquad \square$

This gives a criterion for Gateaux-differentiability of convex functions.

Corollary 3.8. *Let $f \in \Gamma_0(E)$, $x \in E$ such that f is finite and continuous at x.*

1) *If $\partial f(x) = \{p\}$, f is Gateaux-differentiable at x and $D_G f(x) = p$.*

2) If f is Gateaux-differentiable at x, then $\partial f(x)$ is a singleton $\partial f(x) = \{D_G f(x)\}$.

Proof. The first claim is a direct consequence of (3.35). Now assume that f is Gateaux-differentiable at x and set $q := D_G f(x)$ and let $p \in \partial f(x)$, for $h \in E$ and $t > 0$ we thus have

$$\frac{f(x + th) - f(x)}{t} \geq p(h)$$

letting $t \to 0^+$ gives $q(h) \geq p(h)$ and h being arbitrary this gives $p = q$ so that $\partial f(x)$ consists of the single point q. $\square$

The next corollary expresses a simple link between differentiability and strict convexity of the Legendre transform.

Corollary 3.9. *Let $f \in \Gamma_0(E)$ be such that f^* is strictly convex. If $x \in E$ is such that f is finite and continuous at x, then f is Gateaux-differentiable at x.*

Proof. Let $(p_1, p_2) \in (\partial f(x))^2$, i.e., $f^*(p_i) = p_i(x) - f(x)$ for $i = 1, 2$. For $t \in (0, 1)$ we have by Young–Fenchel inequality $f^*((1 - t)p_1 + tp_2) \geq ((1 - t)p_1 + tp_2)(x) - f(x) = (1 - t)f^*(p_1) + tf^*(p_2)$ but if $p_1 \neq p_2$, strict convexity of f^* gives $f^*((1 - t)p_1 + tp_2) < (1 - t)f^*(p_1) + tf^*(p_2)$. This shows that $p_1 = p_2$. Therefore, $\partial f(x)$ is a singleton and f is is Gateaux-differentiable at x. $\square$

The previous result suggests a strategy to approximate convex functions by regular ones: add a small strictly convex term εg to f^* and then consider $f_\varepsilon = (f^* + \varepsilon g)^*$. The Moreau–Yosida regularization is a powerful regularization technique based on this idea (see Exercise 3.28).

We end this section with some subdifferential calculus rules. One should indeed be cautious when manipulating subgradients of sums of convex functions and subgradients of convex functions which are obtained via composition with a linear map. Let us start with the subdifferential of a sum.

Theorem 3.11. *Let f_1 and $f_2 \in \Gamma_0(E)$ and let $x \in E$. Then*

$$\partial f_1(x) + \partial f_2(x) \subset \partial(f_1 + f_2)(x). \tag{3.37}$$

If there exists $x_0 \in \text{dom}(f_1) \cap \text{dom}(f_2)$ such that f_2 is continuous at x_0 then

$$\partial f_1(x) + \partial f_2(x) = \partial(f_1 + f_2)(x), \ \forall x \in E. \tag{3.38}$$

Proof. Let $p_i \in \partial f_i(x)$ for $i = 1$, 2, summing the inequalities

$$f_i(y) \geq f_i(x) + p_i(y - x), \ \forall y \in E$$

immediately gives (3.37). Assume now that there exists $x_0 \in \mathrm{dom}(f_1) \cap \mathrm{dom}(f_2)$ such that f_2 is continuous at x_0, let $x \in E$ and $p \in \partial(f_1 + f_2)(x)$. Our goal is to show that p can be decomposed as $p = p_1 + p_2$ with $(p_1, p_2) \in \partial f_1(x) \times \partial f_2(x)$. For every $y \in E$, one has

$$f_1(y) + f_2(y) \geq f_1(x) + f_2(x) + p(y - x). \tag{3.39}$$

Note first that both $f_1(x)$ and $f_2(x)$ should be finite and also that it is impossible to have at the same time

$$\lambda > f_2(y) \text{ and } -\lambda \geq f_1(y) - p(y - x) - f_1(x) - f_2(x),$$

which means that the open convex set $\mathrm{int}(\mathrm{epi}(f_2))$ (non-empty thanks to our assumption) can be separated from the convex set

$$B := \{(\mu, z) \in \mathbb{R} \times E : -\mu \geq f_1(z) - p(z - x) - f_1(x) - f_2(x)\}.$$

Hence there exists $(\alpha, q) \in \mathbb{R} \times E^* \setminus \{0, 0\}$ such that

$$q(y) + \alpha\lambda \geq \alpha\mu + q(z), \ \forall(\lambda, y) \in \mathrm{int}(\mathrm{epi}(f_2)), \ \forall(\mu, z) \in B. \tag{3.40}$$

Arguing as in the proof of Theorem 3.10, we find that $\alpha > 0$ so, up to changing q into $\frac{q}{\alpha}$ and using the density of $\mathrm{int}(\mathrm{epi}(f_2))$ in $\mathrm{epi}(f_2)$, (3.40) becomes

$$q(y) + \lambda \geq \mu + q(z), \ \forall(\lambda, y) \in \mathrm{epi}(f_2), \ \forall(\mu, z) \in B,$$

which by the very definition of B is equivalent to

$$f_1(z) + f_2(y) \geq q(z - y) + p(z - x) + f_1(x) + f_2(x), \ \forall(y, z) \in E^2. \tag{3.41}$$

Taking $z = x$ in (3.41) thus gives $-q \in \partial f_2(x)$ and likewise taking $y = x$ we deduce $(p + q) \in \partial f_1(x)$. This shows that $p \in \partial f_1(x) + \partial f_2(x)$. $\square$

    One does not always have (3.38) (see Exercise 3.6) and this is a manifestation of the fact that even though the subdifferential is a powerful and flexible notion it is very nonlinear. However the (qualification like) condition of Theorem 3.11 is not the optimal one that guarantees that the subdifferential of the sum is the sum of the subdifferentials, a celebrated theorem by Attouch and Brezis [7] indeed gives a sharper condition.

Let us now consider the composition of a convex function with a linear map. Let E and F be two normed spaces, $\Lambda \in \mathcal{L}_c(E, F)$, $f \in \Gamma_0(F)$ and $g := f \circ \Lambda$ so that obviously $g \in \Gamma_0(E)$. If f was differentiable, the usual chain rule would give $g'(x)(h) = f'(\Lambda(x))(\Lambda(h))$ for every $(x, h) \in E \times E$. The adjoint of Λ, denoted Λ^*, is the map $\Lambda^* : F^* \to E^*$, defined by

$$\Lambda^*(q)(h) = q(\Lambda(h)), \ \forall (q, h) \in F^* \times E.$$

Λ^* is obviously linear and since $|\Lambda^*(q)(h)| \leq \|q\|_{F^*} \|\Lambda\|_{\mathcal{L}_c(E,F)} \|h\|_E$ we have $\Lambda^* \in \mathcal{L}_c(F^*, E^*)$; finally, the chain rule can be rewritten as $g'(x) = \Lambda^*(f'(\Lambda(x)))$. The analogous chain rule for subdifferentials is as follows.

Theorem 3.12. *Let $g = f \circ \Lambda$ with $f \in \Gamma_0(F)$ and $\Lambda \in \mathcal{L}_c(E, F)$ as above and let $x \in E$. Then*

$$\Lambda^*(\partial f(\Lambda(x))) \subset \partial g(x). \tag{3.42}$$

If there exists $x_0 \in E$ such that f is finite and continuous at $\Lambda(x_0)$, then

$$\Lambda^*(\partial f(\Lambda(x))) = \partial g(x), \quad \forall x \in E. \tag{3.43}$$

Proof. Let $q \in \partial f(\Lambda(x))$. Then for every $y \in F$ one has

$$f(y) \geq f(\Lambda(x)) + q(y - \Lambda(x)).$$

For $y = \Lambda(z)$ this rewrites as $g(z) \geq g(x) + q(\Lambda(z-x)) = g(x) + (\Lambda^* q)(z - x)$ so that $\Lambda^*(q) \in \partial g(x)$. Now assume that f is finite and continuous at $\Lambda(x_0)$ for some $x_0 \in E$ and let $p \in \partial g(x)$ that is

$$f(\Lambda(z)) \geq f(\Lambda(x)) + p(z - x), \quad \forall z \in E.$$

The affine subspace of $\mathbb{R} \times F$, $B := \{(f(\Lambda(x)) + p(z - x), \Lambda(z)), z \in E\}$ is therefore disjoint from int(epi(f)). Arguing as in the proofs of Theorems 3.10 and 3.11, we find $q \in F^*$ such that $f(y) + q(y) \geq \mu + q(\xi)$ for every $y \in F$ and every $(\mu, \xi) \in B$, i.e.,

$$f(y) + q(y) \geq f(\Lambda(x)) + p(z - x) + q(\Lambda(z)), \quad \forall (y, z) \in F \times E. \tag{3.44}$$

Taking $y = \Lambda(x_0)$, we deduce that the linear form $z \in E \mapsto (p + \Lambda^*(q))(z)$ is bounded from above so that $p = \Lambda^*(-q)$. Then (3.44) rewrites as

$$f(y) \geq f(\Lambda(x)) + q(\Lambda(x) - y), \ \forall y \in F$$

hence $-q \in \partial f(\Lambda(x))$ so that $p = \Lambda^*(-q) \in \Lambda^*(\partial f(\Lambda(x)))$. $\qquad \square$

Remark 3.2. In the finite-dimensional case where $E = \mathbb{R}^d$ and $F = \mathbb{R}^m$ and $\Lambda \in \mathcal{L}(\mathbb{R}^d, \mathbb{R}^m) = \mathcal{L}_c(\mathbb{R}^d, \mathbb{R}^m)$, it is convenient to represent Λ by its matrix $M \in \mathcal{M}_{m \times d}(\mathbb{R})$ in some bases of E and F. We identify E and F with their duals using their respective inner products. Given $q \in \mathbb{R}^m$ and $x \in \mathbb{R}^d$ observe that

$$q(\Lambda(x)) = q \cdot Mx = M^T q \cdot x = \Lambda^*(q)(x).$$

The matrix of Λ^* is M^T: it is simply the transpose of the matrix of Λ.

Exercise 3.6. *Let E be a normed space, f_1 and f_2 be in $\Gamma_0(E)$ and $x \in E$.*

1) *Show that if $\lambda > 0$ and $x \in E$ then $\partial(\lambda f_1)(x) = \lambda \partial f_1(x)$.*
2) *Consider the case $E = \mathbb{R}^2$, f_1 (respectively, f_2) is the characteristic function of the closed ball with center $(1, 0)$ (respectively, $(-1, 0)$) and radius 1, compute $\partial f_1(0)$, $\partial f_2(0)$, $\partial f_1(0) + \partial f_2(0)$, $\partial(f_1 + f_2)(0)$ and conclude.*
3) *Let $\Lambda \in \mathcal{L}(\mathbb{R}, \mathbb{R}^2)$ defined by $\Lambda(x) := (0, x)$ for every $x \in \mathbb{R}$, $g(x) := f_1(\Lambda(x))$ where f_1 is as in the previous question, compute Λ^*, $\partial g(0)$, $\Lambda^*(\partial f_1(0, 0))$ and conclude.*

3.5. Exercises

Exercise 3.7. *Let $E := C_b(\mathbb{R}, \mathbb{R})$ (equipped with the uniform norm). Show that there exists $l \in E^*$ with $\|l\|_{E^*} = 1$ and such that*

$$\liminf_{t \to +\infty} f(t) \leq l(f) \leq \limsup_{t \to +\infty} f(t), \quad \forall f \in E.$$

Exercise 3.8. *In the case of a Hilbert space, give an alternative proof of Theorem 3.3 based on the projection onto a convex set theorem.*

Exercise 3.9. *Let $f: \mathbb{R}^d \to \mathbb{R} \cup \{+\infty\}$ be a convex lsc function, let $(\Omega, \mathcal{F}, \mu)$ be a probability space and let $u: \Omega \to \mathbb{R}^d$ be measurable. Show Jensen's inequality:*

$$f\left(\int_\Omega u(\omega)\,d\mu(\omega)\right) \leq \int_\Omega f(u(\omega))\,d\mu(\omega). \tag{3.45}$$

Exercise 3.10. *Let $f \in C(\mathbb{R}^d, \mathbb{R})$ be such that*

$$\frac{f(x)}{|x|} \to +\infty \text{ as } |x| \to \infty.$$

1) *Show that* $f^{**} \in C(\mathbb{R}^d, \mathbb{R})$.
2) *Show that*

$$\frac{f^{**}(x)}{|x|} \to +\infty \text{ as } |x| \to \infty.$$

3) *Show that both f and f^{**} attain their minimum and that*

$$\inf_{\mathbb{R}^d} f = \inf_{\mathbb{R}^d} f^{**}.$$

Exercise 3.11 (Polar cones and the bipolar theorem). *Let E be a Hilbert space (identified with its dual). A non-empty subset C of E is called a cone if $\lambda x \in C$ whenever $\lambda \geq 0$ and $x \in C$. Let A be a non-empty subset of E, the polar cone of A, denoted A^*, is defined by*

$$A^* := \{x \in E \ : \ a \cdot x \geq 0, \ \forall a \in A\}.$$

The bipolar cone A^{**} is the polar of A^*.

1) *Show that A^* is a closed convex cone.*
2) *Show that $A \subset A^{**}$.*
3) *Show that if A is a vector subspace then $A^* = A^{\perp}$.*
4) *In the case $E = \mathbb{R}^d$ and $A = \mathbb{R}^d_+$ identify A^* and A^{**}.*
5) *Show that A^{**} is the smallest closed convex cone containing A and deduce that $A = A^{**}$ if and only if A is a closed convex cone (this is the so-called bipolar theorem).*

Exercise 3.12 (Soft maximum). *For $x \in \mathbb{R}^d$ and $\lambda > 0$, define:*

$$f_\infty(x) := \max(x_1, \ldots, x_d), \ f_\lambda(x) := \frac{1}{\lambda} \log \Big(\sum_{i=1}^{d} \exp(\lambda x_i) \Big).$$

1) *Show that f_∞ is convex and that $f_\infty^* = \chi_\Delta$ where $\Delta := \{p \in \mathbb{R}^d_+, \sum_{i=1}^d p_i = 1\}$.*
2) *Show that*

$$f_\infty \leq f_\lambda \leq f_\infty + \frac{1}{\lambda} \log(d).$$

3) *Show that f_λ is of class C^2 (this, together with the previous inequality, explains why it is called a soft maximum function), compute ∇f_λ and $D^2 f_\lambda$.*
4) *Show that f_λ is convex, is it strictly convex?*

5) *Show that*

$$f_\lambda^*(p) = \begin{cases} \dfrac{1}{\lambda} \sum_{i=1}^{d} p_i \log(p_i) & \text{if } p \in \Delta \\ +\infty & \text{otherwise.} \end{cases}$$

Exercise 3.13. *Let $C_1, \ldots, C_n$ be non-empty subsets of $\mathbb{R}^d$ show that*

$$\max_{i=1,\ldots,n} \sigma_{C_i} = \sigma_{\mathrm{co}(\bigcup_{i=1}^n C_i)}.$$

Now let $(C_i)_{i \in I}$ be a family of non-empty subsets of $\mathbb{R}^d$ show that

$$\sup_{i \in I} \sigma_{C_i} = \sigma_{\overline{\mathrm{co}}(\bigcup_{i \in I} C_i)},$$

where $\overline{\mathrm{co}}(\bigcup_{i \in I} C_i)$ is the closed convex hull of $\bigcup_{i \in I} C_i$ that is the smallest closed convex set containing $\bigcup_{i \in I} C_i$ which is also the closure (prove it) of $\mathrm{co}(\bigcup_{i \in I} C_i)$.

Exercise 3.14. *Equip $\mathcal{M}_d(\mathbb{R})$ with its usual scalar product (see Exercise 1.16) and for $M \in \mathcal{S}_d^{++}(\mathbb{R})$ (see Exercise 1.17) set*

$$\varphi(M) := -\log(\det(M)).$$

1) *Let $M \in \mathcal{S}_d^{++}(\mathbb{R})$. Show that φ is differentiable at M and show that $\nabla\varphi(M) = -M^{-1}$.*
2) *Let $(M_1, M_2) \in (\mathcal{S}_d^{++}(\mathbb{R}))^2$. Compute $\langle \nabla\varphi(M_1) - \nabla\varphi(M_2), M_1 - M_2 \rangle$.*
3) *Use Exercise 1.18 to prove that φ is convex on $\mathcal{S}_d^{++}(\mathbb{R})$, is it strictly convex?*

Exercise 3.15. *Let E be a normed space, $f \in \Gamma_0(E)$ such that $f \geq 0$ and $f(0) = 0$. Define inductively the sequence $(g_n)_{n \geq 1}$ by*

$$g_1 = f, \quad g_{n+1} := g_n \square f.$$

1) *Study the sequence (g_n) when $E = \mathbb{R}$ when $f(x) = \frac{x^2}{2}$ and when $f(x) = |x|$.*
2) *Show that $0 \leq g_{n+1} \leq g_n \leq f$.*
3) *Deduce that g_n converges pointwise.*
4) *Prove that g_n^* converges to $\chi_{\partial f(0)}$.*
5) *What is the pointwise limit of g_n?*

Exercise 3.16. *Let C be a convex compact subset of $\mathbb{R}^2$. Prove that* $\text{Ext}(C)$ *is closed. Give a counterexample in dimension 3. (Hint: consider the convex hull of $(0,0,1)$, $(0,0,-1)$ and the horizontal disk $\{(1 + \cos(\theta), \sin(\theta), 0),\ \theta \in \mathbb{R}\}$.)*

Exercise 3.17. *Let A be a non-empty compact subset of $\mathbb{R}^d$. Show that* $\text{co}(A)$ *is compact and that* $\text{Ext}(\text{co}(A)) \subset A$.

Exercise 3.18. *Let $p_1, \ldots, p_m$ be m vectors in $\mathbb{R}^d$, $b \in \mathbb{R}^m$, and*

$$C := \{x \in \mathbb{R}^d \ : \ p_i \cdot x \le b_i,\ i = 1, \ldots, m\}$$

for $x \in C$ denote by $I(x)$ the set of i's for which $p_i \cdot x = b_i$ and by $A_{I(x)}$ the matrix whose rows are $\{p_i\}_{i \in I(x)}$. Then prove that for $x \in C$ one has

$$x \in \text{Ext}(C) \iff \text{the rank of } A_{I(x)} \text{ is } d.$$

Deduce that $\text{Ext}(C)$ (if non-empty) is finite and give a bound on its cardinality. Give a necessary condition for $\text{Ext}(C)$ to be non-empty.

Exercise 3.19. *Show that $x \in \mathbb{R} \mapsto |x|$ is convex, compute its Legendre transform and its subdifferential at each point. Same question with $x \in \mathbb{R} \mapsto \max(0, x^2)$ and $x \in \mathbb{R} \mapsto \max(x, 2x - 1)$.*

Exercise 3.20. *For $x \in \mathbb{R}$, define*

$$\text{Ent}(x) = \begin{cases} x \log x & \text{if } x > 0, \\ 0 & \text{if } x = 0, \\ +\infty & \text{if } x < 0, \end{cases}$$

show that Ent is convex and lsc, compute its Legendre transform and its subdifferential at each point where it is non-empty.

Exercise 3.21. *For $(x, m) \in \mathbb{R}^d \times \mathbb{R}$, define*

$$\phi(x, m) := \begin{cases} \dfrac{|x|^2}{m} & \text{if } m > 0, \\ 0 & \text{if } x = 0 \text{ and } m = 0, \\ +\infty & \text{otherwise}, \end{cases}$$

show that ϕ is convex and lsc, compute its Legendre transform and its subdifferential at each point where it is non-empty.

Exercise 3.22. *Let E be a Hilbert space (so that we identify E with E^*), and $f : E \to \mathbb{R} \cup \{+\infty\}$. Show that $f = f^*$ if and only if $f = \frac{1}{2}\| \cdot \|^2$.*

Exercise 3.23. *Let E be a normed space, f and g be two convex functions $E \to \mathbb{R}$ such that $f \geq g$. Show that if $f(x_0) = g(x_0)$ then $\partial g(x_0) \subset \partial f(x_0)$, and deduce that if, in addition, f is Gateaux-differentiable at x_0 then so is g.*

Exercise 3.24. *Let E be a normed space, $f : E \mapsto \mathbb{R} \cup \{+\infty\}$ be convex and $g : E \to \mathbb{R}$ be convex and differentiable. Let $x \in E$ and $q \in \partial(f+g)(x)$. Show directly (i.e., without using Theorem 3.11) that $(q - g'(x)) \in \partial f(x)$. (Hint: use the fact that $f(y) - f(x) \geq \frac{1}{t}(f(x+t(y-x)) - f(x))$ for $t \in (0, 1]$, use then q and g to have a lower bound on the previous difference quotient and then let $t \to 0^+$.) Deduce that $\partial(f + g)(x) = g'(x) + \partial f(x)$.*

Exercise 3.25. *The aim of this exercise is to prove Theorem 2.17 when F is a general Banach space.*

1) *Let $(F, \| \cdot \|_F)$ be a normed space and $y \in F$. Show that*

$$\partial(\| \cdot \|)(y) = \{p \in F^* \ : \ \|p\|_{F^*} \leq 1, \ \|y\|_F = p(y)\}.$$

2) *Use the previous question together with Proposition 3.13 to prove Theorem 2.17 when F and E are Banach spaces.*

Exercise 3.26. *Let E be a Banach space and $f \in \Gamma_0(E)$. Assume that (x_n), (p_n) are sequences in E and E^*, respectively, such that $p_n \in \partial f(x_n)$ for every n, that x_n converges strongly to x in E and that p_n converges weakly-* (see Exercise 1.22) to p in E^*. Show that $p \in \partial f(x)$.*

Exercise 3.27. *Let $f : \mathbb{R}^d \to \mathbb{R}$ be a convex function. Show that for every $R > 0$, $\sup\{|p|, \ p \in \partial f(x), \ |x| \leq R\}$ is finite. Show that if f is Gateaux-differentiable on $\mathbb{R}^d$ then it is of class C^1. (Hint: take x_n converging to some x, show that $\nabla f(x_n)$ converges to $\nabla f(x)$ and use Theorem 2.1.)*

Exercise 3.28 (Moreau–Yosida regularization). *This exercise is devoted to a well-known regularization technique for convex functions on a Hilbert space called Moreau–Yosida approximation. Let E be a Hilbert space (so E^* is identified with E) and $f \in \Gamma_0(E)$. For $\varepsilon > 0$ define*

$$f_\varepsilon := \left(f^* + \frac{\varepsilon}{2}\| \cdot \|^2 \right)^*$$

and

$$g_\varepsilon(x) := \inf_{y \in E} \left\{ \frac{1}{2\varepsilon} \|x - y\|^2 + f(y) \right\}, \quad \forall x \in E.$$

1) Show that f_ε is finite, continuous and Gateaux-differentiable.
2) Show that g_ε is finite, convex and continuous.
3) Compute g_ε^* and show that $f_\varepsilon = g_\varepsilon$.
4) Show that f_ε converges pointwise to f as $\varepsilon \to 0^+$.
5) Show that ∇f_ε is $\frac{1}{\varepsilon}$-Lipschitz. (Hint: use Fenchel reciprocity formula, apply Exercise 3.24 to $f^* + \frac{\varepsilon}{2} \|\cdot\|^2$ and use the monotonicity of ∂f.)

Chapter 4

Optimality Conditions for Differentiable Optimization

Let E be a normed space, A be a non-empty subset of E and $f : A \to \mathbb{R}$ be a given function. The aim of this chapter is to establish, under suitable (differentiability) assumptions on f and A, necessary conditions for a point x to be a solution of

$$\inf_{x \in A} f(x). \tag{4.1}$$

As already mentioned, there is no loss of generality in considering only minimization problems since maximizing a function g is the same as minimizing the function $-g$. The function f appearing in (4.1) is called the objective function (or also the cost, the criterion, or the energy depending on the context) and A is the set of admissible points.

Definition 4.1. Let $x^* \in A$. One says that

- x^* is a solution of (4.1) (or a global minimizer of f on A) if $f(x^*) \leq f(x)$ for every $x \in A$;
- x^* is a global strict minimizer of f on A if $f(x^*) < f(x)$ for every $x \in A \setminus \{x^*\}$;
- x^* is a local minimizer of f on A if there exists $r > 0$ such that $f(x^*) \leq f(x)$ for every $x \in B(x^*, r) \cap A$;
- x^* is a local strict minimizer of f on A if there exists $r > 0$ such that $f(x^*) < f(x)$ for every $x \in (B(x^*, r) \cap A) \setminus \{x^*\}$.

Throughout this chapter, we will make some differentiability assumptions on f and the optimality conditions we are looking for a (local or global) minimizer x^* of f on A will involve the first and second derivatives of f. Of course, these conditions will very much depend on the *local structure of A around x^**. Indeed, the only information we have concerning a local minimizer x^* is that $f(x^*) \leq f(x)$ for points x in A close to x^*. To translate this inequality into a (hopefully useful) differential condition, we have to understand how we can move in a differentiable way from x^* to nearby admissible points (or competitors). If, for instance, x^* is an isolated point of A, i.e., for some $r > 0$, $A \cap B(x^*, r) = \{x^*\}$, x^* is automatically a local strict minimizer of f and no differential condition can be hoped for. On the contrary, if x^* is an interior point of A (unconstrained case), then for every direction h, $x^* + th \in A$ for small enough t and then, the inequality $f(x^* \pm th) - f(x^*) \geq 0$ implies that, if f is differentiable at x^*, it has to be a critical point of f, i.e., $f'(x^*) = 0$. The constrained situation where A is defined by both equality and inequality constraints is more involved and will be investigated in details in this chapter leading to the classic multiplier rules of Lagrange (equality constraints) and Karush–Kuhn–Tucker (equality and inequality constraints). For more on nonlinear programming (i.e., constrained optimization in finite dimensions), the book [16] by Bertsekas is a very good and complete reference.

The basic idea behind any first-order optimality condition is given by the following elementary argument.

Proposition 4.1. *Let $x^* \in A$ and $h \in E$ such that for some $t_0 > 0$ the segment $[x^*, x^* + t_0 h]$ lies inside A. If x^* is a local minimizer of f on A and if the right-directional derivative*

$$D^+ f(x; h) = \lim_{t \to 0^+} \frac{f(x^* + th) - f(x^*)}{t}$$

exists, it is non-negative.

Proof. Since x^* is a local minimizer of f on A and $[x^*, x^* + t_0 h]$ lies inside A for small enough $t > 0$, $\frac{f(x^* + th) - f(x^*)}{t} \geq 0$ passing to the limit $t \to 0^+$ gives the result. $\square$

4.1. Unconstrained optimization

In this section, we consider the simplest unconstrained situation where x^* is an interior point of the admissible set A. It is called unconstrained because *any* small perturbation of x^* remains admissible. The standard first-order condition for an interior local minimizer reads as follows.

Proposition 4.2 (Fermat's rule). *Let x^* be an interior point of A. If x^* is a local minimizer of f on A and f is Gateaux-differentiable at x^*, then x^* is a critical point of f, i.e.,*

$$D_G f(x^*) = 0.$$

In the convex case where, in addition, both f and A are convex, any critical point of f is a global minimizer of f on A.

Proof. Let $h \neq 0$ for $t > 0$ small enough $f(x^* + th) - f(x^*) \geq 0$, dividing by t and letting t tend to 0^+ gives $D_G f(x^*)(h) \geq 0$ but since h is arbitrary and $D_G f(x^*)$ is linear this implies $D_G f(x^*) = 0$. The statement in the convex case follows from Proposition 3.9. □

The second-order necessary optimality condition for a local minimizer is expressed as follows.

Proposition 4.3. *Let x^* be an interior point of A. If x^* is a local minimizer of f on A and f is twice-differentiable at x^*, then*

$$f'(x^*) = 0 \text{ and } f''(x^*)(h, h) \geq 0, \quad \forall h \in E,$$

i.e., x^* is a critical point of f and the symmetric bilinear form $f''(x^*)$ is positive semidefinite.

Proof. We already know that $f'(x^*) = D_G f(x^*) = 0$. Now for $h \in E$ and t sufficiently close to 0, $f(x^* + th) - f(x^*) \geq 0$. Taylor's formula thus gives

$$0 \leq f(x^* + th) - f(x^*) = \frac{t^2}{2} f''(x^*)(h, h) + o(t^2),$$

dividing by t^2 and letting t tends to 0, we get $f''(x^*)(h, h) \geq 0$. □

Let us also mention a sufficient condition for a local strict minimizer.

Proposition 4.4. *Let x^* be an interior point of A. If f is twice-differentiable at x^* and*

$$f'(x^*) = 0 \text{ and there exists } \lambda > 0 \text{ such that } f''(x^*)(h, h) \geq \lambda \|h\|^2, \forall h \in E,$$

then x^* is a local strict minimizer of f on A.

Proof. The second-order Taylor expansion of f around x^* reads as follows:

$$f(x^* + h) - f(x^*) = \frac{1}{2} f''(x^*)(h, h) + \|h\|^2 \varepsilon(h) \geq \left(\frac{\lambda}{2} + \varepsilon(h) \right) \|h\|^2,$$

where $\varepsilon(h)$ tending to 0 as $h \to 0$. In particular, there exists $r > 0$ such that $B(x, r) \subset A$ and $\frac{\lambda}{2} + \varepsilon(h) > 0$ for every $h \in B(0, r)$, enabling us to conclude that $f(y) > f(x^*)$ for every $y \in B(x^*, r) \setminus \{x^*\}$. $\qquad\square$

Recall that a symmetric $d \times d$ matrix S is called positive definite when

$$Sh \cdot h > 0, \ \forall h \in \mathbb{R}^d \setminus \{0\}.$$

A direct consequence of the compactness of the unit sphere of $\mathbb{R}^d$ is as follows.

Lemma 4.1. *Let S be a symmetric positive definite $d \times d$ matrix. Then there exists $\lambda > 0$ such that*

$$Sh \cdot h \geq \lambda \|h\|^2, \quad \forall h \in \mathbb{R}^d.$$

Proof. Let $\lambda := \inf\{Sh \cdot h : \|h\| = 1\}$. By compactness of the unit sphere, this infimum is attained and, by positive definiteness of S, we thus have $\lambda > 0$. Hence, for every $h \in \mathbb{R}^d \setminus \{0\}$,

$$S\frac{h}{\|h\|} \cdot \frac{h}{\|h\|} \geq \lambda,$$

which gives the desired result by homogeneity. $\qquad\square$

Proposition 4.5. *Let A be a non-empty subset of $\mathbb{R}^d$, $f : A \to \mathbb{R}$, x^* be an interior of A such that f is twice differentiable at x^* and*

$$\nabla f(x^*) = 0, \ \text{and} \ D^2 f(x^*) \ \text{is positive definite}.$$

Then x^* is a local strict minimizer of f on A.

4.2.   Equality constraints

We now consider the case where the admissible set is defined by equality constraints. We shall, throughout this section, consider the case where E is a normed space. The objective function f is defined on a non-empty open subset Ω of E. The constraints on the admissible points for the minimization of the objective f will then be given by $g(x) = 0$ where g is a map defined on Ω with values in a certain normed vector space F. We will consider several cases by increasing order of difficulty:

- the case of finitely many affine constraints which corresponds to $F = \mathbb{R}^m$ and $g(x) = (g_1(x), \ldots, g_m(x)) = (p_1(x) - c_1, \ldots, p_m(x) - c_m)$ where the c_i's are constant and the p_i are linear forms; optimality conditions will then be obtained by an elementary linear algebra result stated in Lemma 4.2;
- the finite-dimensional case $E = \mathbb{R}^d$ and $g = (g_1, \ldots, g_m)$ where the g_i's are real-valued but nonlinear. In this case, we will mimic the linear case by linearizing locally around x^* the constraints, which will be done by using the implicit function theorem. To do so, we will have to assume that $g'(x^*)$ is surjective (equivalently, see Lemma 4.3, the linear forms $g_1'(x^*), \ldots, g_m'(x^*)$ are linearly independent);
- the general case where both E and F might be infinite-dimensional Banach spaces (infinitely many constraints), where we will need further functional analytic arguments and assumptions (existence of a topological complement, see Section 4.2.1 for details) for making the implicit function machinery work.

4.2.1. *Algebraic and topological preliminaries*

In this section, we gather some elementary algebraic and topological results which will be useful in establishing first-order optimality conditions for minimization under equality constraints. The following result is classic (it may be viewed as a consequence of the Minkowski–Farkas theorem (Theorem 3.5) but can be proved directly by elementary linear algebra).

Lemma 4.2. *Let E be a real vector space and $(p_1, \ldots, p_m, p)$ be $m + 1$ linear forms on E. Then, the following conditions are equivalent:*

1) $\bigcap_{i=1}^m \ker(p_i) \subset \ker(p)$;
2) *there exists* $(\lambda_1, \ldots, \lambda_m) \in \mathbb{R}^m$ *such that* $p = \sum_{i=1}^m \lambda_i p_i$.

Proof. The implication 2) $\Rightarrow$ 1) is obvious. To show the converse implication, assume 1) and consider the linear subspace of $\mathbb{R}^{m+1}$, $G := \{(p_1(x), \ldots, p_m(x), p(x)), \ x \in E\}$ then $\alpha := (0, \ldots, 0, 1) \notin G$, hence G is included in an hyperplane H of $\mathbb{R}^{m+1}$ not containing α. We can write H in the form $H = \{z \in \mathbb{R}^{m+1} : \sum_{i=1} \mu_i z_i + \mu z = 0\}$ for $(\mu_1, \ldots, \mu_m, \mu) \in \mathbb{R}^{m+1} \setminus \{0\}$. Thus, we have $\sum_{i=1} \mu_i p_i + \mu p = 0$ on E. Now, since $\alpha \notin H$, we have $\mu \neq 0$ hence $p = \sum_{i=1}^m \lambda_i p_i$ with $\lambda_i = -\frac{\mu_i}{\mu}$. $\square$

Lemma 4.3. *Let E be a real vector space and $(p_1, \ldots, p_m)$ be m linear forms on E. Then, the following conditions are equivalent:*

1) $\{p_1, \ldots, p_m\}$ are *linearly independent*;
2) *the linear map* $g \in \mathcal{L}(E, \mathbb{R}^m)$ *defined by*

$$g(x) := (p_1(x), \ldots, p_m(x)), \ \forall x \in E$$

   is surjective.

Proof. The linear forms $\{p_1, \ldots, p_m\}$ are linearly dependent if and only if there is a non-zero vector $\lambda \in \mathbb{R}^m$ such that $\sum_{i=1}^m \lambda_i p_i = 0$ which is equivalent to the existence of an hyperplane of $\mathbb{R}^m$ containing the range of g which is the same as the non-surjectivity of g. $\square$

Let us now recall that in a real vector space E two linear subspaces E_1 and E_2 are called algebraic complements, which is denoted by

$$E_1 \oplus E_2 = E$$

if

$$E_1 \cap E_2 = \{0\} \quad \text{and} \quad E_1 + E_2 = E,$$

where $E_1 + E_2 := \{x_1 + x_2, \ (x_1, x_2) \in E_1 \times E_2\}$. In other words, E_1 and E_2 are algebraic complements if the linear map

$$\begin{cases} E_1 \times E_2 \to E, \\ (x_1, x_2) \mapsto x_1 + x_2 \end{cases}$$

is an isomorphism, the inverse of this isomorphism is the map

$$\begin{cases} E \to E_1 \times E_2, \\ x \mapsto (\pi_1(x), \pi_2(x)), \end{cases}$$

where $(\pi_1(x), \pi_1(x)) \in E_1 \times E_2$ is such that $x = \pi_1(x) + \pi_2(x)$, $\pi_1(x)$ (respectively, $\pi_2(x)$) is called the projection of x onto E_1 (respectively, E_2) along E_2 (respectively, E_1). When $E = E_1 \oplus E_2$, one can (and usually does) identify E with $E_1 \times E_2$ through the projection maps (π_1, π_2).

Any linear subspace E_1 of E admits algebraic complements. Indeed, taking a basis B_1 of E_1, we can find a basis (it is easy to see in finite dimensions, in infinite dimensions, one has to use Zorn's lemma) of E containing B_1, i.e., of the form $B_1 \cup B_2$ — the linear subspace E_2 spanned by B_2 then is an algebraic complement of E_1. If E is a Hilbert space (in particular, this covers the case of finite dimensions), if E_1 is a *closed* subspace of E, then the fact that the orthogonal of E_1, $E_2 := E_1^\perp$ is a complement of E_1 directly follows from Proposition 1.16.

The following result is classic (and actually a standard tool to prove the rank-nullity theorem).

Lemma 4.4. *Let E and F be two real vector spaces, $u \in \mathcal{L}(E, F)$ be surjective, $E_1 := \ker(u)$ and E_2 be a complement of E_1. Then, the restriction $u_{|E_2}$ is an isomorphism between E_2 and F.*

Proof. Since u is surjective and $E_1 \oplus E_2 = E$, any $y \in F$ can be written as $y = u(x_1 + x_2)$ for some $(x_1, x_2) \in E_1 \times E_2$ hence $y = u(x_1) + u(x_2) = u(x_2)$ which shows that $u_{|E_2}$ is surjective. Now if $x \in \ker(u_{|E_2})$ then $x \in E_1 \cap E_2 = \{0\}$ which shows that $u_{|E_2}$ is injective. $\qquad\square$

If E is finite dimensional and $E = E_1 \oplus E_2$, then E_1 and E_2 are automatically closed and the projections (π_1, π_2) are automatically continuous. In infinite dimensions, this needs not to be the case in general. This leads to the notion of topological complements.

Definition 4.2. Let E be a Banach space and E_1 be a closed subspace of E. Then a linear subspace E_2 of E is called a topological complement of E_1 if it is closed and $E_1 \oplus E_2 = E$. A closed subspace of E is called topologically complemented if it admits a topological complement. $\qquad\cdot$

Proposition 4.6. *Let E be a Banach space, E_1 be a closed subspace of E and E_2 be a topological complement of E_1. Then, the corresponding projections (π_1, π_2) are continuous. In other words, not only every $x \in E$ can be decomposed in a unique way as $x = x_1 + x_2$ with $(x_1, x_2) \in E_1 \times E_2$ but one also has the existence of a constant C such that the estimates $\|x_1\| \leq C\|x\|$, $\|x_2\| \leq C\|x\|$ hold for every $x \in E$.*

Proof. Since E_1 and E_2 are closed, equipped with the norm of E, they are Banach spaces and the linear map

$$\begin{cases} E_1 \times E_2 \to E, \\ (x_1, x_2) \mapsto x_1 + x_2 \end{cases}$$

is continuous (by the triangle inequality!) and invertible, it thus follows from Theorem 1.23 that its inverse is continuous, which is exactly the desired continuity of the projections: $(\pi_1, \pi_2) \in \mathcal{L}_c(E, E_1) \times \mathcal{L}_c(E, E_2)$. $\qquad\square$

The existence of a topological complement to a given closed subspace of a Banach space cannot be taken for granted in general, however it holds in the following situations.

Proposition 4.7. *Let E be a Banach space and E_1 be a closed subspace of E. Then E_1 is topologically complemented:*

1) *when E_1 has finite codimension;*
2) *when E_1 has finite dimension;*
3) *when E is a Hilbert space.*

Proof. 1) If E_1 has finite codimension it has a finite-dimensional algebraic complement which, as such, is closed. 2) If E_1 is finite dimensional, let $\{e_1, \ldots, e_n\}$ be a basis of E and let $(g_1, \ldots, g_n) \in (E_1^*)^n$ be such that $g_j(e_i) = \delta_{ij}$, thanks to Corollary 3.1, each g_i can be extended to E by a continuous linear form still denoted g_i, then $E_2 := \bigcap_{i=1}^n \ker(g_i)$ is closed and is an algebraic complement of E_1. 3) If E is a Hilbert space, as already noted as a consequence of Proposition 1.16, the orthogonal of E_1 is a closed algebraic complement of E_1. □

4.2.2. *Lagrange rule in the case of affine constraints*

Let us start with the simplest case of constrained minimization, namely minimization under linear equality constraints

$$\inf\{f(x), \ x \in \Omega, \ p_i(x) = c_i, \ i = 1, \ldots, m\}, \qquad (4.2)$$

where Ω is a non-empty open subset of the real normed space E, $f : \Omega \to \mathbb{R}$, $p_1, \ldots, p_m$ are m-linear forms on E and $(c_1, \ldots, c_m)$ are m real constants.

The derivation of the first-order condition for (4.2) is based on a simple application of Lemma 4.2. Indeed, from Proposition 4.1 and Lemma 4.2, we have the following proposition.

Proposition 4.8. *Let x^* be a local minimizer of (4.2). If f is Gateaux-differentiable at x^*, then there exists $(\lambda_1, \ldots, \lambda_m) \in \mathbb{R}^m$ such that*

$$D_G f(x^*) = \sum_{i=1}^m \lambda_i p_i. \qquad (4.3)$$

Proof. Let $h \in \bigcap_{i=1}^m \ker(p_i)$. Then, for every $t \in \mathbb{R}$, $p_i(x^* + th) = c_i$ so for $t > 0$ small enough $\frac{f(x^* + th) - f(x^*)}{t} \geq 0$ yielding $D_G f(x^*)(h) \geq 0$. Using the same argument with $-h$ instead of h, we deduce that $D_G f(x^*)(h) = 0$. This means that $\bigcap_{i=1}^m \ker(p_i) \subset \ker(D_G f(x^*))$ so that the desired result directly follows from Lemma 4.2. □

The reals $(\lambda_1, \ldots, \lambda_m) \in \mathbb{R}^m$ which appear in (4.3) are called *Lagrange multipliers*. In the convex case where f is convex and Ω is convex, condition (4.3) is also sufficient for a global minimum. Indeed, assume (4.3) and let $y \in \Omega$ such that $p_i(y) = c_i$ for $i = 1, \ldots, m$; since $p_i(y - x^*) = 0$, using (4.3) and the *above the tangent* characterization of the convexity of f, we get

$$f(y) \geq f(x^*) + D_G f(x^*)(y - x^*) = f(x^*) + \sum_{i=1}^{m} p_i(y - x^*) = f(x^*).$$

4.2.3. *Lagrange rule in the finite-dimensional case*

Let us consider now the case where $E = \mathbb{R}^d$, $F = \mathbb{R}^m$, f is defined on a nonempty open set Ω of $\mathbb{R}^d$, $g = (g_1, \ldots, g_m)$ with $g_i : \Omega \to \mathbb{R}$. Consider then the constrained minimization problem:

$$\inf_{x \in A} f(x), \quad \text{with } A := \{x \in \Omega, \ g_i(x) = 0, \ i = 1, \ldots, m\}. \quad (4.4)$$

Before we state and prove an optimality condition in the form of a suitable multiplier rule for (4.4), we will need some preliminaries. Our goal is to linearize the constraint $g(x) = 0$ near a point $x^* \in A$ of interest thanks to the implicit function theorem. For this purpose, we will assume that

$$g \in C^1(\Omega, \mathbb{R}^m), \ \{\nabla g_1(x^*), \ldots, \nabla g_m(x^*)\} \text{ are linearly independent.} \quad (4.5)$$

Note that assumption (4.5) imposes that $m \leq d$ so that there are less constraints than degrees of freedom for x.

The assumption that $\{\nabla g_1(x^*), \ldots, \nabla g_m(x^*)\}$ are linearly independent is equivalent, thanks to Lemma 4.3, to the fact that the linear map $g'(x^*)$: $\mathbb{R}^d \mapsto \mathbb{R}^m$ is surjective. Let us then define

$$E_1 := \ker g'(x^*) = \{h \in \mathbb{R}^d : \nabla g_i(x^*) \cdot h = 0, \ i = 1, \ldots, m\} \quad (4.6)$$

and consider E_2 a complement of E_1. Note that the dimension of $E_1 = \ker g'(x^*)$ is $d - m$ and that the dimension of E_2 is m.

We then identify $\mathbb{R}^d$ with $E_1 \times E_2$, that is, slightly abusing notations write elements of $\mathbb{R}^d$ as $x = (\pi_1(x), \pi_2(x)) = (x_1, x_2) \in E_1 \times E_2$. Likewise, we will denote $x^* = (x_1^*, x_2^*)$ and will view g as a map from an open subset of $E_1 \times E_2$ to $\mathbb{R}^m$ that is we will write $g(x_1, x_2)$ instead of $g(x_1 + x_2)$, which amounts to compose the initial map g with the isomorphism $(x_1, x_2) \in E_1 \times E_2 \mapsto x_1 + x_2 \in E$. Since $E = \mathbb{R}^d$ and $F = \mathbb{R}^m$ are finite dimensional, this isomorphism is continuous, so this composition does not affect the fact that g is C^1. Denoting by $\partial_1 g$ and $\partial_2 g$ the corresponding partial derivatives of g, note that since $E_1 = \ker(g'(x^*))$ and $\partial_1 g(x^*)$ is the restriction of $g'(x^*)$ to E_1, by construction, we have

$$\partial_1 g(x^*) = 0. \tag{4.7}$$

It now follows from Lemma 4.4 that $\partial_2 g(x^*) = g'(x^*)|_{E_2}$ is an isomorphism between E_2 and $\mathbb{R}^m$. The implicit function theorem thus enables us to find open neighborhoods U_1 and U_2 of x_1^* and x_2^* with $U_1 \times U_2 \subset \Omega$ and $\varphi \in C^1(U_1, U_2)$ such that

$$\{(x_1, x_2) \in U_1 \times U_2 : g(x_1, x_2) = 0\} = \{(x_1, \varphi(x_1)), \; x_1 \in U_1\}. \tag{4.8}$$

In other words, in a neighborhood of x^*, A can be represented as the graph of φ. Keeping this construction in mind, it is easy to prove the following Lagrange multiplier rule for (4.4).

Theorem 4.1. *Let x^* be a local minimizer of (4.4). If f is Fréchet-differentiable at x^* and (4.5) holds, then there exist Lagrange multipliers $(\lambda_1, \ldots, \lambda_m) \in \mathbb{R}^m$ such that*

$$\nabla f(x^*) + \sum_{i=1}^m \lambda_i \nabla g_i(x^*) = 0. \tag{4.9}$$

Proof. Let $h_1 \in E_1$. For $t \in \mathbb{R}$ small enough (so that $x_1^* + th_1 \in U_1$) define $\gamma(t) := (x_1^* + th_1, \varphi(x_1^* + th_1))$. Since $\gamma(t) \in A$, $\gamma(0) = x^*$ and x^* is a local minimizer of f in A, the function $t \mapsto f(\gamma(t))$ has a local minimum at 0, hence

$$0 = \frac{d}{dt}(f \circ \gamma)(0) = \partial_1 f(x^*)(h_1) + (\partial_2 f(x^*) \circ \varphi'(x_1^*))(h_1)$$

but differentiating the relation $g(x_1, \varphi(x_1)) = 0$ at x_1^* yields

$$0 = \partial_1 g(x^*) + \partial_2 g(x^*) \circ \varphi'(x_1^*).$$

Since $\partial_1 g(x^*) = 0$ and $\partial_2 g(x^*)$ is an isomorphism, we deduce that $\varphi'(x_1^*) = 0$ hence

$$\partial_1 f(x^*)(h_1) = f'(x^*)(h_1) = \nabla f(x^*) \cdot h_1 = 0, \quad \forall h_1 \in E_1.$$

Since $E_1 = \bigcap_{i=1}^m \ker(g_i'(x^*))$, Lemma 4.2 enables us to conclude that $f'(x^*)$ is a linear combination of $g_1'(x^*), \ldots, g_m'(x^*)$ which is the same as (4.9). $\qquad\square$

Note that, by linear independence of $\{\nabla g_1(x^*), \ldots, \nabla g_m(x^*)\}$, the Lagrange multipliers λ_i in (4.3) are unique. In practice, Theorem 4.1 gives a necessary optimality condition where we look for $d + m$ unknowns (the d components of x^* and the multipliers λ_i). The constraints on x^* give us m equations and the optimality condition (4.3) gives us d extra equations. This can be conveniently expressed by means of the Lagrangian of problem (4.4):

$$\mathscr{L}(x, \lambda) := f(x) + \sum_{i=1}^{m} \lambda_i g_i(x), \ x \in \Omega \subset \mathbb{R}^d, \ \lambda = (\lambda_1, \ldots, \lambda_m) \in \mathbb{R}^m.$$

$$(4.10)$$

Indeed, the Lagrange multiplier rule (4.9) can be expressed as

$$\nabla \mathscr{L}(x^*, \lambda) = 0, \text{ i.e., } \nabla_x \mathscr{L}(x^*, \lambda) = 0 \text{ and } \nabla_\lambda \mathscr{L}(x^*, \lambda) = 0, \qquad (4.11)$$

where $\nabla_\lambda \mathscr{L}$ stands for the gradient of the Lagrangian with respect to λ so that the condition $\nabla_\lambda \mathscr{L}(x^*, \lambda) = 0$ is just a convenient way to write the constraint $0 = g_1(x) = \cdots = g_m(x)$.

Let us consider an elementary exercise as a direct application.

Exercise 4.1. *We wish to solve*

$$\inf\{x + y + 2z^2 : x^2 + y^2 + z^2 = 1\}.$$

This consists in minimizing a polynomial on the unit sphere S which is compact, so existence of a minimizer (as well as of a maximizer) of $f(x, y, z) = x + y + 2z^2$ on S can be taken for granted. There is a single equality constraint $g = 0$ with $g(x, y, z) := x^2 + y^2 + z^2 - 1$ and $\nabla g(x, y, z) = 2(x, y, z)$ which does not vanish on S. So, thanks to Theorem 4.1, if (x, y, z) is a solution there exists a $\lambda \in \mathbb{R}$ such that

$$1 = 2\lambda x = 2\lambda y, \ 4z = 2\lambda z \text{ and } x^2 + y^2 + z^2 = 1.$$

This gives $x = y$ and either $\lambda = 2$ or $z = 0$. These two cases lead to $(x, y, z) \in \{(1/4, 1/4, (7/8)^{1/2}), \ (1/4, 1/4, -(7/8)^{1/2}), (1/\sqrt{2}, 1/\sqrt{2}, 0), (-1/\sqrt{2}, -1/\sqrt{2}, 0)\}$. A direct evaluation of f enables us to conclude that $(-1/\sqrt{2}, -1/\sqrt{2}, 0)$ is the only point where f achieves its minimum on S (similarly, $(1/4, 1/4, (7/8)^{1/2})$ and $(1/4, 1/4, -(7/8)^{1/2})$ are the maximizers of f on S).

4.2.4. *Lagrange rule in Banach spaces*

We now consider the case where E is a Banach space, Ω is a non-empty open subset of E, and the constraint reads as $g(x) = 0$ where $g : \Omega \to F$ and F is another Banach space. The problem we consider now is

$$\inf_{x \in A} f(x), \quad \text{with } A := \{x \in \Omega, \ g(x) = 0\}. \tag{4.12}$$

Note that, in (4.12), we allow both the optimization variable x and the constraint $g(x) = 0$ to be infinite dimensional (so that existence of a solution has to be carefully studied).

We shall use exactly the same strategy based on the implicit function theorem as in the finite-dimensional case, but we have to pay attention to the fact that in infinite dimensions, the implicit function theorem for a map defined on a product $E_1 \times E_2$ with values in F requires E_1 and E_2 to be Banach spaces, so if E_1 and E_2 are linear subspaces of E, they should be closed. This is why, given $x^* \in A$, we shall not only assume

$$g \in C^1(\Omega, F), \ g'(x^*) \text{ is surjective} \tag{4.13}$$

but also

$$\ker(g'(x^*)) \text{ is topologically complemented.} \tag{4.14}$$

As in the finite-dimensional case, we then define $E_1 := \ker(g'(x^*))$ (which is a closed subspace of E since $g'(x^*) \in \mathcal{L}_c(E, F)$) and let E_2 be a linear subspace of E such that

$$E_2 \text{ is closed}, \quad E_1 \oplus E_2 = E. \tag{4.15}$$

We can again identify the Banach spaces E and $E_1 \times E_2$ through linear continuous isomorphisms (Proposition 4.6), view Ω as an open subset of $E_1 \times E_2$ and g as a C^1 map of two variables $(x_1, x_2) \in \Omega$. Denoting by $\partial_i g(x^*)$ the partial derivatives of g with respect to x_i at x^* we thus have $\partial_1 g(x^*) = 0$ and $\partial_2 g(x^*)$ is a continuous isomorphism between E_2 and F. The implicit function theorem (recall that Theorem 2.16 was stated in a general Banach spaces framework) thus enables us to find open neighborhoods U_1 and U_2 of x_1^* and x_2^* with $U_1 \times U_2 \subset \Omega$ and $\varphi \in C^1(U_1, U_2)$ such that (4.8) holds.

Theorem 4.2. *Let x^* be a local minimizer of (4.12). If f is Fréchet-differentiable at x^* and (4.13)–(4.14) hold, then there exists $\Lambda \in F^*$ such that*

$$f'(x^*) = \Lambda \circ g'(x^*). \tag{4.16}$$

Proof. Arguing exactly as in the proof of Theorem 4.1, we find that $f'(x^*)(h_1) = 0$ for every $h_1 \in E_1 := \ker(g'(x^*))$. Since both $g'(x^*)$ and $f'(x^*)$ vanish on E_1, one has $f'(x^*) = f'(x^*) \circ \pi_2$ and $g'(x^*) = g'(x^*) \circ \pi_2$ where π_2 is the (continuous) projection onto E_2 along E_1. Denoting by $\Psi \in \mathcal{L}_c(F, E_2)$ the inverse of $\partial_2 g(x^*) = g'(x^*)_{|E_2}$ we have $\pi_2 = \Psi \circ g'(x^*) \circ \pi_2 = \Psi \circ g'(x^*)$ so that $f'(x^*) = f'(x^*) \circ \Psi \circ g'(x^*) = \Lambda \circ g'(x^*)$ with $\Lambda := f'(x^*) \circ \Psi \in F^*$. □

Note that since $g'(x^*)$ is surjective the linear form Λ in (4.16) is unique.

4.2.5. *Second-order conditions*

We consider again (4.12) where E and F are Banach spaces, assuming further differentiability conditions we obtain the following second-order optimality condition.

Theorem 4.3. *Let x^* be a local minimizer of (4.12). If f is twice differentiable at x^*, $g \in C^2(\Omega, F)$, (4.13)–(4.14) hold and $\Lambda \in F^*$ is such that (4.16) holds, then*

$$f''(x^*)(h, h) - \Lambda(g''(x^*)(h, h)) \geq 0, \quad \forall h \in \ker(g'(x^*)). \qquad (4.17)$$

Proof. We adopt the same notations as in the previous section, in particular set $E_1 = \ker(g'(x^*))$, E_2 is a topological complement of E_1, and express locally around $x^* = (x_1^*, x_2^*)$ the constraint $g(x_1, x_2) = 0$ in the form (4.8), i.e., as the graph of a map $\varphi : U_1 \to U_2$ where U_1 and U_2 are open neighborhoods of x_1^* in E_1 and x_2^* in E_2, respectively. Since g is assumed to be C^2, $\varphi \in C^2(U_1, U_2)$ (see Remark 2.9). We then define for all $x_1 \in U_1$, $F(x_1) := g(x_1, \varphi(x_1))$ and note that F has a local minimum at x_1^* and since F is twice differentiable at x_1^* we have

$$F'(x_1^*) = 0, \quad F''(x_1^*)(h, h) \geq 0, \quad \forall h \in E_1. \qquad (4.18)$$

Differentiating the identity $g(x_1, \varphi(x_1)) = 0$ which holds on U_1, we first obtain

$$\partial_1 g(x_1, \varphi(x_1)) + \partial_2 g(x_1, \varphi(x_1)) \circ \varphi'(x_1) = 0, \quad \forall x_1 \in U_1, \qquad (4.19)$$

at $x_1 = x_1^*$, since $\partial_1 g(x^*) = 0$ and $\partial_2 g(x_1^*, \varphi(x_1^*)) = \partial_2 g(x^*)$ is invertible, we deduce that

$$\varphi'(x_1^*) = 0. \qquad (4.20)$$

Differentiating (4.19), we then have

$$\partial_{11}^2 g(x_1, \varphi(x_1)) + 2\partial_{12}^2 g(x_1, \varphi(x_1)) \circ \varphi'(x_1)$$
$$+ \partial_{22}^2 g(x_1, \varphi(x_1))(\varphi'(x_1), \varphi'(x_1)) + \partial_2 g(x_1, \varphi(x_1)) \circ \varphi''(x_1) = 0. \tag{4.21}$$

For $x_1 = x_1^*$, using (4.20), this yields

$$\partial_{11}^2 g(x^*) + \partial_2 g(x^*) \circ \varphi''(x_1^*) = 0. \tag{4.22}$$

Performing similar computations as in (4.21) and (4.22) leads to

$$F''(x_1^*)(h, h) = \partial_{11}^2 f(x^*)(h, h) + \partial_2 f(x^*)(\varphi''(x_1^*)(h, h)), \quad \forall h \in E_1.$$

Recalling the Lagrange relation $\partial_2 f(x^*) = \Lambda \circ \partial_2 g(x^*)$ and (4.22) gives

$$F''(x_1^*)(h, h) = \partial_{11}^2 f(x^*)(h, h) + \Lambda(\partial_2 g(x^*) \circ \varphi''(x_1^*)(h, h))$$
$$= \partial_{11}^2 f(x^*)(h, h) - \Lambda(\partial_{11}^2 g(x^*)(h, h))$$
$$= f''(x^*)(h, h) - \Lambda(g''(x^*)(h, h)).$$

The desired result thus follows from (4.18). $\qquad\square$

Let us finish this section with a sufficient condition for local strict minimizers of (4.12).

Theorem 4.4. *Let $x^* \in A$. If f is twice Fréchet-differentiable at x^*, $g \in C^2(\Omega, F)$, (4.13)–(4.14) hold, $\Lambda \in F^*$ is such that (4.16) holds and there is a $\lambda > 0$ such that*

$$f''(x^*)(h, h) - \Lambda(g''(x^*)(h, h)) \geq \lambda \|h\|^2, \quad \forall h \in \ker(g'(x^*)), \tag{4.23}$$

then x^* is a local strict minimizer of f on A.

Proof. Introducing the same function F ($F(x_1) := f(x_1, \varphi(x_1))$) as in the proof of Theorem 4.3, we deduce from Proposition 4.4 that x_1^* is a local strict minimizer of F on U_1 that is x^* a local strict minimizer of f on A since, around x^*, A is the graph of φ. $\qquad\square$

4.3. Equality and inequality constraints

In this section, we shall consider both equality and inequality constraints. Let us consider an elementary example to have a guess of what the multiplier rule should look like when there are inequality constraints. Let $f : [0,1] \to \mathbb{R}$ be a differentiable function and suppose that x^* is a local minimizer of f on $[0,1]$ which corresponds to the simple constraint $0 \leq x \leq 1$. Then, there are three cases:

- $x^* \in (0,1)$, then for $h \in \mathbb{R}$ small enough both $x^* + h$ and $x^* - h$ remain in the admissible set $[0,1]$ (one says that the constraints are not binding at x^*) so that $f'(x^*) = 0$;
- $x^* = 0$ in which case one can only compare $f(0)$ with $f(h)$ when $h \geq 0$ yielding $f'(0) = \lim_{h \to 0+} \frac{f(h) - f(0)}{h} \geq 0$;
- $x^* = 1$, then one can only say that $f(1) - f(1-h) \leq 0$ yielding $f'(1) \leq 0$.

One can unify these three cases by saying that any local minimizer x^* of f under the constraints $-x \leq 0$ and $x \leq 1$ satisfies

$$f'(x^*) = \mu_0 - \mu_1 \tag{4.24}$$

with

$$\mu_0 \geq 0, \ \mu_1 \geq 0, \ \mu_0 x^* = 0, \ \mu_1(x^* - 1) = 0, \tag{4.25}$$

μ_0 and μ_1 are called Karush–Kuhn–Tucker (in short KKT) multipliers associated with the constraints $-x \leq 0$ and $x \leq 1$. Note that the main difference with Lagrange multipliers associated with equality constraints is that they have a sign and that they are non-zero only when the corresponding constraint is binding, as expressed in the *complementarity slackness condition* (4.25).

4.3.1. *Karush–Kuhn–Tucker conditions for affine constraints*

Again, we start with the simplest case of affine constraints, namely minimization under linear equality and inequality constraints:

$$\inf\{f(x), \ x \in \Omega, \ p_i(x) = c_i, \ i = 1, \ldots, m, \ q_j(x) \leq d_j, j = 1, \ldots, p\}, \tag{4.26}$$

where Ω is a non-empty open subset of the real normed space E, $f : \Omega \to \mathbb{R}$, $(p_1, \ldots, p_m, q_1, \ldots, q_p) \in (E')^m \times (E^*)^p$ are linear form on E, $(c_1, \ldots, c_m)$

and $(d_1, \ldots, d_p)$ are $m + p$ real constants. The admissible set is then given by

$$A := \{x \in \Omega,\ p_i(x) = c_i,\ i = 1, \ldots, m,\ q_j(x) \le d_j, j = 1, \ldots, p\}.$$

For $x \in A$, we denote by $J(x)$ the set of binding (one also says active) inequality constraints at x, i.e.,

$$J(x) := \{j \in \{1, \ldots, p\}\ :\ q_j(x) = d_j\}. \tag{4.27}$$

The derivation of the KKT theorem for minimization subject to linear constraints as in (4.26) easily follows from the Minkowski–Farkas theorem.

Proposition 4.9. *Let x^* be a local minimizer of (4.2). If f is Gateaux-differentiable at x^*, then there exist Lagrange multipliers $(\lambda_1, \ldots, \lambda_m) \in \mathbb{R}^m$ and KKT multipliers $(\mu_j)_{j \in J(x^*)} \in \mathbb{R}_+^{J(x^*)}$ such that*

$$D_G f(x^*) = \sum_{i=1}^m \lambda_i p_i - \sum_{j \in J(x^*)} \mu_j q_j. \tag{4.28}$$

Proof. Let $h \in E$ such that

$$p_i(h) = 0, i = 1, \ldots, m,\ q_j(x)h \le 0,\ \forall j \in J(x^*).$$

Since $q_j(x^*) < d_j$ for $j \in \{1, \ldots, p\} \setminus J(x^*)$, it follows from the continuity of q_j that for small enough $t > 0$ one has $q_j(x^* + th) \le d_j$ for $j \in \{1, \ldots, p\} \setminus J(x^*)$. Therefore, $x^* + th \in A$ for small enough $t > 0$. Passing to the limit in $t^{-1}[f(x^* + th) - f(x^*)] \ge 0$ gives

$$D_G f(x^*)(h) \ge 0.$$

The Minkowski–Farkas theorem (Theorem 3.5) therefore implies that there exist $(\lambda_1, \ldots, \lambda_m) \in \mathbb{R}^m$ and KKT multipliers $(\mu_j)_{j \in J(x^*)} \in \mathbb{R}_+^{J(x^*)}$ such that (4.28) is satisfied. $\qquad \square$

The fact that the only KKT multipliers which appear in (4.28) correspond to active constraints can also be rewritten by writing a multiplier for each constraint, i.e., rewrite (4.28) as

$$D_G f(x^*) = \sum_{i=1}^m \lambda_i p_i - \sum_{j=1}^p \mu_j q_j \tag{4.29}$$

but it has to be supplemented with the non-negativity and complementarity slackness conditions:

$$\mu_j \ge 0,\ \mu_j(d_j - q_j(x^*)) = 0;\ \forall j = 1, \ldots, p. \tag{4.30}$$

We let the reader check that the conditions above are sufficient in the convex case.

Exercise 4.2. *Under the assumptions of Proposition 4.9, further assume that Ω is convex and f is convex in Ω. Show that (4.29)–(4.30) are necessary and sufficient conditions for x^* to be a global minimizer of f on A.*

Let us illustrate Proposition 4.9 with the case of a quadratic minimization with affine constraints.

Example 4.1. Let A be a symmetric positive definite $d \times d$ matrix, p_1, p_2 be linearly independent vectors of $\mathbb{R}^d$ and consider

$$\inf_{x \in \mathbb{R}^d} \frac{1}{2} Ax \cdot x \; : \; p_1 \cdot x_1 = 1, \; p_2 \cdot x \geq 1. \tag{4.31}$$

Let us first observe that (4.31) consists of minimizing a strictly convex and coercive objective function over a (non-empty since p_1 and p_2 are linearly independent) closed convex subset of $\mathbb{R}^d$ so that it admits a unique solution that we denote by x^*. Applying Proposition 4.9, we find that there exist $\lambda \in \mathbb{R}$ and $\mu \geq 0$ with $\mu(p_2 \cdot x - 1) = 0$ (complementarity slackness) such that

$$x^* = \lambda A^{-1} p_1 + \mu A^{-1} p_2. \tag{4.32}$$

Since $p_2 \cdot x^* = 1$, we have

$$1 = \lambda A^{-1} p_1 \cdot p_1 + \mu A^{-1} p_1 \cdot p_2. \tag{4.33}$$

Let us then distinguish two cases:

1) the inequality constraint is not binding, i.e., $p_2 \cdot x^* > 1$ then $\mu = 0$ and one directly deduces from (4.32)–(4.33)

$$x^* = \frac{A^{-1} p_1}{A^{-1} p_1 \cdot p_1};$$

this is indeed the minimizer whenever

$$\frac{A^{-1} p_1 \cdot p_2}{A^{-1} p_1 \cdot p_1} \geq 1, \quad \text{i.e.,} \quad A^{-1} p_1 \cdot p_2 \geq A^{-1} p_1 \cdot p_1.$$

2) $A^{-1} p_1 \cdot p_2 < A^{-1} p_1 \cdot p_1$; in this case, the inequality constraint is binding; $p_2 \cdot x^* = 1$, this gives the extra linear relation

$$1 = \lambda A^{-1} p_1 \cdot p_2 + \mu A^{-1} p_2 \cdot p_2. \tag{4.34}$$

We solve the linear system (4.33)–(4.34) whose determinant:

$$m := (A^{-1} p_1 \cdot p_1)(A^{-1} p_2 \cdot p_2) - (A^{-1} p_1 \cdot p_2)^2$$

is positive (strict inequality in Cauchy–Schwarz since p_1 and p_2 are linearly independent) so that

$$x^* := \frac{A^{-1}p_2 \cdot p_2 - A^{-1}p_1 \cdot p_2}{m} A^{-1}p_1 + \frac{A^{-1}p_1 \cdot p_1 - A^{-1}p_1 \cdot p_2}{m} A^{-1}p_2.$$

4.3.2. Karush–Kuhn–Tucker conditions in the general case

Let us now consider two Banach spaces E and F, Ω an open subset of E, $f : \Omega \to \mathbb{R}$ (the cost function), $g : \Omega \to F$ (defining the equality constraints), $p \in \mathbb{N}^*$, and for $j = 1, \ldots, p$, $k_j : \Omega \to \mathbb{R}$ (defining the inequality constraints). Consider then:

$$\inf_{x \in A} f(x), \quad A := \{x \in \Omega : g(x) = 0, \ k_j(x) \leq 0, j = 1, \ldots, p\}. \tag{4.35}$$

For $x \in A$, as we did in the affine case, let us denote by $J(x)$ set of binding constraints at x, i.e.,

$$J(x) := \{j \in \{1, \ldots, p\} : k_j(x) = 0\}.$$

We are interested in optimality conditions for a local solution $x^* \in A$ of (4.35). We already saw that conditions (4.13)–(4.14) permit to treat the equality constraints by means of the implicit function theorem. To deal with the inequality conditions, we will assume that

$$k_j \text{ is continuous at } x^* \text{ for } j = 1, \ldots, p,$$

$$k_j \text{ is differentiable at } x^* \text{ for } j \in J(x^*), \tag{4.36}$$

as well as the so-called Mangasarian–Fromowitz condition:

$$\exists h_0 \in \ker(g'(x^*)) : k_j'(x^*)(h_0) < 0, \ \forall j \in J(x^*). \tag{4.37}$$

The Karush–Kuhn–Tucker theorem then reads as follows.

Theorem 4.5 (Karush–Kuhn–Tucker). *Let x^* be a local minimizer of (4.12). If f is Fréchet-differentiable at x^* and (4.13)–(4.14)–(4.36)–(4.37) hold, then there exist $\Lambda \in F^*$ and $(\mu_j)_{j \in J(x^*)} \in \mathbb{R}_+^{J(x^*)}$ such that*

$$f'(x^*) = \Lambda \circ g'(x^*) - \sum_{j \in J(x^*)} \mu_j k_j'(x^*). \tag{4.38}$$

Proof. We treat the equality constraint exactly as we did in the proof of Theorem 4.2, and write locally near x^* the constraint $g(x) = 0$ as $x = (x_1, \varphi(x_1)) \in E_1 \times E_2$ with $E_1 := \ker(g'(x^*))$ and E_2 a topological complement of E_1. Let $h \in E_1$ such that $k'_j(x^*)(h) \leq 0$ for every $j \in J(x^*)$. Let $h_0 \in E_1$ be as in the Mangasarian–Fromowitz condition (4.37); for $\varepsilon > 0$ set $h_\varepsilon := h + \varepsilon h_0$. For small $t \geq 0$, then define

$$\gamma_\varepsilon(t) := (x_1^* + th_\varepsilon, \varphi(x_1^* + th_\varepsilon)).$$

By construction $g(\gamma_\varepsilon(t)) = 0$. Differentiating γ_ε at $t = 0^+$, using the fact that $\varphi'(x_1^*) = 0$ (see the proof of Theorem 4.1) we first obtain

$$\gamma_\varepsilon(t) = x^* + th_\varepsilon + o(t). \tag{4.39}$$

If $j \notin J(x^*)$, since $k_j(x^*) < 0$ and k_j is continuous at $x^* = \gamma_\varepsilon(0)$, we have $k_j(\gamma_\varepsilon(t)) \leq 0$ for $t \geq 0$ small enough. Finally, when $j \in J(x^*)$, $k_j(x^*) = 0$, since $(k_j \circ \gamma_\varepsilon)'(0) = k'_j(x^*)(h_\varepsilon)$, we have

$$k_j(\gamma_\varepsilon(t)) = tk'_j(x^*)h_\varepsilon + o(t)$$
$$= t(k'_j(x^*)(h) + \varepsilon k'_j(x^*)(h_0) + o(1)) \leq t(\varepsilon k'_j(x^*)(h_0) + o(1))$$

and since $k'_j(x^*)(h_0) < 0$, we also have $k_j(\gamma_\varepsilon(t)) \leq 0$ for small enough $t \geq 0$. For $t \geq 0$ small enough, we therefore have $\gamma_\varepsilon(t) \in A$ so that $f(\gamma_\varepsilon(t)) - f(\gamma(0)) \geq 0$ for small $t \geq 0$. This yields

$$0 \leq \frac{d}{dt}(f \circ \gamma_\varepsilon)(0) = f'(x^*)(h_\varepsilon) = f'(x^*)(h) + \varepsilon f'(x^*)(h_0).$$

Letting $\varepsilon \to 0^+$ we get

$$f'(x^*)(h) \geq 0, \forall h \in E_1 \ : \ k'_j(x^*)(h) \leq 0, \quad \forall j \in J(x^*).$$

We thus deduce from Theorem 3.5 (applied to linear forms on E_1) that there exists $(\mu_j)_{j \in J(x^*)} \in \mathbb{R}_+^{J(x^*)}$ such that

$$\left(f'(x^*) + \sum_{j \in J(x^*)} \mu_j k'_j(x^*)\right)(h) = 0, \quad \forall h \in E_1. \tag{4.40}$$

We then conclude as in the proof of Theorem 4.2 by writing π_2 (the projection onto E_2 along E_1) as $\pi_2 = \Psi \circ g'(x^*)$ with $\Psi \in \mathcal{L}_c(F, E_2)$ and thus

(4.40) rewrites

$$\left(f'(x^*) + \sum_{j \in J(x^*)} \mu_j k'_j(x^*)\right) = \left(f'(x^*) + \sum_{j \in J(x^*)} \mu_j k'_j(x^*)\right) \circ \pi_2 = \Lambda \circ g'(x^*)$$

with $\Lambda = (f'(x^*) + \sum_{j \in J(x^*)} \mu_j k'_j(x^*)) \circ \Psi \in F^*$.   □

Exercise 4.3. *Solve the minimization problem*

$$\inf\{x_2 \ : x_2^3 \geq x_1^2\}.$$

Does the conclusion of the Karush–Kuhn–Tucker theorem hold at the global minimizer? Conclude.

Remark 4.1. In the optimization literature, there are many technical conditions which ensure the validity of the Karush–Kuhn–Tucker theorem. This is what is generally referred to as the qualification of the constraints at x^*. The Mangasarian–Fromowitz condition is the most commonly used. Let us make some comments on (4.37) and mention some variants:

- As mentioned in Section 4.3.1, if all the constraints are affine (and continuous), there is no need neither for (4.13)–(4.14) nor for (4.37).
- In the case where $g = 0$ (no equality constraint), Ω is convex and the functions k_j are convex and differentiable, then the Slater condition

$$\exists x \in \Omega : k_j(x) < 0, \ \forall j \in J(x^*) \tag{4.41}$$

  implies (4.37). Indeed, since for every $j \in J(x^*)$,

$$0 > k_j(x) - k_j(x^*) \geq \nabla k_j(x^*)(x - x^*)$$

  (4.37) holds with $h_0 = x - x^*$.
- Similarly, when g is affine continuous, Ω is convex, the functions k_j are convex and differentiable and (4.41) holds, the Karush–Kuhn–Tucker conditions are valid at the local minimizer x^*.
- If $E = \mathbb{R}^d$, $g = (g_1, \ldots, g_m) : \Omega \to \mathbb{R}^m$, a sufficient condition for (4.37) to hold is that the family of vectors $\nabla g_1(x^*), \ldots, \nabla g_m(x^*), (\nabla k_j(x^*))_{j \in J(x^*)}$ is linearly independent (use Lemma 4.3).

- In the same case, the linear independence of $\nabla g_1(x^*), \ldots, \nabla g_m(x^*)$ together with (4.37) implies a certain positive independence condition; more precisely, if $(\lambda_1, \ldots, \lambda_m, (\mu_j)_{j \in J(x^*)}) \in \mathbb{R}^m \times \mathbb{R}_+^{J(x^*)}$ are such that

$$\sum_{i=1}^{m} \lambda_i \nabla g_i(x^*) = \sum_{j \in J(x^*)} \mu_j \nabla k_j(x^*),$$

then $\lambda_i = \mu_j = 0$ for every $(i,j) \in \{1, \ldots, m\} \times J(x^*)$ (to see this, take the scalar product of the relation above with h_0 given by (4.37) to deduce that $\mu_j = 0$ and then the linear independence of the gradients of the equality constraints to get $\lambda_i = 0$).

4.4. Exercises

Exercise 4.4. *Does the following problem*

$$\inf\{x^2 + 2y^2 + \sin(xy) : (x,y) \in \mathbb{R}^2\}$$

admit solutions? If so, compute them.

Exercise 4.5. *Let E be a Hilbert space, $f \in C^1(E)$ be convex and C be a closed convex cone of E. Show that x is a minimizer of f on C if and only if $\nabla f(x) \cdot x = 0$ and $\nabla f(x) \in C^*$ where C^* is the polar cone of C (see Exercise 3.11).*

Exercise 4.6. *Let $f \in C^1(\mathbb{R}^d, \mathbb{R})$ such that:*

$$\frac{f(x)}{|x|} \to +\infty \quad as \ |x| \to +\infty.$$

Prove that ∇f is surjective.

Exercise 4.7. *Let E be a Banach space and $f : E \to \mathbb{R}$, lower-semicontinuous, be bounded from below and Gateaux-differentiable. Let $\varepsilon > 0$ and $x_\varepsilon \in E$ such that*

$$f(x_\varepsilon) \leq \inf_E f + \varepsilon.$$

Show that there exists $y_\varepsilon \in E$ which satisfies

$$f(y_\varepsilon) \leq f(x_\varepsilon), \quad \|x_\varepsilon - y_\varepsilon\| \leq \sqrt{\varepsilon}, \quad \|D_G f(y_\varepsilon)\|_{E^*} \leq \sqrt{\varepsilon}.$$

Exercise 4.8. *Let Ω be a non-empty bounded open subset of $\mathbb{R}^d$ and let $f \in C^2(\mathbb{R}^d, \mathbb{R})$ such that*

$$\Delta f + |\nabla f|^2 = 1 \text{ in } \Omega \quad \text{and} \quad f = 0 \text{ on } \partial\Omega$$

($\Delta f := \sum_{i=1}^d \partial_{ii}^2 f = \text{tr}(D^2 f)$ denotes the Laplacian of f). Show that $f \leq 0$ on $\bar{\Omega}$ (consider a maximum point of f on $\overline{\Omega}$ and show that it belongs to $\partial\Omega$).

Exercise 4.9. *Let $f : \mathbb{R}^d \to \mathbb{R} \cup \{+\infty\}$ be convex and proper and $g : \mathbb{R}^d \to \mathbb{R}$ be Gateaux-differentiable. Show that if x^* is a local minimizer of $f + g$ then f is subdifferentiable at x^* and $-\nabla g(x^*) \in \partial f(x^*)$.*

Exercise 4.10. *Let $f, g \in C^1(\mathbb{R}^d, \mathbb{R})$ such that*

$$A := \{x \in \mathbb{R}^d : g(x) \leq 0\} \neq \emptyset, \text{ and } f(x) \to \infty \text{ as } |x| \to \infty. \quad (4.42)$$

We are interested in the constrained problem

$$\inf_{x \in A} f(x). \quad (4.43)$$

Defining $g_+(x) := \max(g(x), 0)$ we consider for $\varepsilon > 0$ the (unconstrained) penalized version of (4.43):

$$\inf_{x \in \mathbb{R}^d} f(x) + \frac{1}{2\varepsilon} g_+(x)^2. \quad (4.44)$$

1) *Show that g_+^2 is of class C^1 and compute its gradient.*
2) *Show that (4.44) admits at least a solution x_ε and give the first optimality condition satisfied by x_ε.*
3) *Show that x_ε is bounded. Let x^* be a cluster point of $(x_\varepsilon)_\varepsilon$. Show that $x^* \in A$ and that x^* solves (4.43).*
4) *Show (using the previous questions and not Lagrange's theorem) that if $\nabla g(x^*) \neq 0$, there exists $\mu \geq 0$ such that $\mu g(x^*) = 0$ and $\nabla f(x^*) + \mu \nabla g(x^*) = 0$.*

Exercise 4.11. *Show that ellipses with prescribed length and maximal area are disks.*

Exercise 4.12. *Rigorously solve the program*

$$\sup\{\sqrt{xyz} : (x, y, z) \in \mathbb{R}_+^3, \ x + y + z = 1\}.$$

Exercise 4.13. *Let A be a $d \times d$ matrix with real entries and S be the Euclidean sphere of $\mathbb{R}^d$. Show that the problem*

$$\inf_{x \in S} Ax \cdot x$$

has at least a solution x_0 and that any such solution is an eigenvector of $\frac{1}{2}(A+A^T)$. Show that $Ax_0 \cdot x_0$ is the smallest eigenvalue of $\frac{1}{2}(A+A^T)$. Using the previous question and an induction argument on the dimension, prove that if $A \in S_d(\mathbb{R})$, then there exists an orthonormal basis of eigenvectors of A (spectral theorem).

Exercise 4.14. *Rigorously solve the program*

$$\inf\{x^3 - y^3 : x^2 + y^2 \le 1\}.$$

Exercise 4.15. *Let $f : [0,1] \to \mathbb{R}$ be strictly convex. Solve*

$$\min\left\{\sum_{i=1}^{n} f(x_i) : x_i \ge 0, \sum_{i=1}^{n} x_i = 1\right\}.$$

Exercise 4.16. *Consider the minimization problem*

$$\inf\left\{\tfrac{1}{2}(x-2)^4 + y^2 : x^2 - y^2 \le 0\right\}. \tag{4.45}$$

1) *Show that (4.45) admits at least a solution.*
2) *Show that if (x,y) solves (4.45) then $x^2 = y^2$.*
3) *Solve (4.45).*

Exercise 4.17. *Let $a \in \mathbb{R}$ be a parameter. According to the value of a, solve the problem*

$$\inf\{x^3 + y^3 + az^3 : x^2 + y^2 + z^2 = 1\}.$$

Exercise 4.18. *For $f \in L^2 := L^2(\mathbb{R}^d)$ (the space of square integrable functions with respect to the Lebesgue measure on $\mathbb{R}^d$) and $K \in L^2(\mathbb{R}^d \times \mathbb{R}^d)$ symmetric (i.e., $K(x,y) = K(y,x)$) and non-zero, i.e., $\|K\|_{L^2} > 0$, we define $T_K(f)$ by*

$$T_K(f)(x) := \int_{\mathbb{R}^d} K(x,y)f(y)\,dy, \quad \forall x \in \mathbb{R}^d.$$

We also define

$$J_K(f) := \frac{1}{2}\int_{\mathbb{R}^d \times \mathbb{R}^d} K(x,y)f(x)f(y)\,dx\,dy.$$

We are interested in the following problem:

$$\inf\{\|f\|_{L^2} : f \in L^2(\mathbb{R}^d), J_K(f) = 1\}. \tag{4.46}$$

1) Show that $T_K \in \mathcal{L}_c(L^2(\mathbb{R}^d))$ and that whenever f_n converges weakly to f in L^2, $T_K(f_n)$ converges strongly to $T_K(f)$ in L^2 (one says that the linear operator T_K is compact in L^2).

2) Show that $J_K(f_n)$ converges to $J_K(f)$ whenever f_n converges weakly to f in L^2 and prove that (4.46) has at least a solution f^*.

3) Show that if f^* solves (4.46) there exists $\lambda \in \mathbb{R}$ such that $T_K(f^*) = \lambda f^*$ (in other words, f^* is an eigenfunction of T_K).

4) Show that T_K is self-adjoint, i.e., $\int_{\mathbb{R}^d} T_K(f)g = \int_{\mathbb{R}^d} f T_K(g)$ for every f and g in L^2.

5) Let us fix a solution f_1 of (4.46). Prove that

$$\inf \left\{ \|f\|_{L^2} \; : \; f \in L^2(\mathbb{R}^d), J_K(f) = 1, \int_{\mathbb{R}^d} f f_1 = 0 \right\} \qquad (4.47)$$

has at least a solution and that if f_2 is such a solution, it is also an eigenfunction of T_K.

Exercise 4.19. Let $A = (a_{ij})$ be an $n \times p$ matrix with real and strictly positive entries and consider the minimization problem

$$\inf_{M = [m_{ij}] \in \mathcal{M}_{n,p}(\mathbb{R})} \sum_{i=1}^{n} \sum_{j=1}^{p} m_{ij} \log \left(\frac{m_{ij}}{a_{ij}} \right)$$

subject to the constraints

$$m_{ij} \geq 0, \; \sum_{i=1}^{n} m_{ij} = 1, \forall j, \; \sum_{j=1}^{p} m_{ij} = 1, \forall i.$$

1) Show that this problem admits a unique minimizer $M = [m_{ij}]$.

2) Show that $m_{ij} > 0$ for every i and j.

3) Establish optimality conditions for M.

4) Deduce that there are two diagonal matrices D (of size n) and Δ (of size p) with strictly positive entries such that $\Delta A D$ has sums and rows whose coefficients all sum to 1. Are the matrices D and Δ unique?

Exercise 4.20. Let us equip $\mathcal{M}_d(\mathbb{R})$ with the scalar product $A \cdot B := \operatorname{tr}(A^T B)$ and the corresponding Hilbertian norm $\|\cdot\|$. Recall that the orthogonal group is

$$\mathcal{O}_d(\mathbb{R}) := \{ U \in \mathcal{M}_d(\mathbb{R}) \; : \; UU^T = \operatorname{id} \}.$$

Given $M \in \mathcal{M}_d(\mathbb{R})$, we are interested in the projection problem:

$$\inf_{U \in \mathcal{O}_d(\mathbb{R})} \|M - U\|. \qquad (4.48)$$

1) *Show that $\mathcal{O}_d(\mathbb{R})$ is compact and deduce that (4.48) has a solution.*
2) *Show that (4.48) is equivalent to*

$$\sup_{U \in \mathcal{O}_d(\mathbb{R})} M \cdot U. \qquad (4.49)$$

3) *Let $g : \mathcal{M}_d(\mathbb{R}) \to \mathcal{S}_d(\mathbb{R})$ be defined by $g(M) := MM^T - \mathrm{id}$ for every $M \in \mathcal{M}_d(\mathbb{R})$. Let $U \in \mathcal{O}_d(\mathbb{R})$. Show that $H \in \ker(g'(U))$ if and only if HU^T is skew-symmetric and deduce that $g'(U)$ is surjective.*
4) *Let U solve (4.49). Show that $M = SU$ with $S \in \mathcal{S}_d(\mathbb{R})$ (Hint: show that $MU^T \cdot A = 0$ for every skew-symmetric A and use Exercise 1.16.)*
5) *Let S be as in the previous question. Show that $\mathrm{tr}(S) \geq \mathrm{tr}(SO)$ for every $O \in \mathcal{O}_d(\mathbb{R})$ and deduce that S is positive semidefinite.*
6) *Show that if M is invertible the (polar factorization) decomposition $M = SU$ with U orthogonal and S positive semidefinite symmetric is unique (Hint: have a look at Exercise 1.17.)*

Exercise 4.21. *We consider the variant of Exercise 4.20 where instead of minimizing the distance to the orthogonal group, we minimize the distance to the special orthogonal group*

$$\mathcal{SO}_d(\mathbb{R}) := \{U \in \mathcal{O}_d(\mathbb{R}) \, : \, \det(U) = 1\},$$

that is, given $M \in \mathcal{M}_d(\mathbb{R})$, we consider

$$\inf_{U \in \mathcal{SO}_d(\mathbb{R})} \|M - U\|. \qquad (4.50)$$

1) *Show that this problem admits a solution U and that, for some neighbourhood V of U, $V \cap \mathcal{SO}_d(\mathbb{R}) = V \cap \mathcal{O}_d(\mathbb{R})$.*
2) *Prove that $M = SU$ with S symmetric and that S has at most one negative eigenvalue λ^- and that, in the case there is such a negative eigenvalue, denoting by λ^+ the smallest non-negative eigenvalue of S, one has $\lambda^+ + \lambda^- \geq 0$.*

Exercise 4.22. *We consider a finite-dimensional optimization problem on $\mathbb{R}^d$ with differentiable equality constraints*

$$g_i(x) = 0, \quad i = 1, \ldots, m$$

and differentiable inequality constraints

$$k_j(x) \leq 0, \quad j = 1, \ldots, p.$$

We consider an admissible point x^* and consider the conditions

$$\nabla g_1(x^*), \ldots, \nabla g_m(x^*) \text{ linearly independent} \tag{4.51}$$

and

$$\exists h_0 \in \mathbb{R}^d, \nabla g_i(x^*) \cdot h_0 = 0, \ \nabla k_j(x^*) \cdot h_0 < 0, \ \forall (i,j) \in \{1, \ldots, m\} \times J(x^*). \tag{4.52}$$

Show that (4.51)–(4.52) is equivalent to the fact that whenever

$$\sum_{i=1}^{m} \lambda_i \nabla g_i(x^*) = \sum_{j \in J(x^*)} \mu_j \nabla k_j(x^*)$$

with $\lambda_i \in \mathbb{R}$ and $\mu_j \geq 0$ then necessarily $\lambda_i = \mu_j = 0$ for every $(i,j) \in \{1, \ldots, m\} \times J(x^*)$ (Hint: separate the space spanned by the $\nabla g_j(x^*)$ from the convex hull of $\{\nabla k_j(x^*)\}_{j \in J(x^*)}$.)

Exercise 4.23. *Let A be a non-empty subset of $\mathbb{R}^d$ and $x \in A$, the contingent cone of A at x, denoted by $T_A(x)$, is defined by*

$$T_A(x) := \Big\{ h \in \mathbb{R}^d : \exists t_k > 0, t_k \to 0^+, x_k \in A$$

$$\text{such that } \frac{x_k - x}{t_k} \to h \text{ as } k \to +\infty \Big\}.$$

1) *Show that $T_A(x)$ is a cone (i.e., if $h \in T_A(x)$, then $\lambda h \in T_A(x)$ for every $\lambda \geq 0$).*
2) *Prove that $T_A(x) = \bigcap_{\alpha > 0} \overline{\bigcup_{\lambda \geq \alpha} \lambda(A - x)}$ and deduce that $T_A(x)$ is closed.*
3) *Let $f : \mathbb{R}^d \to \mathbb{R}$. If $x^* \in A$ is a local minimizer of f on A and f is differentiable at x^*, show that $\nabla f(x^*) \cdot h \geq 0$ for every $h \in T_A(x^*)$.*

4) Let $g_1,\ldots,g_m : \mathbb{R}^d \to \mathbb{R}$ and $k_1,\ldots,k_p : \mathbb{R}^d \to \mathbb{R}$ be differentiable functions and set

$$A := \{x \in \mathbb{R}^d \ : \ g_i(x) = 0, \ i = 1,\ldots,m, \ k_j(x) \le 0, \ j = 1,\ldots,p\}.$$

For $x \in A$ we set $J(x) := \{j \in \{1,\ldots,p\} \ : \ k_j(x) = 0\}$. Show that if $x \in A$, then

$$T_A(x) \subset K_A(x),$$

where $K_A(x)$ is the closed convex cone:

$$K_A(x) := \{h \in \mathbb{R}^d \ : \ \nabla g_i(x) \cdot h = 0, \ \nabla k_j(x) \cdot h \le 0,$$

$$\forall (i,j) \in \{1,\ldots,m\} \times J(x)\}.$$

5) Show that if $\nabla g_1(x),\ldots,\nabla g_m(x)$ are linearly independent and there exists $h_0 \in K_A(x)$ such that $\nabla k_j(x) \cdot h_0 < 0$ for every $j \in J(x)$ then $T_A(x) = K_A(x)$.

Chapter 5

Problems Depending
on a Parameter

Optimization problems that appear in applied contexts (in particular in decision sciences: mathematical economics, game theory, etc.) often involve parameters that can influence the cost, the constraints or both. It is therefore natural to wonder under which conditions the value (i.e., the minimal cost) or the minimizers behave in a stable way (continuous or even better, differentiable) as one varies these parameters. The aim of this chapter is to introduce some tools to study such stability questions. Since the value function plays a key role in this chapter, I thought it was a good occasion to introduce dynamic optimization problems in discrete time and the dynamic programming approach pioneered by Bellman.

5.1. Setting and examples

Throughout this chapter, we are given two metric spaces X and Y (with respective distances denoted d_X and d_Y), a function $\Phi : X \times Y \to \mathbb{R} \cup \{+\infty\}$ and a *set-valued* map[a] $\Gamma : Y \mapsto 2^X$ (i.e., for every $y \in Y$, $\Gamma(y)$ is a-possibly empty-subset of X). We interpret $y \in Y$ as a parameter, $x \in X$ as the decision or control variable (i.e., the one over which we optimize), Φ is the cost function and the set-valued map Γ captures the constraints of the problem. In this abstract setting, we consider for every $y \in Y$

$$v(y) := \inf_{x \in \Gamma(y)} \{\Phi(x,y)\} = \inf_{x \in X} \{\Phi(x,y) + \chi_{\Gamma(y)}(x)\}, \ \forall y \in Y. \qquad (5.1)$$

---

[a] Also often called a *correspondence* in the literature.

As usual, if $\Gamma(y) = \emptyset$ the previous infimum is $+\infty$. The function $v : Y \to \overline{\mathbb{R}}$ is called the value function. Our purpose is to identify conditions under which the value function v as well as the (possibly empty) set of solutions

$$M(y) := \operatorname{argmin}\{\Phi(\cdot, y) + \chi_{\Gamma(y)}\} = \{x \in \Gamma(y) : \Phi(x, y) = v(y)\}$$

behave nicely, in a sense to be made precise, with respect to y.

Before going further, let us consider some examples.

Example 5.1. Take $X = \mathbb{R}^d$, $Y = \mathbb{R}^m$, $\Gamma(y) = \mathbb{R}^d$ for every $y \in \mathbb{R}^m$ (unconstrained case) and assume that for every $y \in \mathbb{R}^m$, $\Phi(\cdot, y)$ is $C^2(\mathbb{R}^d)$ and that there exists $\lambda(y) > 0$ such that

$$D_{xx}^2 \Phi(x, y)(h, h) \geq \lambda(y)|h|^2, \ \forall (x, h) \in \mathbb{R}^d \times \mathbb{R}^d, \tag{5.2}$$

which in particular implies that $\Phi(\cdot, y)$ is strictly convex for every value of the parameter $y \in \mathbb{R}^m$. Moreover, Taylor formula together with (5.2) gives

$$\Phi(x, y) = \Phi(0, y) + \nabla_x \Phi(0, y) \cdot x + \int_0^1 (1-t) D_{xx}^2 \Phi(tx, y)(x, x) dt$$

$$\geq \Phi(0, y) + \nabla_x \Phi(0, y) \cdot x + \frac{\lambda(y)}{2}|x|^2,$$

which implies that $\Phi(\cdot, y)$ is coercive. For each y, there is therefore a unique minimizer of $x \mapsto \Phi(x, y)$ which is characterized by the first-order condition

$$\nabla_x \Phi(x, y) = 0.$$

If we further assume that $(x, y) \mapsto \nabla_x \Phi(x, y)$ is of class C^1, (5.2) and the implicit function theorem imply that the minimizer $\overline{x}(y)$ of $\Phi(\cdot, y)$ is a C^1 function of y. Moreover, the Jacobian of this optimal map is given by

$$J\overline{x}(y) = -[D_{xx}^2 \Phi(\overline{x}(y), y)]^{-1} D_{yx}^2 \Phi(\overline{x}(y), y). \tag{5.3}$$

The value function $v(y) = \inf_{x \in \mathbb{R}^n} \Phi(x, y) = \Phi(\overline{x}(y), y)$ therefore is also of class C^1, more precisely, by the chain-rule and using $0 = \nabla_x \Phi(\overline{x}(y), y)$, we have:

$$\nabla v(y) = J\overline{x}(y)^T \nabla_x \Phi(\overline{x}(y), y) + \nabla_y \Phi(\overline{x}(y), y)$$

$$= \nabla_y \Phi(\overline{x}(y), y).$$

In this very regular situation, the gradient of the value function coincides with the partial gradient of Φ with respect to y evaluated at $(\overline{x}(y), y)$, this kind of result is usually referred to as an envelope theorem. Envelope theorems will be explored in more general situations in Section 5.3.1.

Example 5.2. Take $X = Y = \mathbb{R}$, $\Phi(x, y) = x^2$ and

$$\Gamma(y) := \begin{cases} [-1, 1] & \text{if } y \neq 0, \\ \{1\} & \text{if } y = 0, \end{cases}$$

so that $v(0) = 1$ and $v(y) = 0$ whenever $y \neq 0$. What is causing the discontinuity of v at 0 here is the fact that for y close to 0 elements of $\Gamma(y)$ do not remain close to $\Gamma(0)$. This kind of discontinuity will be ruled out by an assumption on Γ called *upper hemicontinuity*.

Example 5.3. Take $X = Y = \mathbb{R}$, $\Phi(x, y) = x^2$ and $\Gamma(y) := \{x \in \mathbb{R} : (x - 1)y^2 \geq 0\}$ then $v(0) = 0$, $M(0) = \{0\}$ and $v(y) = 1$, $M(y) = \{1\}$ when $y \in \mathbb{R} \setminus \{0\}$ so that v (though lsc) is not continuous at 0. This is due to the fact that the constrained set $\Gamma(y)$ suddenly becomes the whole of $\mathbb{R}$ when $y = 0$. In particular, the minimizer 0 cannot be approximated by points in $\Gamma(y')$ for $y' \neq 0$ close to 0. This kind of discontinuity will be ruled out by an assumption on Γ called *lower hemicontinuity*.

Example 5.4. Take $X = [-1, 1]$, $Y = \mathbb{R}$, $\Phi(x, y) := xy$ and $\Gamma(y) = X$ for every $y \in \mathbb{R}$. Then $v(y) = -|y|$ which is Lipschitz continuous but not differentiable at 0 and

$$M(y) = \begin{cases} \{1\} & \text{if } y < 0, \\ [-1, 1] & \text{if } y = 0, \\ \{-1\} & \text{if } y > 0. \end{cases}$$

When $y < 0$ (respectively, $y > 0$), $M(y)$ is the singleton $\{1\}$ (respectively, $\{-1\}$) and v is differentiable at y with $v'(y) = 1 = \partial_y \Phi(1, y)$, (respectively, $v'(y) = -1 = \partial_y \Phi(-1, y)$)). But when $y = 0$ there are several minimizers and the graph of v is touched from above at 0 by two lines with slopes 1 and -1. This non-uniqueness of the minimizer for the value 0 of the parameter is the very reason why the value function has a corner at $y = 0$.

5.2. Continuous dependence

5.2.1. *Notions of continuity for set-valued maps*

The graph of the set-valued map $\Gamma : Y \to 2^X$ is by definition the subset of $Y \times X$ given by

$$\mathrm{graph}(\Gamma) := \{(y, x) : y \in Y,\ x \in \Gamma(y)\}.$$

One says that Γ is non-empty (respectively, closed, compact) valued if $\Gamma(y)$ is non-empty (respectively, closed, compact) for every $y \in Y$. Of course, a map $y \in Y \mapsto \gamma(y) \in X$ defines a special (single-valued) correspondence $y \mapsto \{\gamma(y)\}$, as we already know continuity of such a single-valued map γ at the point $y \in Y$ is characterized by one of the equivalent statements:

1) if V is an open set containing $\gamma(y)$, then $\gamma(y') \in V$ for y' close enough to y;
2) if y_n converges to y, then $\gamma(y_n)$ converges to $\gamma(y)$.

The generalizations of these two notions to the set-valued case are no longer equivalent and require specific definitions, called upper hemicontinuity and lower hemicontinuity, respectively.

Definition 5.1. Let $y \in Y$. One says that:

1) Γ is upper hemicontinuous (uhc) at y if for every open subset V of X such that $\Gamma(y) \subset V$, there exists $\delta > 0$ such that $\Gamma(y') \subset V$ whenever $d_Y(y', y) \leq \delta$;
2) Γ is lower hemicontinuous (lhc) at y if for every $x \in \Gamma(y)$ and for every open subset V of X containing x, there exists $\delta > 0$ such that $\Gamma(y') \cap V \neq \emptyset$, whenever $d_Y(y', y) \leq \delta$;
3) Γ is hemicontinuous (hc) at y if it is both upper and lower hemicontinuous at y.

Finally, Γ is called uhc (respectively, lhc or hc) on Y if it is uhc (respectively, lhc or hc) at every point of Y.

In the single-valued case where $\Gamma(y) = \{\gamma(y)\}$, we have the obvious equivalence:

$$\gamma \text{ continuous at } y \iff \Gamma \text{ uhc at } y \iff \Gamma \text{ lhc at } y.$$

In the set-valued case, it is always good to have in mind elementary counterexamples, like:

- the correspondence $\Gamma : \mathbb{R} \to 2^{\mathbb{R}}$ defined by

$$\Gamma(y) = \begin{cases} 0 & \text{if } y < 0, \\ [0,1] & \text{if } y = 0, \\ 1 & \text{if } y > 0, \end{cases}$$

which is uhc but not lhc at 0;
- the correspondence $\Gamma : \mathbb{R} \to 2^{\mathbb{R}}$ defined by

$$\Gamma(y) = \begin{cases} 0 & \text{if } y \leq 0, \\ [-1,1] & \text{if } y > 0, \end{cases}$$

which is lhc but not uhc at 0.

By definition, Γ is uhc on Y if and only if whenever V is open in X then the set

$$\{y \in Y : \Gamma(y) \subset V\} \text{ is open in } Y.$$

Passing to complements, upper hemicontinuity can be expressed equivalently as: for every F closed subset of X, then the set

$$\{y \in Y : \Gamma(y) \cap F \neq \emptyset\} \text{ is closed in } Y.$$

In particular, if Γ is uhc $\{y \in Y : \Gamma(y) \neq \emptyset\}$ is closed.

In a similar way, saying that Γ is lhc on Y means that

$$V \text{ is open in } X \Rightarrow \{y \in Y : \Gamma(y) \cap V \neq \emptyset\} \text{ is open in } Y,$$

which is equivalent to

$$F \text{ is closed in } X \Rightarrow \{y \in Y : \Gamma(y) \subset F\} \text{ is closed in } Y.$$

Note that if Γ is lhc then $\{y \in Y : \Gamma(y) \neq \emptyset\}$ is open.

As usual, when dealing with topological notions, it is convenient to have sequential characterizations. Let us start with a connection between upper hemicontinuity and the closed graph property.

Lemma 5.1. *If Γ has closed values and is uhc, then its graph is closed.*

Proof. Let $(y_n, x_n) \in (X \times Y)^{\mathbb{N}}$ with $x_n \in \Gamma(y_n)$ converge to some (y, x). Note that since Γ is uhc and $\Gamma(y_n) \neq \emptyset$ then $\Gamma(y) \neq \emptyset$. If x was not in the closed set $\Gamma(y)$ there would exist $\varepsilon > 0$ such that $\Gamma(y) \subset X \setminus \overline{B}(x, \varepsilon)$ hence by upper hemicontinuity, for large enough n, one would also have $\Gamma(y_n) \subset X \setminus \overline{B}(x, \varepsilon)$ implying that $d_X(x_n, x) > \varepsilon$, which contradicts the convergence of x_n to x. $\square$

Note that Γ being uhc does not necessarily imply that it has closed values (take $X = Y = \mathbb{R}$ and $\Gamma(y) = (0,1)$ for every $y \in \mathbb{R}$). Let us also remark that the fact that Γ has a closed graph is not enough in general to guarantee that it is uhc as the following example shows

$$X = Y = \mathbb{R}, \Gamma(y) = \begin{cases} \left\{ \dfrac{1}{y} \right\} & \text{if } y \neq 0, \\ \{0\} & \text{if } y = 0. \end{cases}$$

Under an additional compactness assumption, a sequential characterization of upper hemicontinuity is as follows.

Lemma 5.2. *If $\Gamma(y)$ is compact, then the following conditions are equivalent:*

1) Γ *is uhc at y;*
2) *for every sequence (y_n) converging to y, for every sequence $(x_n) \in X^{\mathbb{N}}$ such that $x_n \in \Gamma(y_n)$, for every n, (x_n) has a convergent subsequence (x_{n_k}) which has its limit in $\Gamma(y)$.*

Proof. 1) $\Rightarrow$ 2) Since Γ is uhc at y and since for every $\varepsilon > 0$ the open set $V_\varepsilon := \bigcup_{z \in \Gamma(y)} B(z, \varepsilon)$ contains $\Gamma(y)$, for n large enough, $\Gamma(y_n) \subset V_\varepsilon$. If we specify $\varepsilon = 1/k$, we thus can find a subsequence y_{n_k} such that $\Gamma(y_{n_k}) \subset V_{1/k}$ so that there exists $z_k \in \Gamma(y)$ such that $d_X(x_{n_k}, z_k) \leq 1/k$. But since $\Gamma(y)$ is compact, (z_k) has a (not relabeled) subsequence which converges in $\Gamma(y)$ hence so does (x_{n_k}). 2) $\Rightarrow$ 1) Let V be an open set containing the compact set $\Gamma(y)$, if Γ was not uhc at y, we could find a sequence y_n converging to y such that $\Gamma(y_n) \cap (X \setminus V) \neq \emptyset$. In this event, let us choose $x_n \in \Gamma(y_n) \setminus V$. Assumption 2) ensures that (x_n) has a subsequence which converges some $x \in \Gamma(y) \subset V$ but since $(X \setminus V)$ is closed we get $x \notin V$ which contradicts $x \in \Gamma(y)$. $\qquad\square$

Finally, if X is compact, one has the following lemma.

Lemma 5.3. *If X is compact and Γ has closed values, then Γ is uhc if and only if its graph is closed.*

Proof. Thanks to Lemma 5.1, we only have to prove the if part. Assume then that the graph of Γ is closed and take (y_n) converging to y and $x_n \in \Gamma(x_n)$. Since X is compact, some subsequence (x_{n_k}) converges to some $x \in X$ but since the graph of Γ is closed $x \in \Gamma(y)$. This proves that Γ is uhc thanks to Lemma 5.2 and the fact that $\Gamma(y)$ is compact. $\qquad\square$

As for lower hemicontinuity, a general sequential characterization is as follows.

Lemma 5.4. Γ *is lhc at* y *if and only if, for every* $x \in \Gamma(y)$ *and every sequence* (y_n) *converging to* y, *there exists a subsequence* (y_{n_k}) *and* $x_{n_k} \in \Gamma(y_{n_k})$ *such that* (x_{n_k}) *converges to* x.

Proof. Assume that Γ is lhc at y, that $x \in \Gamma(y)$ and that (y_n) converges to y. Then for every $\varepsilon > 0$, $B(x, \varepsilon) \cap \Gamma(y_n) \neq \emptyset$ for n large enough. Taking $\varepsilon = 1/k$ we can then find a subsequence y_{n_k} such that $B(x, 1/k) \cap \Gamma(y_{n_k}) \neq \emptyset$. We can therefore find $x_{n_k} \in B(x, 1/k) \cap \Gamma(y_{n_k})$ and such a sequence has the desired property. Conversely, assume that the sequential statement holds and let us show that Γ is lhc at y. Let $x \in \Gamma(y)$ and let V be an open set containing x, we have to show that there is a $\delta > 0$ such that $\Gamma(y') \cap V \neq \emptyset$ whenever $d_Y(y', y) \leq \delta$. If it was not the case, there would exist a sequence y_n such that $d_Y(y_n, y) \leq 1/n$ with $\Gamma(y_n) \subset (X \setminus V)$ for every n. But, by assumption there exists a subsequence $x_{n_k} \in \Gamma(y_{n_k})$ converging to x which would imply that $x \notin V$ since V is open, we have reached the desired contradiction. $\qquad\square$

5.2.2. *Semicontinuity of values*

Proposition 5.1. *If* Φ *is usc on* $\mathrm{graph}(\Gamma)$ *and* Γ *is lhc at* y, *then the value function* v *is usc at* y.

Proof. Let (y_n) converge to y. Then choose a suitable subsequence in such a way that

$$\limsup_n v(y_n) = \lim_k v(y_{n_k}).$$

If $\Gamma(y) = \emptyset$ then $v(y) = +\infty$ and there is nothing to prove, we may therefore assume that $\Gamma(y) \neq \emptyset$. Let $x \in \Gamma(y)$, since Γ is lhc at y, it follows from Lemma 5.4 that y_{n_k} has a (not relabeled) subsequence such that one can find a sequence $(x_{n_k})_k$ with $x_{n_k} \in \Gamma(y_{n_k})$ converging to x. Since $v(y_{n_k}) \leq \Phi(x_{n_k}, y_{n_k})$ and Φ is usc we get

$$\limsup_n v(y_n) \leq \limsup_k \Phi(x_{n_k}, y_{n_k}) \leq \Phi(x, y),$$

and since this inequality holds for any $x \in \Gamma(y)$ taking the infimum provides the desired inequality

$$\limsup_n v(y_n) \leq v(y).$$

$\qquad\square$

Remark 5.1. The statement of Proposition 5.1 may look a bit asymmetric and weird but this is due to our convention of considering only minimization problems and to have defined the value as an infimum. If we change our convention and consider the (equivalent!) maximization problem

$$w(y) = -v(y) = \sup_{x \in \Gamma(y)} \Psi(x, y) \text{ with } \Psi = -\Phi,$$

we can rephrase Proposition 5.1 in a more symmetric way as follows: w is lsc at y as soon as Ψ is lsc and Γ is lhc at y.

Proposition 5.2. *If X is compact, Φ is lsc on graph(Γ), $\Gamma(y)$ is closed and Γ is uhc at y, then the value function v is lsc at y.*

Proof. Let (y_n) converge to y, let us prove

$$\liminf_n v(y_n) \geq v(y). \tag{5.4}$$

First observe that if $\Gamma(y) = \emptyset$ then since Γ is uhc at y, this implies that $\Gamma(y') = \emptyset$ for every y' in a neighborhood of y so that $v = +\infty$ in this neighborhood, in this case, the desired inequality is obvious. Thus we may assume that $\Gamma(y) \neq \emptyset$ and also pass to a subsequence in such a way that

$$\liminf_n v(y_n) = \lim_k v(y_{n_k}).$$

If $v(y_{n_k}) = +\infty$ for arbitrary large k then $\liminf_n v(y_n) = +\infty$ and there is again nothing to prove. We may therefore assume that $v(y_{n_k}) < +\infty$ (which implies $\Gamma(y_{n_k}) \neq \emptyset$). Now by definition of v there exists $x_{n_k} \in \Gamma(y_{n_k})$ such that

$$v(y_{n_k}) \geq \Phi(x_{n_k}, y_{n_k}) - \frac{1}{k}$$

but since $\Gamma(y)$ is closed and X is compact, $\Gamma(y)$ is compact and Lemma 5.2 implies that (x_{n_k}) has a (not relabeled) subsequence which converges to some $x \in \Gamma(y)$, the lower semicontinuity of Φ thus enables us to conclude:

$$\lim_k v(y_{n_k}) \geq \liminf_k \Phi(x_{n_k}, y_{n_k}) \geq \Phi(x, y) \geq v(y). \qquad \square$$

Combining Propositions 5.1 and 5.2, we obtain the following theorem.

Theorem 5.1. *If X is compact, Φ is continuous on* graph(Γ) *and Γ has non-empty closed values and is hemicontinuous then the value function v is real-valued and continuous on Y. Moreover, the optimal set-valued map given by*

$$M(y) := \{x \in \Gamma(y) : \Phi(x,y) = v(y)\}$$

is uhc with non-empty compact values. In particular, in the uniqueness case where M is single-valued, the optimal solution depends continuously on the parameter $y \in Y$.

Proof. We only have to check that M is uhc, the fact that it has nonempty compact values follows directly from the continuity of Φ, the compactness of X and the fact that Γ has closed and nonempty values. Using Lemma 5.3, we thus have to check that M has a closed graph but if a sequence in the graph of M converges its limit (y,x) is in the graph of Γ and it satisfies $v(y) = \Phi(x,y)$ since this condition is closed due to the continuity of both v and Φ. □

The semicontinuity/continuity results given above are tightly related to the more general theory of Γ-convergence for which we refer the reader to the textbooks of Dal Maso [41] and Braides [21] (also see Exercises 5.4–5.8).

5.3. Parameter-independent constraints, envelope theorems

Now we are interested in differentiating (when it is possible which in particular requires some linear structure of the parameter space), the value of optimization problems depending on a parameter (only in the objective, the case where the parameter enters in the constraints will be addressed in Section 5.4). Envelope theorems basically give conditions that guarantee some differentiability of the value and explicit formulas for the derivatives. These kinds of results are particularly useful in microeconomics. As we have seen in Example 5.4, the smoothness of Φ does not guarantee differentiability of the value functions which typically has kinks when there are several minimizers. In this context, it is more natural to study the local behavior of the value function with non-smooth notions such as subgradients or super-gradients, we will discuss this in Section 5.3.2 and will focus on suprema of convex functions in Section 5.3.3.

5.3.1. *Differentiability under local uniqueness*

We consider the case where the parameter space Y is a normed space, assume that X is a compact metric space[b] and, given $\Phi \in C(X \times Y)$, we consider

$$v(y) := \min_{x \in X} \Phi(x, y), \ \forall y \in Y. \tag{5.5}$$

Consider the following assumptions

$$\Phi(x, \cdot) \text{ is differentiable for every } x \in X \text{ and } \partial_y \Phi \in C(X \times Y, Y^*) \tag{5.6}$$

and given a parameter $y_0 \in Y$, there is uniqueness of the minimizer around y_0, that is there exists $r > 0$ such that

$$\forall y \in B(y_0, r) \text{ there exists a unique } x \in X, \text{ such that } v(y) = \Phi(x, y). \tag{5.7}$$

For $y \in B(y_0, r)$, we shall denote by $x := \overline{x}(y)$ the unique point in X for which $v(y) = \Phi(x, y)$. Note that the optimal map $y \mapsto \overline{x}(y)$ is continuous on $B(y_0, r)$ thanks to Theorem 5.1.

Theorem 5.2. *Under the assumptions (5.6)–(5.7) above, the value function v is of class C^1 on $B(y_0, r)$ and one has*

$$v'(y) = \partial_y \Phi(\overline{x}(y), y), \ \forall y \in B(y_0, r). \tag{5.8}$$

Proof. Let $y \in B(y_0, r)$, $h \in Y$ and choose $t > 0$ small enough so that $y + th$ remains in $B(y_0, r)$. To shorten notation, then set $y_t := y + th$ and $x_t := \overline{x}(y_t)$. We then firstly have

$$v(y_t) \leq \Phi(\overline{x}(y), y_t) = \Phi(\overline{x}(y), y) + t\partial_y \Phi(\overline{x}(y), y)h + o(t)$$
$$= v(y) + t\partial_y \Phi(\overline{x}(y), y)h + o(t)$$

so

$$\limsup_{t \to 0^+} \frac{1}{t}(v(y_t) - v(y)) \leq \partial_y \Phi(\overline{x}(y), y)h. \tag{5.9}$$

Secondly, we observe that

$$v(y) \leq \Phi(x_t, y) = v(y_t) + \Phi(x_t, y) - \Phi(x_t, y_t).$$

---

[b]This assumption can be replaced by suitable coercivity conditions.

With the mean value theorem, there exists $\theta_t \in (0, 1)$ such that

$$\Phi(x_t, y_t) - \Phi(x_t, y) = t\partial_y\Phi(x_t, y + t\theta_t h)h$$

so

$$\liminf_{t \to 0^+} \frac{1}{t}(v(y_t) - v(y)) \geq \liminf_{t \to 0^+} \partial_y\Phi(x_t, y + t\theta_t h)h.$$

Thanks to (5.6) and the fact that $x_t \to \overline{x}(y)$ as $t \to 0^+$ we thus get

$$\liminf_{t \to 0^+} \frac{1}{t}(v(y_t) - v(y)) \geq \partial_y\Phi(\overline{x}(y), y)h. \tag{5.10}$$

Together with (5.9) this implies that v is Gateaux-differentiable at y with Gateaux derivative $\partial_y\Phi(\overline{x}(y), y)$ which is continuous with respect to y so that v is C^1 and the desired result is established. $\qquad\square$

The conclusion of the envelope theorem can easily be reached when the value function is differentiable at the value of interest of the parameter, but again non-uniqueness of a minimizer is the main obstacle for such differentiability, this is clarified in the following proposition.

Proposition 5.3. *Assume (5.6), if v is Gateaux-differentiable at y and $\overline{x}(y) \in \text{argmin}\Phi(\cdot, y)$ then*

$$D_G v(y) = \partial_y\Phi(\overline{x}(y), y).$$

This implies that if $\Phi(\cdot, y)$ has two minimizers on X, x_1 and x_2, with $\partial_y\Phi(x_1, y) \neq \partial_y\Phi(x_2, y)$ then v is not Gateaux-differentiable at y.

Proof. Let $h \in Y$ and $t \in \mathbb{R}$. Then

$$v(y + th) = v(y) + tD_G v(y)h + o(t) \leq \Phi(\overline{x}(y), y + th)$$

$$= \Phi(\overline{x}(y), y) + t\partial_y\Phi(\overline{x}(y), y)h + o(t),$$

so that dividing by $t > 0$ and letting $t \to 0$, we get

$$(D_G v(y) - \partial_y\Phi(\overline{x}(y), y))h \leq 0.$$

Changing h into $-h$, we get an equality in the previous inequality and since h is arbitrary we get $D_G v(y) = \partial_y\Phi(\overline{x}(y), y)$. $\qquad\square$

Functions defined as infima of smooth functions however satisfy quite generally a *superdifferentiability* property, as the following exercise shows.

Exercise 5.1. *Let Y be a normed space, $f \in C(Y, \mathbb{R})$, $y \in Y$. Then $p \in Y^*$ is called a supergradient at y if*

$$f(z) \leq f(y) + p(z - y) + o(\|z - y\|)$$

and f is called superdifferentiable at y if it admits a supergradient at y.

1) *Show that if f and $-f$ are superdifferentiable at y, then f is Fréchet-differentiable at y.*
2) *Let v be the value function defined by (5.5), assume that $\overline{x}(y) \in X$ satisfies $v(y) = \Phi(\overline{x}(y), y)$ and that p is a supergradient of $\Phi(\overline{x}(y), \cdot)$ at y, show that p is a supergradient of v at y.*

5.3.2. Non-smooth cases

Of course, this seems a bit too optimistic to assume *a priori* as in Proposition 5.3 that the value function is differentiable. However, there are many cases where the value function can be shown to be globally or locally Lipschitz hence differentiable almost everywhere thanks to Rademacher's theorem recalled below. A first obvious observation is the following lemma.

Lemma 5.5. *Assume that Φ is locally uniformly continuous with respect to y uniformly in x which means that for every $y_0 \in Y$ there exists a neighborhood U of y_0 and a continuous function $\omega : \mathbb{R}_+ \to \mathbb{R}_+$ such that $\omega(0) = 0$ and*

$$|\Phi(x, y) - \Phi(x, y')| \leq \omega(\|y - y'\|), \quad \forall (x, y, y') \in X \times U^2,$$

then the value function v satisfies

$$|v(y) - v(y')| \leq \omega(\|y - y'\|), \quad \forall (y, y') \in U^2.$$

Proof. By definition for every $(y, y') \in Y^2$ one has

$$\Phi(x, y') \leq \Phi(x, y) + \omega(\|y - y'\|), \quad \forall x \in X$$

taking the minimum with respect to x gives $v(y') \leq v(y) + \omega(\|y - y'\|)$ and reversing the role of y and y' proves the claim. $\qquad \square$

In particular, if for every $y_0 \in Y$ there exists a neighborhood U of y_0 and $M \in \mathbb{R}_+$ such that $\Phi(x, \cdot)$ is M-Lipschitz on U for every $x \in X$ then v is locally-Lipschitz on Y (i.e., each point of Y has a neighborhood on which v is Lipschitz). The following theorem due to Rademacher (see [47], [5] or [6] for a proof) ensures that in finite dimensions, locally Lipschitz functions are differentiable almost everywhere (hence on a dense set). Note that since convex (and concave) functions from $\mathbb{R}^d$ to $\mathbb{R}$ are locally Lipschitz, they are differentiable almost everywhere.

Theorem 5.3 (Rademacher). *Let Ω be a non-empty open set of $\mathbb{R}^d$ and $f : \Omega \to \mathbb{R}$ be locally Lipschitz then f is differentiable Lebesgue-almost everywhere on Ω.*

As a consequence, we have a simple sufficient for a.e. differentiability of the value function in finite dimensions.

Corollary 5.1. *If $Y = \mathbb{R}^d$ and for every $y_0 \in \mathbb{R}^d$, there exists a neighborhood U of y_0 and $M \in \mathbb{R}_+$ such that $\Phi(x, \cdot)$ is M-Lipschitz on U for every $x \in X$, then v is differentiable at y for Lebesgue-almost every $y \in \mathbb{R}^d$.*

Let us mention other situations where the value function has special properties:

- If Y is a vector space and Φ is concave with respect to y for every $x \in X$, then v is concave as an infimum of concave functions. We shall investigate this case in the following section (actually we will deal with suprema of convex functions but this is equivalent after an obvious change of sign).
- If Y is a vector space, X is a convex subset of a vector space and Φ is convex *with respect to both arguments* then v is convex (see Lemma 6.3), this is the situation of interest for convex duality which we will study more in details in Chapter 6.
- If Y is an interval of the real line and $\Phi(x, \cdot)$ is non-increasing (respectively, non-decreasing) for every x then so is v.

5.3.3. *The envelope theorem for suprema of convex functions*

In the case where Φ is concave with respect to the parameter, the value function being concave it is natural to expect envelope theorems in terms of superdifferentials (the superdifferential of a concave function v being by definition $-\partial(-v)$ with $\partial(-v)$ being the subdifferential of the convex function $-v$). To be consistent with our convention of dealing rather with convex than concave functions instead of considering infima of concave functions we shall rather (and this is equivalent up to a minus sign) deal with suprema of convex functions.

Let us first consider the case of the supremum of finitely many convex functions on $\mathbb{R}^d$.

Proposition 5.4. *Let $n \in \mathbb{N}^*$, $f_1, \dots, f_n \colon \mathbb{R}^d \to \mathbb{R}$ be convex functions and $w := \max\{f_1, \dots, f_n\}$ define for each $y \in \mathbb{R}^d$, $I(y) := \{i \in \{1, \dots, n\} : w(y) = f_i(y)\}$. Let $y \in \mathbb{R}^d$. Then*

$$\partial w(y) = \mathrm{co}\left(\bigcup_{i \in I(y)} \partial f_i(y) \right). \tag{5.11}$$

Proof. Note that being finite convex functions on $\mathbb{R}^d$, w and the f_i's are continuous and have a non-empty (and convex compact) subdifferential at every point. Fix $y \in \mathbb{R}^d$, and observe that since $f_j(y) < w(y)$ for $j \notin I(y)$, by continuity, there exists a neighborhood of y on which $w = \max_{i \in I(y)} f_i$. Let $i \in I(y)$ and $q \in \partial f_i(y)$. Then for every $z \in \mathbb{R}^d$ we have

$$w(z) \geq f_i(z) \geq f_i(y) + q \cdot (z - y) = w(y) + q \cdot (z - y),$$

hence $q \in \partial w(y)$. Since $\partial w(y)$ is convex we deduce that

$$\mathrm{co}\left(\bigcup_{i \in I(y)} \partial f_i(y) \right) \subset \partial w(y).$$

To prove the converse inclusion since both $\partial w(y)$ and $\mathrm{co}(\bigcup_{i \in I(y)} \partial f_i(y))$ are convex and closed (actually compact) it is enough, thanks to Theorem 3.4, to prove that the support function of $\partial w(y)$ is smaller than that of $\mathrm{co}(\bigcup_{i \in I(y)} \partial f_i(y))$. We know from Proposition 3.13 that

$$\sigma_{\partial w(y)} = D^+ w(y; \cdot)$$

and (see Exercise 3.13)

$$\sigma_{\mathrm{co}(\bigcup_{i \in I(y)} \partial f_i(y))} = \max_{i \in I(y)} \sigma_{\partial f_i(y)}$$

$$= \max_{i \in I(y)} D^+ f_i(y; \cdot).$$

So we have to show

$$D^+ w(y; \cdot) \leq \max_{i \in I(y)} D^+ f_i(y; \cdot). \tag{5.12}$$

Assume by contradiction that (5.12) is false then there exist an $h \in \mathbb{R}^d$ and an $\varepsilon > 0$ such that for $t > 0$ small enough one has

$$w(y+th) - w(y) \geq \varepsilon t + \max_{i \in I(y)} (f_i(y+th) - f_i(y)) > \max_{i \in I(y)} (f_i(y+th) - w(y))$$

so that $w(y+th) > \max_{i \in I(y)} f_i(y+th)$ but for t small enough $w(y+th) = \max_{i \in I(y)} f_i(y+th)$. This shows (5.12) and ends the proof of (5.11). $\square$

The case of more general suprema of convex functions, indexed by a compact set, was pioneered by Valadier [94] (see [61] and the references therein for interesting recent extensions). A generalization of the finite case is as follows.

Theorem 5.4. *Let X be a compact metric space, $\Psi \in C(X \times \mathbb{R}^d)$ such that for each $x \in X$, $\Psi_x := \Psi(x, \cdot)$ is convex on $\mathbb{R}^d$. Defining for every $y \in \mathbb{R}^d$,*

$$w(y) := \max_{x \in X} \Psi_x(y), \quad M(y) := \{x \in X : \Psi_x(y) = w(y)\}$$

we have

$$\partial w(y) = \mathrm{co}\left(\bigcup_{x \in M(y)} \partial \Psi_x(y)\right).$$

Proof. Let $y \in \mathbb{R}^d$. The inclusion $\mathrm{co}(\bigcup_{x \in M(y)} \partial \Psi_x(y)) \subset \partial w(y)$ is easy and can be shown arguing as in the proof of Proposition 5.4. The converse inclusion requires some extra work. We first claim that, for every K closed in X, the set $\bigcup_{x \in K} \partial \Psi_x(y)$ is compact. Indeed if $x_n \in K$ and $q_n \in \partial \Psi_{x_n}(y)$ we have

$$\Psi_{x_n}\left(y + \frac{q_n}{|q_n| + 1}\right) - \Psi_{x_n}(y) \geq \frac{|q_n|^2}{|q_n| + 1}$$

but the left-hand side being bounded uniformly in n, this gives a bound on q_n. By compactness of K, this means that, up to subsequences, we may assume that x_n and q_n converge to some x and q in X and $\mathbb{R}^d$, respectively. Since Ψ_{x_n} converges pointwise to Ψ_x, passing to the limit $n \to \infty$ in the subgradient inequality $\Psi_{x_n}(z) - \Psi_{x_n}(y) \geq q_n \cdot (z - y)$ we deduce that $q \in \partial \Psi_x(y)$. This shows that $\bigcup_{x \in K} \partial_x \Psi_x(y)$ is compact. Since $M(y)$ is compact, we deduce that both $\bigcup_{x \in M(y)} \partial \Psi_x(y)$ and its convex hull are compact. Our goal now, as in the proof of Proposition 5.4, is to show that

$$D^+ w(y; h) \leq \sup_{x \in M(y)} D^+ \Psi_x(y; h), \quad \forall h \in \mathbb{R}^d.$$

Let $h \in \mathbb{R}^d$, fix $s > 0$, take $t \in (0, s]$ and $x_t \in M(y + th)$ then

$$\frac{1}{t}(w(y+th) - w(y)) \leq \frac{1}{t}(\Psi_{x_t}(y+th) - \Psi_{x_t}(y)) \leq \frac{1}{s}(\Psi_{x_t}(y+sh) - \Psi_{x_t}(y)),$$

where we have used the monotonicity of the slope for the convex function Ψ_{x_t}. Since X is compact, there is a vanishing sequence $t_n \to 0$ such that x_{t_n} converges to some $\overline{x}$ and it is immediate to check that $\overline{x} \in M(y)$. We then pass to the limit $n \to \infty$ in

$$\frac{1}{t_n}(w(y+t_n h) - w(y)) \leq \frac{1}{s}(\Psi_{x_{t_n}}(y+sh) - \Psi_{x_{t_n}}(y))$$

to obtain

$$D^+ w(y; h) \leq \frac{1}{s}(\Psi_{\overline{x}}(y+sh) - \Psi_{\overline{x}}(y))$$

and then we let $s \to 0^+$, together with $\overline{x} \in M(y)$, to conclude

$$D^+ w(y; h) = \sigma_{\partial w(y)}(h) \leq D^+ \Psi_{\overline{x}}(y; h) = \sigma_{\partial \Psi_{\overline{x}}(y)}(h)$$

$$\leq \sigma_{\text{co}(\bigcup_{x \in M(y)} \partial \Psi_x(y))}(h)$$

which, since $\text{co}(\bigcup_{x \in M(y)} \partial \Psi_x(y))$ is convex and closed, enables us to deduce that

$$\partial w(y) \subset \text{co}\left(\bigcup_{x \in M(y)} \partial \Psi_x(y) \right). \qquad \square$$

5.4. Parameter-dependent constraints

5.4.1. *Smoothness of Lagrange points*

We now address constrained minimization problems depending on a parameter both in the objective and in the constraints. To make things simple, we will assume that both the parameter and the control variables are finite dimensional and will deal only with equality constraints. We refer the interested reader to the book of Bonnans and Shapiro [18] for a more general and detailed analysis.

We are given a cost function $f : (x, y) \in \mathbb{R}^n \times \mathbb{R}^d \mapsto f(x, y) \in \mathbb{R}$ and m constraints given by a function $g = (g_1, \ldots, g_m) : \mathbb{R}^n \times \mathbb{R}^d \to \mathbb{R}^m$. Given a value $y \in \mathbb{R}^d$ of the parameter, we thus consider

$$(\mathcal{P}_y) \ \inf\{f(x, y) : x \in \mathbb{R}^n, \ g(x, y) = 0\}, \ v(y) := \inf(\mathcal{P}_y). \tag{5.13}$$

Since our analysis will rely very much on the Lagrange optimality conditions from Chapter 4 and the implicit function theorem, we will make the following smoothness assumption on the data f and g:

$$f \in C^1(\mathbb{R}^n \times \mathbb{R}^d, \mathbb{R}), \ g \in C^1(\mathbb{R}^n \times \mathbb{R}^d, \mathbb{R}^m), \tag{5.14}$$

and

$$\nabla_x f \in C^1(\mathbb{R}^n \times \mathbb{R}^d, \mathbb{R}^n), \ J_x g \in C^1(\mathbb{R}^n \times \mathbb{R}^d, \mathcal{M}_{m \times n}(\mathbb{R})), \tag{5.15}$$

where $\nabla_x f$ denotes the gradient of f with respect to x and $J_x g$ denotes the partial Jacobian matrix of g with respect to x, i.e.,

$$J_x g(x, y) = \begin{pmatrix} \nabla_x g_1(x, y)^T \\ \vdots \\ \nabla_x g_m(x, y)^T \end{pmatrix}.$$

Note in particular that for every y, $f(\cdot, y)$ and $g_i(\cdot, y)$ are of class C^2 on $\mathbb{R}^n$ and the Hessians with respect to x, $D^2 f_{xx}$ and $D^2_{xx} g_i$, depend continuously on (x, y). We know from Lagrange's theorem that if $x \in \mathbb{R}^n$ solves $(\mathcal{P}_y)$ and

$$\nabla_x g_1(x, y), \ldots, \nabla_x g_m(x, y) \ \text{are linearly independent}, \tag{5.16}$$

then there exists a (unique) vector of Lagrange multipliers $\lambda := (\lambda_1, \ldots, \lambda_m)^T \in \mathbb{R}^m$ such that

$$\nabla_x f(x, y) - \sum_{i=1}^m \lambda_i \nabla_x g_i(x, y) = 0, \ \text{i.e.,} \ \nabla_x f(x, y) = J_x g(x, y)^T \lambda. \tag{5.17}$$

Moreover, recall that the second-order optimality necessary condition for $(\mathcal{P}_y)$ from Chapter 4 reads

$$\left(D^2_{xx} f(x, y) - \sum_{i=1}^m \lambda_i D^2_{xx} g_i(x, y) \right)(h, h) \geq 0, \forall h \in \ker(J_x g(x, y)). \tag{5.18}$$

We have also seen in Chapter 4 that the stronger condition

$$\left(D_{xx}^2 f(x,y) - \sum_{i=1}^m \lambda_i D_{xx}^2 g_i(x,y) \right)(h,h) > 0, \forall h \in \ker(J_x g(x,y)) \setminus \{0\},$$

(5.19)

together with (5.17), implies that x is a strict local minimizer for problem $(\mathcal{P}_y)$.

Given a parameter $y \in \mathbb{R}^d$, we call $(\mathcal{L}_y)$ the corresponding Lagrange problem which consists in finding $(x, \lambda) \in \mathbb{R}^n \times \mathbb{R}^m$ such that

$$(\mathcal{L}_y) \quad g(x,y) = 0, \ \nabla_x f(x,y) = J_x g(x,y)^T \lambda. \tag{5.20}$$

We rewrite this problem in a more synthetic way as follows:

$$\text{find } (x,\lambda) \in \mathbb{R}^n \times \mathbb{R}^m \text{ s.t. } \Phi(x,\lambda,y) = 0,$$

where

$$\Phi(x,\lambda,y) := \begin{pmatrix} \nabla_x f(x,y) - \sum_{i=1}^m \lambda_i \nabla_x g_i(x,y) \\ g(x,y) \end{pmatrix}. \tag{5.21}$$

Note that $\Phi \in C^1(\mathbb{R}^n \times \mathbb{R}^m \times \mathbb{R}^d, \mathbb{R}^{n+m})$ and that the partial Jacobian matrix of Φ with respect to (x,λ) is the $(n+m) \times (n+m)$ matrix which can be written blockwise as follows:

$$J_{x,\lambda}\Phi(x,\lambda,y) = \begin{pmatrix} D_{xx}^2 f(x,y) - \sum_{i=1}^m \lambda_i D_{xx}^2 g_i(x,y) & -J_x g(x,y)^T \\ -J_x g(x,y) & 0 \end{pmatrix}. \tag{5.22}$$

Lemma 5.6. *If $(x,\lambda,y) \in \mathbb{R}^n \times \mathbb{R}^m \times \mathbb{R}^d$ is such that (5.16) and (5.19) hold, then $J_{x,\lambda}\Phi(x,\lambda,y)$ is invertible.*

Proof. Let $(h,\mu) \in \mathbb{R}^n \times \mathbb{R}^m$ and assume $(h,\mu) \in \ker(J_{x,\lambda}\Phi(x,\lambda,y))$ then using (5.22) we have

$$\left(D_{xx}^2 f(x,y) - \sum_{i=1}^m \lambda_i D_{xx}^2 g_i(x,y) \right) h = J_x g(x,y)^T \mu \tag{5.23}$$

and

$$J_x g(x, y)h = 0. \tag{5.24}$$

Taking the scalar product of (5.23) with h and using (5.24) we thus get

$$\left(D_{xx}^2 f(x, y) - \sum_{i=1}^m \lambda_i D_{xx}^2 g_i(x, y) \right)(h, h) = \mu^T J_x g(x, y)h = 0$$

but since $h \in \ker(J_x g(x, y))$, (5.19) implies that $h = 0$ and $\mu \in \ker(J_x g(x, y))^T$. Now note that (5.16) means that $J_x g(x, y)$ is surjective, i.e., $\ker(J_x g(x, y)^T) = \{0\}$ hence $\mu = 0$. This proves the invertibility of $J_{x, \lambda} \Phi(x, \lambda, y)$. $\qquad \square$

We thus deduce from the implicit function theorem the following proposition.

Proposition 5.5. *Let $y_0 \in \mathbb{R}^d$ and let $(x_0, \lambda_0) \in \mathbb{R}^n \times \mathbb{R}^m$ be a solution of $(\mathcal{L}_{y_0})$, assume that $J_x g(x_0, y_0)$ is surjective and*

$$\left(D_{xx}^2 f(x_0, y_0) - \sum_{i=1}^m \lambda_{0,i} D_{xx}^2 g_i(x_0, y_0) \right)(h, h) > 0,$$

$$\forall h \in \ker(J_x g(x_0, y_0)) \setminus \{0\}.$$

There exist an open neighborhood U of y_0, an open neighborhood V of (x_0, λ_0) and a map $(\overline{x}, \overline{\lambda}) \in C^1(U, V)$ such that $(\overline{x}(y_0), \overline{\lambda}(y_0)) = (x_0, \lambda_0)$ and for every $y \in U$, the Lagrange problem $(\mathcal{L}_y)$ admits $(\overline{x}(y), \overline{\lambda}(y))$ as unique solution in V.

Proof. This follows directly from Lemma 5.6 and the implicit function theorem. $\qquad \square$

Regarding the continuity of solutions and multipliers, under a local uniqueness assumption, we have:

Lemma 5.7. *Let $y_0 \in \mathbb{R}^d$, assume that there exist $r > 0$ and a compact subset K of $\mathbb{R}^n$ such that*

$$\forall y \in B(y_0, r), \quad (\mathcal{P}_y) \text{ has a unique solution in } K. \tag{5.25}$$

Let us denote by $x^*(y)$ the solution of $(\mathcal{P}_y)$ in K and $x_0 := x^*(y_0)$. If $J_x g(x_0, y_0)$ is surjective then x^* is continuous at y_0. Moreover, there exists $r' \in (0, r]$ such that for every $y \in B(y_0, r')$, there exists a unique $\lambda^*(y) \in \mathbb{R}^m$ such that $(x^*(y), \lambda^*(y))$ solves $(\mathcal{L}_y)$. Finally, $y \mapsto \lambda^*(y)$ is continuous at y_0.

Proof. First observe that since $J_x g(x_0, y_0)$ is surjective, by the implicit function theorem one can express near (x_0, y_0) the constraint $g(x, y) = 0$ as $x_2 = \varphi(x_1, y)$ where $x = x_1 + x_2$ with $x_1 \in \ker(J_x g(x_0, y_0))$, $x_2 \in \ker(J_x g(x_0, y_0))^{\perp}$ and φ is of class C^1. Note in particular that $x_{0,1} + \varphi(x_{0,1}, y_0) = x_0$ where $x_{0,1}$ denotes the orthogonal projection of x_0 onto $\ker(J_x g(x_0, y_0))$. Now let $y_n \in B(y_0, r)$ converge to y_0. Our aim is to show that $x_n := x^*(y_n)$ converges to x_0. Since K is compact, taking a subsequence if necessary, we may assume that x_n converges to some $x \in K$. The continuity of g implies that $g(x, y_0) = 0$. Note that $f(x_n, y_n) \le f(x_{0,1} + \varphi(x_{0,1}, y_n), y_n)$ which passing to the limit gives $f(x, y_0) \le f(x_0, y_0)$. But since x_0 is the only solution of $(\mathcal{P}_{y_0})$ in K this gives $x = x_0$. In fact, the sequence (x_n) takes its values in the compact set K and admits x_0 as only cluster point so it converges to x_0.

Since $J_x g(x_0, y_0)$ has rank m there is an $m \times m$ submatrix of $J_x g(x_0, y_0)$ which is invertible. Using the continuity of the determinant and the continuity of x^*, we deduce that there exists $r' \in (0, r]$ such that for every $y \in B(y_0, r')$, the matrix $J_x g(x^*(y), y)$ has rank m. Therefore, $x^*(y)$ being a solution of $(\mathcal{P}_y)$, by Lagrange's theorem, there exists a unique $\lambda^*(y)$ such that $(x^*(y), \lambda^*(y))$ solves $(\mathcal{L}_y)$. The continuity of λ^* is obtained by remarking that for every $y \in B(y_0, r')$, the symmetric $m \times m$ matrix

$$J_x g(x^*(y), y) J_x g(x^*(y), y)^T$$

is invertible so that the Lagrange relation $\nabla_x f(x^*(y), y) = J_x g(x^*(y), y)^T \lambda^*(y)$ gives

$$\lambda^*(y) = \left(J_x g(x^*(y), y) J_x g(x^*(y), y)^T \right)^{-1} J_x g(x^*(y), y) \nabla_x f(x^*(y), y),$$

which depends continuously on y. $\qquad\square$

Combining Proposition 5.5 with Lemma 5.7, we obtain a C^1 regularity result.

Theorem 5.5. *Let $y_0 \in \mathbb{R}^d$, assume that there exist $r > 0$ and a compact subset K of $\mathbb{R}^n$ such that (5.25) holds, and denote by $x^*(y)$ the solution of $(\mathcal{P}_y)$ in K, $x_0 := x^*(y_0)$. Further assume that $J_x g(x_0, y_0)$ has rank m, define λ^* as in Lemma 5.7 and $\lambda_0 := \lambda^*(y_0)$. If*

$$\left(D_{xx}^2 f(x_0, y_0) - \sum_{i=1}^{m} \lambda_{0,i} D_{xx}^2 g_i(x_0, y_0) \right)(h, h) > 0,$$

$$\forall h \in \ker(J_x g(x_0, y_0)) \setminus \{0\},$$

then $y \mapsto (x^*(y), \lambda^*(y))$ is C^1 in a neighborhood of y_0.

Exercise 5.2. *Under the assumptions of Theorem 5.5, show that for y close enough to y_0 and for every $h \in \ker(J_x g(x^*(y), y)) \setminus \{0\}$ one has*

$$D_{xx}^2 f(x^*(y), y) - \sum_{i=1}^{m} \lambda_i^*(y) D_{xx}^2 g_i(x^*(y), y))(h, h) > 0,$$

for every $h \in h \in \ker(J_x g(x^*(y), y) \setminus \{0\}$. (Hint: set

$$M(x, y, \lambda) := D_{xx}^2 f(x, y) - \sum_{i=1}^{m} \lambda_i D_{xx}^2 g_i(x, y),$$

and

$$m(x, y, \lambda) := \min\{M(x, y, \lambda)(h, h), \ |h| = 1, \ h \in \ker(J_x g(x, y)\}$$

and deduce from Proposition 5.2 that m is lsc.)

5.4.2. *Multipliers and the marginal price of constraints*

If there exists a solution $x^*(y)$ of $(\mathcal{P}_y)$ which depends in a C^1 way on y (and we gave sufficient conditions for this in Theorem 5.5) then the value function being given by $v(y) = f(x^*(y), y)$ it is C^1 itself and the chain rule reveals that the gradient of v is related to the Lagrange multipliers as follows.

Theorem 5.6. *Let $y_0 \in \mathbb{R}^d$, assume that there is a C^1 map x^* defined in a neighborhood of y_0 such that $x^*(y)$ solves $(\mathcal{P}_y)$ for every y in this neighborhood. Set $x_0 := x^*(y_0)$ and assume that $J_x g(x_0, y_0)$ has rank m. Let $\lambda_0 \in \mathbb{R}^m$ be such that $\nabla_x f(x_0, y_0) = J_x g(x_0, y_0)^T \lambda_0$. Then the value function*

$$v(y) := \min_{x \in \mathbb{R}^d} \{f(x, y) : g(x, y) = 0\} = f(x^*(y), y)$$

is of class C^1 and

$$\nabla v(y_0) = \nabla_y f(x_0, y_0) - \sum_{i=1}^{m} \lambda_{0,i} \nabla_y g_i(x_0, y_0).$$

Proof. We differentiate $v(y) = f(x^*(y), y)$ at $y = y_0$ to get first

$$\nabla v(y_0) = Jx^*(y_0)^T \nabla_x f(x_0, y_0) + \nabla_y f(x_0, y_0).$$

By the Lagrange relation $\nabla_x f(x_0, y_0) = \sum_{i=1}^{m} \lambda_{0,i} \nabla_x g_i(x_0, y_0)$ we then have

$$\nabla v(y_0) = \sum_{i=1}^{m} \lambda_{0,i} Jx^*(y_0)^T \nabla_x g_i(x_0, y_0) + \nabla_y f(x_0, y_0). \qquad (5.26)$$

But differentiating, the condition $g_i(x^*(y), y) = 0$ at $y = y_0$ gives

$$Jx^*(y_0)^T \nabla_x g_i(x_0, y_0) + \nabla_y g_i(x_0, y_0) = 0.$$

Replacing in (5.26) yields

$$\nabla v(y_0) = -\sum_{i=1}^{m} \lambda_{0,i} \nabla_y g_i(x_0, y_0) + \nabla_y f(x_0, y_0).$$

$\square$

In the special case $f(x, y) = f(x)$ and $g(x, y) = g(x) - y$ (think of budget constraints), the previous result simply yields $\nabla v(y) = \lambda(y)$. In other words, the multiplier gives the marginal impact of y (an increase on the budget, say) on the value, this is why multipliers are sometimes called shadow prices or prices of the constraints. The theory of convex duality developed in Chapter 6 gives more general results of this kind in the framework of convex programming.

5.5.　Discrete-time dynamic programming

We end this chapter with a short presentation of dynamic optimization in discrete time (for continuous time, we refer the reader to Chapter 9 devoted to the calculus of variations). Such problems arise in various settings, including shortest paths on graphs and growth models in economics (see [67, 93]).

5.5.1.　*Finite horizon*

Given $T \in \mathbb{N}^*$, a non-empty set X, a family of set-valued maps $\Gamma_t : X \to 2^X$ and costs $f_t : X \times X \to \mathbb{R}$ for $t = 0, \ldots, T-1$, consider the dynamic optimization problem with finite horizon T, starting with initial condition $x \in X$:

$$v(0, x) := \inf \left\{ \sum_{s=0}^{T-1} f_s(x_s, x_{s+1}) : x_0 = x, x_{s+1} \in \Gamma_s(x_s), s = 0, \ldots, T-1 \right\}.$$

The function $x \in X \mapsto v(0, x) \in \overline{\mathbb{R}}$ is called the Bellman value function at date 0. One can also consider the value function $v(t, \cdot)$ for $t \in \{0, \ldots, T-1\}$

by considering the problem which starts at time t from $x \in X$:

$$v(t,x) := \inf \left\{ \sum_{s=t}^{T-1} f_s(x_s, x_{s+1}) : x_t = x, x_{s+1} \in \Gamma_s(x_s), \ s = 0, \dots, T-1 \right\},$$

so that, in particular

$$v(T-1, x) = \inf_{y \in \Gamma_{T-1}(x)} f_{T-1}(x, y). \tag{5.27}$$

The key idea of dynamic programming, due to Bellman, is based on the elementary but powerful observation that the value functions satisfy recursive relations called dynamic programming equations. Fix $x \in X$ and let $t \in \{0, \dots, T-2\}$, then

$$v(t,x) = \inf_{y \in \Gamma_t(x)} \left\{ f_t(x,y) + \inf_{y_{s+1} \in \Gamma_s(y_s), \ s=t+1, \dots, T-1, y_{t+1}=y} \sum_{s=t+1}^{T-1} f_s(y_s, y_{s+1}) \right\}$$

so that $v(t, \cdot)$ can be deduced from $v(t+1, \cdot)$ by Bellman's dynamic programming equation

$$v(t, x) = \inf_{y \in \Gamma_t(x)} \{ f_t(x, y) + v(t+1, y) \}. \tag{5.28}$$

But since $v(T-1, \cdot)$ is given by (5.27), one can compute the value functions at the different dates by backward induction that is deducing $v(t, \cdot)$ from $v(t+1, \cdot)$ by (5.28) until $t = 0$. Once the values at the different dates have been computed, finding an optimal path for $v(0, x)$ can be brought down to a succession of static problems. We start with $x_0 = x$ and then find x_1 by minimizing over $\Gamma_0(x_0)$, $f_0(x, x_1) + v(1, x_1)$, then x_2 is obtained by minimizing over $\Gamma_1(x_1)$, $f_1(x_1, x_2) + v(2, x_2)$, etc. This also shows that knowing the Bellman functions, not only enables us to solve the single dynamic minimization problem starting from x at time 0 but all the dynamic minimization problems starting from an arbitrary state y at an arbitrary intermediate time t. The tools from this chapter can of course be used to deduce regularity properties of the Bellman functions under suitable assumptions on the costs f_t and the constraints given by the set-valued maps Γ_t.

5.5.2. *Infinite horizon*

We now consider the infinite horizon case:

$$v(x) := \inf \left\{ \sum_{t=0}^{+\infty} \beta^t f(x_t, x_{t+1}) : x_0 = x, \ x_{t+1} \in \Gamma(x_t), \ \forall t \in \mathbb{N} \right\}, \ x \in X,$$

$$\tag{5.29}$$

where:

- $\beta \in (0,1)$ is a certain discount factor (that captures how the costs in the future are discounted seen from today);
- the state space X is assumed to be a compact metric space;
- the set-valued map capturing the constraints $\Gamma : X \mapsto 2^X$ is assumed to have nonempty closed values and a closed graph (so that it is uhc by Lemma 5.3);
- the running cost f is a bounded lsc function, $f : \text{graph}(\Gamma) \to \mathbb{R}$.

These assumptions, in particular the boundedness f, are quite strong but one really has to be cautious due to the fact that we have to deal with an infinite sum.[c] Boundedness of f and $\beta \in (0,1)$ ensures that the series is converging, as well as the obvious bound

$$\frac{1}{1-\beta} \inf_{\text{graph}(\Gamma)} f \leq v(x) \leq \frac{1}{1-\beta} \sup_{\text{graph}(\Gamma)} f. \tag{5.30}$$

In other words, the value function v belongs to $B(X, \mathbb{R})$ the space of bounded real-functions on X (which equipped with the uniform norm $\|\cdot\|_\infty$ is a Banach space). Then observe that given $x \in X$, one has

$$v(x) := \inf_{y \in \Gamma(x)} \left\{ f(x,y) + \inf_{y_{t+1} \in \Gamma(y_t), y_1 = y} \left\{ \sum_{t=1}^{+\infty} \beta^t f(y_t, y_{t+1}) \right\} \right\}.$$

Now remark that a simple reindexing of the time index $t \mapsto t - 1$ gives

$$\inf_{y_{t+1} \in \Gamma(y_t), y_1 = y} \left\{ \sum_{t=1}^{+\infty} \beta^t f(y_t, y_{t+1}) \right\} = \beta v(y),$$

which reveals that v solves the stationary Bellman equation:

$$v(x) = \inf_{y \in \Gamma(x)} \{ f(x,y) + \beta v(y) \}, \quad \forall x \in X. \tag{5.31}$$

---

[c]Take the case where $X = \mathbb{R}$, $\beta = \frac{1}{2}$, $\Gamma(x) = [-2|x|, 2|x|]$ and $f(x,y) = xy$ then $v(0) = 0$ and $v(x) = -\infty$ for $x \neq 0$. Note also that for a sequence starting from $x \neq 0$ of the form $x_{t+1} = 2\varepsilon_t x_t$ with $\varepsilon_t \in \{-1, 1\}$, the series $\sum_t \varepsilon_t \beta^t x_t x_{t+1}$ does not converge, even in $\overline{\mathbb{R}}$ as soon as ε_t takes infinitely often both values 1 and -1. We shall see in Exercises 5.3 and 5.20 some examples where one slightly relaxes these boundedness and compactness assumption.

Note that (5.31) can be written as a fixed-point problem $v = Tv$ where $T : B(X, \mathbb{R}) \to B(X, \mathbb{R})$ is the Bellman operator given by

$$Tw(x) := \inf_{y \in \Gamma(x)} \{f(x, y) + \beta w(y)\}, \quad \forall x \in X.$$

Now it is easy to see that the Bellman operator T has the following properties:

- it is monotone in the sense that $v_1 \leq v_2$ implies that $Tv_1 \leq Tv_2$;
- if $v \in B(X, \mathbb{R})$ and λ is a constant, then $T(v + \lambda) = T(v) + \beta\lambda$.

Combining these two basic observations, one deduces that T is a contraction, and more precisely it is β-Lipschitz on $(B(X, \mathbb{R}), \|\cdot\|_\infty)$. Indeed if v_1 and v_2 are in $B(X, \mathbb{R})$, since $v_2 \leq v_1 + \|v_1 - v_2\|_\infty$ then monotonicity implies that $Tv_2 \leq T(v_1 + \|v_1 - v_2\|_\infty) = Tv_1 + \beta\|v_1 - v_2\|_\infty$. Reversing the role of v_1 and v_2, we thus get

$$Tv_2 \leq Tv_1 + \beta\|v_1 - v_2\|_\infty, \; Tv_1 \leq Tv_2 + \beta\|v_1 - v_2\|_\infty,$$

i.e., the desired contraction estimate

$$\|Tv_1 - Tv_2\|_\infty \leq \beta\|v_1 - v_2\|_\infty. \tag{5.32}$$

We then immediately deduce from (5.32) and the Banach fixed-point theorem (Theorem 1.4) the following characterization of the value function.

Theorem 5.7. *The value function v defined in (5.29) is the only solution in $B(X, \mathbb{R})$ of the Bellman equation (5.31). Moreover, for every $w \in B(X, \mathbb{R})$ the sequence defined inductively by $w_0 = w$, $w_{n+1} = Tw_n$ converges uniformly to v.*

Corollary 5.2. *Assume that f is lsc on $\mathrm{graph}(\Gamma)$, then the value function v is lsc on X. If, in addition, $f \in C(\mathrm{graph}(\Gamma))$ and Γ is lhc, then v is continuous.*

Proof. Let $w \in B(X, \mathbb{R})$ be lsc. Then it follows from Proposition 5.2 that Tw is lsc and then defining inductively the sequence w_n by $w_0 = w$, $w_{n+1} = Tw_n$, each w_n is lsc, but since w_n converges uniformly to v, v is lsc (see Exercise 1.4). If we also assume that Γ is lhc then Tw is continuous as soon as w is thanks to Proposition 5.1, so by the same argument, we deduce that v is continuous since it is the uniform limit of a sequence of continuous functions. $\square$

Note that if v is lsc as ensured by Corollary 5.2 when f is lsc, then given $x \in X$ the one-shot problem

$$\inf_{y \in \Gamma(x)} \{f(x,y) + \beta v(y)\} \qquad (5.33)$$

has solutions so that if we denote by $M(x)$ the (non-empty compact) set of solutions

$$M(x) := \{y \in \Gamma(x) : v(x) = f(x,y) + \beta v(y)\} \qquad (5.34)$$

and define recursively the sequence x_t by

$$x_0 = x, \ x_{t+1} \in M(x_t). \qquad (5.35)$$

Then by construction

$$v(x) = f(x, x_1) + \beta v(x_1) = f(x, x_1) + \beta f(x_1, x_2) + \beta^2 v(x_2)$$

$$= \sum_{t=0}^{T} \beta^t f(x_t, x_{t+1}) + \beta^{T+1} v(x_{T+1})$$

but since $\beta < 1$ and v is bounded, we get

$$v(x) = \sum_{t=0}^{+\infty} \beta^t f(x_t, x_{t+1}),$$

which means that the sequence defined by (5.35) solves the infinite-horizon minimization problem defining $v(x)$.

An application of the ideas above is given by the following exercise.

Exercise 5.3. *Consider the problem:*

$$v(x) := \sup\left\{\sum_{t=0}^{\infty} \beta^t \sqrt{x_t - x_{t+1}} : x_0 = x, \ x_{t+1} \in [0, x_t], t \in \mathbb{N}\right\}, \qquad (5.36)$$

with $x \geq 0$ given and $\beta \in (0,1)$.

1) *Give a Bellman equation satisfied by (5.36).*
2) *Show that $v(x) = c\sqrt{x}$, for $x \in \mathbb{R}_+$ where $c \geq 0$ is a constant to be determined.*
3) *What are the solutions of (5.36)?*

5.6. Exercises

Exercise 5.4. *Let X be a metric space, let $f_n : X \to \mathbb{R} \cup \{+\infty\}$ and $f : X \to \mathbb{R} \cup \{+\infty\}$. One says that the sequence f_n Γ-converges to f if the following two conditions hold:*

- *whenever x_n converges to x in X one has $\liminf_n f_n(x_n) \geq f(x)$ (Γ-liminf inequality);*
- *for every $x \in X$ there exists a sequence (x_n) converging to x such that $\limsup f_n(x_n) \leq f(x)$ (Γ-limsup inequality).*

1) *Show that if X is compact and f_n Γ-converges to f, then $\inf_X f_n$ converges to $\inf_X f$. Give a counterexample when X is not compact.*
2) *Show that if f_n converges uniformly to f and f is lsc, then f_n Γ-converges to f.*
3) *Take $X = \mathbb{R}$ and $f_n(x) := \sin(nx)$, show that f_n Γ-converges to the constant function equal to -1.*
4) *Assume that f_n Γ-converges to f and that there exists a compact subset K of X such that $\inf_X f_n = \inf_K f_n$ for every n. Show that $\inf_X f_n$ converges to $\inf_X f$. Show that if x_n is a minimizer of f_n and x_n converges to some x, then x is a minimizer of f.*
5) *Let us assume that X is compact, let Y be another metric space, $\Gamma : Y \to 2^X$ be a hemicontinuous set-valued-map with non-empty and closed values and $\Phi \in C(X \times Y)$. Let y_n converge to y in Y, define*

$$f_n := \Phi(\cdot, y_n) + \chi_{\Gamma(y_n)}, \ f := \Phi(\cdot, y) + \chi_{\Gamma(y)}$$

show that f_n Γ-converges to f on X.

Exercise 5.5. *Let $X = [0,1]^d$, $f_0 : X \to [0, +\infty]$ be lsc and define*

$$f = f_0 + \chi_{\{0,1\}^d}, \ f_n(x) := f_0(x) + n \sum_{i=1}^{d} x_i(1 - x_i).$$

Show that (f_n) Γ-converges to f on X.

Exercise 5.6. *Let E be a normed space, f_n be a sequence of convex functions $E \to \mathbb{R} \cup \{+\infty\}$. Show that if f_n Γ-converges to f then f is convex.*

Exercise 5.7. *Let E be a Hilbert space, $f \in \Gamma_0(E)$, define the Moreau–Yosida regularizations of f:*

$$f_n(x) := \inf_{y \in E} \left\{ \frac{n}{2} \|x - y\|^2 + f(y) \right\}.$$

Show that f_n Γ-converges to f.

Exercise 5.8. *Let X be a metric space, $f_0, f_1, \ldots, f_m$ be lsc functions with f_0 bounded from below. Define for every $x \in X$*

$$g(x) := \begin{cases} f_0(x) & \text{if } \max_{i=1,\ldots,m} f_i(x) \leq 0, \\ +\infty & \text{otherwise} \end{cases}$$

and for $x \in X$ and $n \in \mathbb{N}^*$:

$$g_n(x) := f_0(x) + \frac{1}{n} \sum_{i=1}^{m} \exp(n f_i(x)).$$

Show that g_n Γ-converges to g.

Exercise 5.9. *Let (X, d) be a compact metric space. The distance from a point $x \in X$ to a subset A of X is given by*

$$\text{dist}(x, A) := \inf_{a \in A} d(x, a).$$

The Hausdorff distance between two nonempty closed (hence compact) subsets of X, A and B is given by

$$d_H(A, B) := \max\{\max_{a \in A} \text{dist}(a, B), \ \max_{b \in B} \text{dist}(b, A)\}.$$

Finally we denote by $K(X)$ the set of nonempty closed subsets of X.

1) *Show that the Hausdorff distance d_H is a distance on $K(X)$.*
2) *Show that if A is a non-empty subset of X, $\text{dist}(\cdot, A)$ is 1-Lipschitz.*
3) *Let (A_n) be a sequence of non-empty compact subsets of X. Show that $\text{dist}(\cdot, A_n)$ has a subsequence $\text{dist}(\cdot, A_{n_k})$ which converges uniformly to some function f.*
4) *Show that A_{n_k} converges for the Hausdorff distance to the set $A := \{x \in X : f(x) = 0\}$.*
5) *Deduce that $(K(X), d_H)$ is a compact metric space.*

Exercise 5.10. *Let X and Y be two metric spaces with X compact and Γ be a set-valued map $Y \to 2^X$ with non-empty closed values. Let $y \in Y$. Show that Γ is hemicontinuous at y if and only if Γ (viewed as a single-valued map from Y to $K(X)$) is continuous at y for the Hausdorff distance.*

Exercise 5.11. *Let X and Y be two metric spaces with X compact and Γ be a set-valued map $Y \to 2^X$ with non-empty closed values, $\Phi\colon X \times Y \to \mathbb{R}$ Lipschitz. Assume that Γ is Lipschitz for the Hausdorff distance and define the value function*

$$v(y) := \min_{x \in \Gamma(y)} \Phi(x, y), \quad \forall y \in Y.$$

Show that v is Lipschitz on Y.

Exercise 5.12. *Let X be a metric space, (A_n) be a sequence of non-empty closed subsets of X and A be non-empty closed subset of X.*

1) *Show that if $d_H(A_n, A) \to 0$ as $n \to \infty$ then χ_{A_n} Γ-converges to χ_A.*
2) *If X is compact and χ_{A_n} Γ-converges to χ_A show that A_n converges to A in the Hausdorff distance.*
3) *Give a counterexample to the previous implication if X is not compact.*

Exercise 5.13. *Given $A \in \mathcal{S}_d(\mathbb{R})$, denote by $\lambda_{\max}(A)$ the largest eigenvalue of A.*

1) *Show that $A \in \mathcal{S}_d(\mathbb{R}) \mapsto \lambda_{\max}(A)$ is convex and compute $\partial \lambda_{\max}(A)$.*
2) *Show that the minimization problem*

$$\inf_{A \in \mathcal{S}_d(\mathbb{R})} \left\{ \frac{1}{2} \|A\|^2 + \lambda_{\max}(A) \right\} \quad \text{where } \|A\|^2 = \operatorname{tr}(AA^T) \qquad (5.37)$$

has a unique solution.
3) *Show that the solution of (5.37) is diagonal (Hint: use uniqueness and the invariance of the problem by $A \mapsto U^T A U$ for U orthogonal), compute the solution of (5.37).*

Exercise 5.14. *Let $A_1, \ldots, A_d$ be d $n \times n$ matrices, for $y \in \mathbb{R}^d$ set $A(y) := \sum_{i=1}^d y_i A_i$ denote by $w(y)$ the largest eigenvalue of $A(y)$.*

1) *Show that*

$$w(y) := \max_{x \in \mathbb{R}^n, \, \|x\| \leq 1} A(y)x \cdot x.$$

2) *Compute $\partial w(y)$.*

Exercise 5.15. *Let $L \in C^1(\mathbb{R}^d \times \mathbb{R}^d, \mathbb{R})$ satisfy*

$$v \in \mathbb{R}^d \mapsto L(x,v) \text{ is strictly convex for every } x \in \mathbb{R}^d \qquad (5.38)$$

and

$$\frac{L(x,v)}{|v|} \to +\infty \text{ as } |v| \to \infty, \text{ for every } x \in \mathbb{R}^d. \qquad (5.39)$$

The function L is called a Lagrangian. The Hamiltonian associated to the Lagrangian L is then defined by

$$H(x,p) := \sup_{v \in \mathbb{R}^d} \{-L(x,v) - p \cdot v\}.$$

1) *Show that for every $(x,p) \in \mathbb{R}^d \times \mathbb{R}^p$ there exists a unique $v = V(x,p) \in \mathbb{R}^d$ such that $H(x,p) + L(x,v) + p \cdot v = 0$ and that $v = V(x,p)$ if and only if $p + \nabla_v L(x,v) = 0$.*
2) *Show that if (x_n, p_n) is a bounded sequence in $\mathbb{R}^d \times \mathbb{R}^d$ then $v_n = V(x_n, p_n)$ is bounded (Hint: show that $\frac{L(x_n, v_n)}{1+|v_n|}$ is bounded, prove the lower semicontinuity of $x \mapsto \inf_{v:|v| \geq r} \frac{L(x,v)}{1+|v|}$ and use (5.39).)*
3) *Show that $V \in C(\mathbb{R}^d \times \mathbb{R}^d, \mathbb{R}^d)$.*
4) *Show that $H \in C^1(\mathbb{R}^d \times \mathbb{R}^d, \mathbb{R})$ and that*

$$\nabla_x H(x,p) = -\nabla_x L(x, V(x,p)),$$

$$\nabla_p H(x,p) = -V(x,p), \ \forall(x,p) \in \mathbb{R}^d \times \mathbb{R}^d.$$

5) *Consider the (second-order) Euler–Lagrange ODE:*

$$\frac{d}{dt} \nabla_v L(x(t), \dot{x}(t)) = \nabla_x L(x(t), \dot{x}(t)), \ t \geq 0. \qquad (5.40)$$

Show that $t \mapsto x(t)$ solves (5.40) if and only if the pair of functions $t \mapsto (x(t), p(t)) := (x(t), -\nabla_v L(x(t), \dot{x}(t)))$ solves the (first-order) Hamiltonian system

$$\dot{x}(t) = -\nabla_p H(x(t), p(t)), \ \dot{p}(t) = \nabla_x H(x(t), p(t)), \ t \geq 0. \qquad (5.41)$$

Exercise 5.16. *Setting* $\mathbb{R}_+^* := (0, +\infty)$, *let* $U \in C^2((\mathbb{R}_+^*)^d, \mathbb{R}) \cap C(\mathbb{R}_+^d, \mathbb{R})$
satisfy

$$U(x) = 0 \text{ whenever } x \in \partial \mathbb{R}_+^d \text{ and}$$

$$\nabla U(x) \in (\mathbb{R}_+^*)^d, D^2 U(x)(h, h) < 0, \forall x \in (\mathbb{R}_+^*)^d, \ \forall h \in \mathbb{R}^d \setminus \{0\}.$$

Given $p \in (\mathbb{R}_+^*)^d$ *and* $w \in \mathbb{R}_+^*$, *consider*

$$v(p, w) = \sup_{x \in \mathbb{R}_+^d} \{U(x) : p \cdot x \leq w\}. \tag{5.42}$$

*The microeconomic interpretation of this problem is that a consumer with
utility function* U *and revenue* w, *given the price vector* p *determines her
consumption by maximizing her utility subject to her budget constraint.*

1) *Show that (5.42) has a unique solution* $x(p, w)$, *that* $x(p, w) \in (\mathbb{R}_+^*)^d$
   and that the budget constraint is binding, i.e., $p \cdot x(p, w) = w$. *Show that
   there is a unique* $\lambda = \lambda(p, w)$ *such that* $\nabla U(x(p, w)) = \lambda p$.
2) *Show that* $(p, w) \in (\mathbb{R}_+^*)^d \times \mathbb{R}_+^* \mapsto (x(p, w), \lambda(p, w))$ *is of class* C^1.
3) *Show that* $v \in C^1((\mathbb{R}_+^*)^d \times \mathbb{R}_+^*)$ *and compute* $\nabla_p v$ *and* $\partial_w v$.

Exercise 5.17. *Consider*

$$\sup_{(x_1, x_2, x_3)} \{f(x_1, x_0) + g(x_2, x_1) + h(x_3, x_2) : x_i \in \Gamma_{i-1}(x_{i-1}), \ i = 1, 2, 3\} \tag{5.43}$$

with $x_0 \geq 0$ *given and*

$$\Gamma_0(x_0) := [0, x_0^4 + 2x_0 + 3], \ \Gamma_1(x_1) := \left[\frac{x_1}{2}, x_1^2 + x_1\right], \ \Gamma_2(x_2) := \left[0, \frac{x_2^2 + 4}{x_2^2}\right];$$

$$f(x_1, x_0) := 2x_1 x_0 - x_1^2 + x_1, \ g(x_2, x_1) = -\frac{1}{2x_2} + x_2 x_1 - \frac{1}{2}x_2^2$$

and:

$$h(x_3, x_2) := \sqrt{x_3} - \frac{1}{2}x_3 x_2.$$

1) *Introduce the value functions of the problem and give the Bellman equations for these value functions. Compute these value functions.*
2) *Solve (5.43).*

Exercise 5.18. *For $x \geq 0$ and $N \in \mathbb{N}^*$, define:*

$$V_N(x) := \sup\left\{ x_1 \times \cdots \times x_N : x_i \geq 0, \ \sum_{i=1}^{N} x_i = x \right\}.$$

1) *Compute V_1 and show that:*

$$V_N(x) = \sup\{yV_{N-1}(x - y) : y \in [0, x]\}.$$

2) *Show that:*

$$V_N(x) = \frac{x^N}{N^N}$$

and deduce the arithmetico-geometric inequality:

$$(|x_1| \cdots |x_N|)^{1/N} \leq \frac{|x_1| + \cdots + |x_N|}{N}.$$

What are the equality cases?

Exercise 5.19. *Set $X := \{1, 2, 3, 4\}$ and, for $i \in X$, $\Gamma(i) := X \setminus \{i\}$. For $i \in X$, consider:*

$$v_i := \sup\left\{ \sum_{t=0}^{\infty} \beta^t |x_{t+1} - x_t| : x_{t+1} \in \Gamma(x_t), \ x_0 = i \right\}, \tag{5.44}$$

where $\beta \in (0, 1)$ is a discount factor.

1) *Write a system of equations solved by v_1, v_2, v_3, v_4.*
2) *Show that $v_1 = v_4$ and $v_2 = v_3$.*
3) *Compute v_1, v_2, v_3, v_4.*
4) *Solve (5.44) (according to the initial condition $x_0 = i$).*

Exercise 5.20. *Optimal growth theory leads to problems of the form:*

$$\sup_{(k_t)} \sum_{t=0}^{\infty} \beta^t \log(k_t^\alpha - k_{t+1}) \tag{5.45}$$

subject to the constraints $k_0 = k > 0$ (given) and $k_{t+1} \in [0, k_t^\alpha]$ for every $t \in \mathbb{N}$, where α and β are constant parameters in $(0, 1)$.

1) *Define the value function w of the problem (5.45) and show that it is finite on $\mathbb{R}_+^*$.*

2) *Define for $k > 0$:*

$$v(k) := \frac{\alpha \log(k)}{1 - \alpha\beta}.$$

Show that $w \le v$ on $\mathbb{R}_+^*$.

3) *Show that w solves the Bellman equation $f = Tf$ where T is defined by*

$$Tf(x) := \sup_{y \in [0, x^\alpha]} \{\log(x^\alpha - y) + \beta f(y)\}$$

for every $x > 0$.

4) *Can one deduce from what has been seen in this chapter that v is the unique solution of the Bellman equation above?*

5) *Show that $Tv = v + c$ where c is a constant to be determined.*

6) *Compute $T^n v$ for $n \in \mathbb{N}$. Show that $T^n v$ converges (pointwise) to a limit v_∞ to be determined. Show that $Tv_\infty = v_\infty$.*

7) *Show that $w \le v_\infty$.*

8) *Show that $w \ge v_\infty$ (this is slightly more difficult) and conclude.*

9) *Show that (5.45) has a unique solution and compute it.*

Chapter 6

Convex Duality and Applications

It should come as no surprise, in particular in view of Chapter 3, that convexity plays a key role in optimization. The aim of this chapter is to present one of the main tools for convex optimization, namely duality theory. We follow very closely here the perturbation viewpoint of the reference textbook by Ekeland and Temam [46] which unifies in an elegant and powerful way the most important aspects of convex duality both in finite and infinite dimensions. For finite dimensions, we warmly recommend Rockafellar's classic book [87] and for more on convex analysis in infinite-dimensional spaces, we advise the reader to consult the seminal contributions of Moreau [75, 76].

6.1. Generalities

Let us first briefly recall why convexity is important for existence, uniqueness and characterization of global minimizers. Firstly, in terms of existence of combining Theorems 1.29 and 3.8 we have the following general result.

Theorem 6.1. Let E be a reflexive Banach space and $f \in \Gamma_0(E)$ be coercive, i.e.,

$$f(x) \to +\infty \quad as \ \|x\| \to +\infty.$$

Then f admits a (global) minimizer on E.

As for uniqueness, a trivial way to guarantee it is by strict convexity:

Lemma 6.1. Let E be a vector space, C be a non-empty convex subset of E and $f : C \to \mathbb{R} \cup \{+\infty\}$ be strictly convex and finite at at least one point of C. Then f admits at most one global minimizer on C.

195

Proof. If x and y were two distinct minimizers of f on C, by strict convexity we would have

$$\frac{x+y}{2} \in C \quad \text{and} \quad f\left(\frac{x+y}{2}\right) < \frac{1}{2}f(x) + \frac{1}{2}f(y) = \inf_{z \in C} f(z),$$

which is absurd. $\qquad\qquad\qquad\qquad\qquad\qquad\qquad\qquad\qquad\qquad\qquad\square$

Finally, as already noticed, a way to characterize global minimizers of convex functions is by the following subgradient criterion.

Proposition 6.1. *Let E be a normed space and $f : E \to \mathbb{R} \cup \{+\infty\}$ be convex and proper. Then $x \in E$ is a minimizer of f on E if and only if*

$$0 \in \partial f(x).$$

In the differentiable case, we have the following characterization, which easily follows from the above tangent property and is therefore left to the reader.

Lemma 6.2. *Let E be a normed space, C be a non-empty convex subset of E and $f : E \to \mathbb{R} \cup \{+\infty\}$ be convex. If f is Gateaux-differentiable at $x \in C$ then there is equivalence between the two following statements:*

- $f(x) \leq f(y)$ *for every $y \in C$;*
- $D_G f(x)(y - x) \geq 0$ *for every $y \in C$.*

The following observation, left as an exercise, is straightforward but sometimes useful in practice.

Exercise 6.1. *Let E be a normed space and $f : E \to \mathbb{R} \cup \{+\infty\}$ be convex and proper. If x is a local minimizer of f then it is also a global minimizer (Hint: use the fact that if f is convex $\frac{1}{t}(f(x+th) - f(x))$ is a non-decreasing function of t.)*

6.2. Convex duality with respect to a perturbation

6.2.1. Setting

In the sequel, E and F will be two given normed spaces with respective topological duals E^* and F^*, we are also given $f \in \Gamma_0(E)$. Our starting point is the convex minimization that we will call the *primal*:

$$\inf_{x \in E} f(x). \tag{6.1}$$

We will assume that there exists $\Phi \in \Gamma_0(E \times F)$ (called a perturbation of f) such that

$$f(x) = \Phi(x, 0), \quad \forall x \in E. \tag{6.2}$$

In other words, the primal problem (6.1) can be seen as a special instance of the minimization parameterized by $y \in F$:

$$\inf_{x \in E} \Phi(x, y) \tag{6.3}$$

corresponding to the special value $y = 0$. The value function of the family of minimization problems (6.3) is defined by

$$v(y) := \inf_{x \in E} \Phi(x, y). \tag{6.4}$$

Identifying $(E \times F)^*$ with $E^* \times F^*$ through

$$(p, q)(x, y) = p(x) + q(y), \quad \forall (x, y, p, q) \in E \times F \times E^* \times F^*.$$

By Young–Fenchel inequality, we have

$$\Phi(x, 0) + \Phi^*(0, q) \geq (0, q)(x, 0) = 0, \quad \forall (x, q) \in E \times F^*$$

so that

$$\Phi(x, 0) = f(x) \geq -\Phi^*(0, q). \tag{6.5}$$

The dual of (6.1) with respect to the perturbation function Φ is then

$$\sup_{q \in F^*} -\Phi^*(0, q). \tag{6.6}$$

Because of (6.5), we see that there is an obvious inequality between the value of the dual and the primal, namely

$$\inf_{x \in E} \Phi(x, 0) \geq \sup_{q \in F^*} -\Phi^*(0, q). \tag{6.7}$$

Hence, if we are lucky enough to find $x \in E$ and $q \in F^*$ such that

$$\Phi(x, 0) + \Phi^*(0, q) = 0, \tag{6.8}$$

then x solves the primal (6.1) and q solves the dual (6.6). Note that (6.8) can equivalently be written in subdifferential terms as

$$(0, q) \in \partial \Phi(x, 0)$$

or equivalently, by Fenchel's reciprocity formula

$$(x, 0) \in \partial \Phi^*(0, q).$$

The natural questions we wish to address now are

- when is (6.7) an equality (absence of duality gap)?
- when does (6.6) admit solutions?
- what do solutions of (6.6) represent (what is, in particular, the connection with the Lagrange and KKT theories we have seen in Chapter 4?).

6.2.2. *A general duality result*

Lemma 6.3. *The value function $v : F \to \overline{\mathbb{R}}$ defined in (6.4) is convex. Its Legendre transform is given by*

$$v^*(q) = \Phi^*(0, q), \quad \forall q \in F^*.$$

Proof. Let $(y_1, y_2) \in F^2$ and $\lambda \in [0, 1]$. Then using the definition of v and the convexity of Φ we have

$$
\begin{aligned}
v((1 - \lambda)y_1 + \lambda y_2) &= \inf_{x \in E} \Phi(x, (1 - \lambda)y_1 + \lambda y_2) \\
&= \inf_{(x_1, x_2) \in E} \Phi((1 - \lambda)x_1 + \lambda x_2, (1 - \lambda)y_1 + \lambda y_2) \\
&\leq \inf_{(x_1, x_2) \in E} ((1 - \lambda)\Phi(x_1, y_1) + \lambda \Phi(x_2, y_2)) \\
&= (1 - \lambda)v(y_1) + \lambda v(y_2).
\end{aligned}
$$

For $q \in F^*$ we have

$$
\begin{aligned}
v^*(q) &= \sup_{y \in F}\{q(y) - \inf_{x \in E} \Phi(x, y)\} \\
&= \sup_{(x,y) \in E \times F} \{q(y) - \Phi(x, y)\} \\
&= \sup_{(x,y) \in E \times F} \{(0, q)(x, y) - \Phi(x, y)\} = \Phi^*(0, q).
\end{aligned}
$$

$\square$

Note that by definition

$$\inf(6.1) = v(0)$$

and by Lemma 6.3, there holds

$$\sup(6.6) = \sup_{q \in F^*} -\Phi^*(0, q) = \sup_{q \in F^*} -v^*(q) = v^{**}(0).$$

So the absence of duality gap amounts to the fact that v and v^{**} agree at 0. One should be careful since v can achieve the value $-\infty$ but, as seen in

Lemma 3.6, if $v(0)$ is finite and v is bounded from above in a neighborhood of 0, then v never takes the value $-\infty$ and is actually continuous hence subdifferentiable at 0. Now we have the following proposition.

Proposition 6.2. *If v is subdifferentiable at 0, then there is no duality gap, i.e.,*

$$\inf(6.1) = v(0) = v^{**}(0) = \sup(6.6)$$

Moreover, $\partial v(0)$ is the set of solutions of (6.6); in particular, (6.6) admits solutions.

Proof. If $q \in \partial v(0)$, then

$$0 = v(0) + v^*(q)$$

but $v^{**}(0) \geq -v^*(q) = v(0)$ by the Young–Fenchel inequality, which, since $v \geq v^{**}$, shows that $v(0) = v^{**}(0)$. Now the fact that $q \in \partial v(0)$ means that

$$-v^*(q) = -\Phi^*(0, q) = v(0) = v^{**}(0) = \sup(6.6).$$

$\square$

We are now in position to prove a general duality result.

Theorem 6.2. *If $v(0)$ is finite and there exists $x_0 \in E$ such that*

$$y \mapsto \Phi(x_0, y) \text{ is finite and continuous at } y = 0, \tag{6.9}$$

then

$$\inf_{x \in E} \Phi(x, 0) = \max_{q \in F^*} -\Phi^*(0, q) \tag{6.10}$$

(where we have written a max in the right-hand side, precisely because it is attained).

Proof. By assumption, the convex function v is finite at 0 and by definition it is smaller than $y \mapsto \Phi(x_0, y)$ which is bounded in a neighborhood of 0, by (6.9); v is therefore continuous at 0 hence subdifferentiable at 0 (see Theorem 3.10). The conclusion then follows from Proposition 6.2. $\square$

Remark 6.1. It is important to note that, given a cost function f, there are many ways to find perturbations so that (6.2) holds. Each of them possibly leads to another dual problem.

Remark 6.2. Theorem 6.2 gives sufficient conditions for the absence of duality gap and also for dual attainment, i.e., the fact that (6.6) admits solutions. This existence result only relies on a separation argument, not on compactness considerations. It also gives a first interpretation of what solutions of (6.6) represent: they are subgradients of the value function at 0 (the value of interest of the parameter). Solutions of the dual therefore capture how the value varies as one slightly perturbs the parameter. This is useful in terms of sensitivity analysis with respect to the parameter (which can influence the cost, the constraints, or both) in the convex framework (see Chapter 5 for more on problems depending on a parameter).

6.3. Applications

As indicated in Remark 6.1, there are many ways to perturb a given minimization problem. There are however some systematic perturbation strategies which may be useful in specific (convex) contexts. We will first investigate the perturbation of linear operators which leads to the Fenchel–Rockafellar duality. Next, we will consider problems with inequality constraints and we will see that perturbing the right-hand side of these constraints will enable us to recover KKT type of results as well as Lagrangian duality.

6.3.1. *The Fenchel–Rockafellar theorem*

Let E and F be two normed spaces, $\Lambda \in \mathcal{L}_c(E, F)$. Recall that the adjoint of Λ is the map $\Lambda^* \in \mathcal{L}_c(F^*, E^*)$ defined by

$$\Lambda^*(q)(x) = q(\Lambda(x)), \quad \forall (q, x) \in F^* \times E. \tag{6.11}$$

Let $J \in \Gamma_0(E \times F)$ and consider

$$\inf_{x \in E} J(x, \Lambda(x)). \tag{6.12}$$

Problem (6.12) is the minimization of the cost J over the graph of Λ which is obviously a closed subspace of $E \times F$. The perturbation we will consider precisely consists in relaxing the constraint of lying in this graph and is given by

$$\Phi(x, y) := J(x, \Lambda(x) - y), \quad \forall (x, y) \in E \times F. \tag{6.13}$$

Indeed, $(x, y) \mapsto (x, \Lambda(x) - y)$ is an isomorphism (it is even an involution) of $E \times F$. To identify the corresponding dual problem, for $q \in F^*$, we then compute $\Phi^*(0, q)$:

$$\Phi^*(0, q) = \sup_{(x,y)\in E\times F} \{q(y) - J(x, \Lambda(x) - y))\}$$

$$= \sup_{(x,z)\in E\times F} \{q(\Lambda(x) - z) - J(x, z)\}$$

$$= \sup_{(x,z)\in E\times F} \{(\Lambda^*(q), -q)(x, z) - J(x, z)\}$$

$$= J^*(\Lambda^*(q), -q).$$

So that the dual of (6.12) (with respect to the perturbation (6.13)) reads

$$\sup_{q\in F^*} -J^*(\Lambda^*(q), -q). \tag{6.14}$$

Applying Theorem 6.2 to the above setting gives the following theorem.

Theorem 6.3. *If* inf(6.12) *is finite and there exists some* $x_0 \in E$ *such that*

$$y \mapsto J(x_0, \Lambda(x_0) - y) \text{ is finite and continuous at } y = 0, \tag{6.15}$$

then

$$\inf_{x\in E} J(x, \Lambda(x)) = \max_{q\in F^*} -J^*(\Lambda^*(q), -q). \tag{6.16}$$

The second condition in the previous theorem is in nature a qualification condition. Convex duality is often a convenient way to write optimality conditions.

Proposition 6.3. *Under the assumptions of Theorem 6.3, given* $x \in E$ *and* $q \in F^*$ *we have an equivalence between the following statements:*

1) x *solves (6.12) and* q *solves (6.14)*;
2) $J(x, \Lambda(x)) + J^*(\Lambda^*(q), -q) = 0$;
3) $(\Lambda^*(q), -q) \in \partial J(x, \Lambda(x))$ *(extremality relation in primal form)*;
4) $(x, \Lambda(x)) \in \partial J^*(\Lambda^*(q), -q)$ *(extremality relation in dual form)*.

Proof. The fact that 1) implies 2) follows directly from (6.16). Now observe that

$$(\Lambda^*(q), -q)(x, \Lambda(x)) = \Lambda^*(q)(x) - q(\Lambda(x)) = 0$$

so if 2) holds we have

$$J(x, \Lambda(x)) + J^*(\Lambda^*(q), -q) = (\Lambda^*(q), -q)(x, \Lambda(x))$$

which is equivalent to 3) (and thus 4)) thanks to Fenchel reciprocity formula). Finally if 4) holds one has

$$\inf(6.12) \leq J(x, \Lambda(x)) = -J^*(\Lambda^*(q), -q) \leq \sup(6.14),$$

with (6.16) this gives the optimality of both x and q. □

If we consider the special case of a separable functional J, i.e.,

$$J(x, y) := G(x) + H(y), \quad \text{with} \quad (G, H) \in \Gamma_0(E) \times \Gamma_0(F),$$

we have the following theorem.

Theorem 6.4 (Fenchel–Rockafellar). *Let* $(G, H) \in \Gamma_0(E) \times \Gamma_0(F)$, $\Lambda \in \mathcal{L}_c(E, F)$. *If the infimum*

$$\inf_{x \in E} \{G(x) + H(\Lambda(x))\} \tag{6.17}$$

is finite and there exists $x_0 \in E$ *such that*

$$G(x_0) < +\infty \quad \text{and } H \text{ is finite and continuous at } \Lambda(x_0),$$

then

$$\inf_{x \in E} \{G(x) + H(\Lambda(x))\} = \max_{q \in F^*} \{-G^*(\Lambda^*(q)) - H^*(-q)\}.$$

The extremality relation obtained by duality then read as follows:

$$G(x) + G^*(\Lambda^*(q)) + H(\Lambda(x)) + H^*(-q) = 0 = \Lambda^*(q)(x) + (-q)(\Lambda(x)).$$

Since by Young–Fenchel inequality, we have

$$G(x) + G^*(\Lambda^*(q)) \geq \Lambda^*(q)(x) \quad \text{and} \quad H(\Lambda(x)) + H^*(-q) \geq (-q)(\Lambda(x)).$$

These two inequalities should in fact be equalities, i.e.,

$$\Lambda^*(q) \in \partial G(x) \quad \text{and} \quad -q \in \partial H(\Lambda(x)), \tag{6.18}$$

or equivalently

$$x \in \partial G^*(\Lambda^*(q)) \quad \text{and} \quad \Lambda(x) \in \partial H^*(-q). \tag{6.19}$$

6.3.2. *Linear programming*

Let A be an $n \times d$ matrix, $b \in \mathbb{R}^n$ and $c \in \mathbb{R}^d$. Let us consider

$$\alpha := \inf_{x \in \mathbb{R}^d} \{c \cdot x \; : \; x \geq 0, \; Ax \leq b\} \tag{6.20}$$

(where $x \geq 0$ and $Ax \leq b$ are shortcut notations that mean $x \in \mathbb{R}^d_+$ and $b - Ax \in \mathbb{R}^n_+$, respectively). This problem will be referred to as primal linear-programming (LP). It is a special instance of (6.17) with $E = \mathbb{R}^d$, $F = \mathbb{R}^n$, $\Lambda(x) = Ax$ (so that $\Lambda^*(q) = A^T q$) and

$$G(x) := \begin{cases} c \cdot x & \text{if } x \geq 0, \\ +\infty & \text{otherwise} \end{cases}$$

and

$$H(y) := \begin{cases} 0 & \text{if } y \leq b, \\ +\infty & \text{otherwise.} \end{cases}$$

Its Fenchel–Rockafellar dual thus reads as

$$\sup_{q \in \mathbb{R}^n} \{-G^*(A^T q) - H^*(-q)\}$$

and direct computations give

$$G^*(A^T q) = \begin{cases} 0 & \text{if } A^T q \leq c, \\ +\infty & \text{otherwise} \end{cases}$$

and

$$-H^*(-q) = \begin{cases} b \cdot q & \text{if } q \leq 0, \\ -\infty & \text{otherwise} \end{cases}$$

so that the dual of (6.20) is another linear-programming problem that takes the form

$$\beta := \sup_{q \in \mathbb{R}^n} \{b \cdot q \; : \; q \leq 0, \; A^T q \leq c\}. \tag{6.21}$$

Of course, one always have $\alpha \geq \beta$ (with $\inf \emptyset = +\infty$ and $\sup \emptyset = -\infty$ so as to cover the case where either the dual or the primal may have no admissible point). Due to the linearity of the problem, one can directly prove the following result usually referred to as the strong linear programming (LP) duality theorem without invoking Theorem 6.4 (note that we do not impose here any qualification condition).

Theorem 6.5. *If (6.20) admits a solution x, then (6.21) admits a solution q and $\alpha = \beta$ (no duality gap). Moreover, primal solutions x and dual solutions q are related by the complementarity slackness conditions*

$$q \cdot (Ax - b) = 0, \quad x \cdot (A^T q - c) = 0. \tag{6.22}$$

Proof. Let x be a solution of (6.20) and define

$$I := \{i \in \{1, \ldots, d\} : x_i = 0\}, \quad J := \{j \in \{1, \ldots, n\} : l_j \cdot x = b_j\},$$

where $(l_1, \ldots, l_n)$ denote the rows of A. If $h \in \mathbb{R}^d$ satisfies

$$h_i \geq 0, \quad l_j \cdot h \leq 0, \ \forall (i, j) \in I \times J, \tag{6.23}$$

then for small enough $t > 0$ the vector $x + th$ is admissible for (6.20) hence by optimality $c \cdot (x + th) \geq c \cdot x$. Thus $c \cdot h \geq 0$ whenever h satisfies (6.23). Thanks to the Minkowski–Farkas theorem we deduce that there exists $(z_i)_{i \in I} \in \mathbb{R}_+^I$ and $(q_j)_{j \in J} \in \mathbb{R}_-^J$ such that

$$c = \sum_{i \in I} z_i e_i + \sum_{j \in J} q_j l_j \tag{6.24}$$

(denoting by $(e_1, \ldots, e_d)$ the canonical basis of $\mathbb{R}^d$). For $i \in \{1, \ldots, d\} \setminus I$ we set $z_i = 0$ and likewise for $j \in \{1, \ldots, n\} \setminus J$ we set $q_j = 0$, so that (6.24) rewrites as

$$c = z + A^T q, \quad q \leq 0, \quad z \geq 0 \text{ so that } A^T q \leq c.$$

In particular q is admissible for (6.21) hence $b \cdot q \leq \beta$. By construction we also have the complementarity slackness conditions $x \cdot z = 0$, i.e., $c \cdot x = q \cdot Ax$ and $q \cdot (b - Ax) = 0$ this gives

$$\beta \leq \alpha = c \cdot x = q \cdot Ax = b \cdot q,$$

which shows that $\alpha = \beta$ and that q solves (6.21). $\qquad\square$

Note that in (6.22), the dual solution q is naturally interpreted as a vector of multipliers associated to the constraints $Ax \geq b : q$ has signed components which are zero whenever the corresponding constraint is not binding. In a symmetric way, x appears as a vector of multipliers for the dual.

Exercise 6.2. *Write (6.21) as (minus) a minimization problem and show that the dual of this minimization problem is nothing but (6.20) (up to a minus sign again of course). Deduce that whenever (6.21) has a solution then so does (6.20) and that $\alpha = \beta$ in this case as well.*

6.3.3. *Semidefinite programming*

In a nutshell, semidefinite programming (SDP) consists of minimizing a linear cost under linear constraints over positive semidefinite matrices. In linear programming, there are finitely many linear constraints whereas SDP involves infinitely many constraints. SDP contains as particular cases some eigenvalue problems (see Exercise 6.3), convex quadratic programming (see Exercise 6.18) and many others (cone programming, volume optimization, relaxation of combinatorial or nonconvex problems, problems on graphs, etc.) which play an important role in various applied settings. We refer the reader to [95] and the references therein for more on SDP. Here, our focus is on duality.

Recall that $\mathcal{S}_d(\mathbb{R})$ is the subspace of $\mathcal{M}_d(\mathbb{R})$ formed by symmetric matrices and $\mathcal{S}_d^+(\mathbb{R})$ consists of positive semidefinite symmetric matrices, i.e., $S \in \mathcal{M}_d(\mathbb{R})$ belongs to $\mathcal{S}_d^+(\mathbb{R})$ if and only if

$$S = S^T \quad \text{and} \quad Sh \cdot h \geq 0, \ \forall h \in \mathbb{R}^d.$$

The SDP cone $\mathcal{S}_d^+(\mathbb{R})$ is obviously a closed convex cone of $\mathcal{S}_d(\mathbb{R})$. For a symmetric $d \times d$ matrix S we write

$$S \succeq 0$$

to express that $S \in \mathcal{S}_d^+$, i.e., S is SDP. Likewise if S_1 and S_2 are symmetric matrices we shall write

$$S_1 \succeq S_2 \iff S_2 \preceq S_1 \iff S_1 - S_2 \succeq 0.$$

One also says that $\succeq$ is the order relation associated with the SDP cone $\mathcal{S}_d^+(\mathbb{R})$. The usual scalar product on $\mathcal{M}_d(\mathbb{R})$ will be denoted with a $\cdot$ (hoping it will not create any confusion with the usual scalar product of $\mathbb{R}^m$ which we also denote with a $\cdot$) and it is given by

$$A \cdot B := \text{tr}(A^T B) = \sum_{i=1}^{d} \sum_{j=1}^{d} A_{ij} B_{ij},$$

which is simply the usual scalar product one gets by identifying $d \times d$ matrices with vectors with d^2 components.

By the spectral theorem (see Exercise 4.13), symmetric matrices admit an orthogonal basis of eigenvectors, i.e., any $S \in \mathcal{S}_d(\mathbb{R})$ can be written as

$$S = U^T \Delta U, \text{ with } U \text{ orthogonal, i.e., } U^T U = \text{id and } \Delta \text{ is diagonal,}$$
$$(6.25)$$

where $\Delta = \text{diag}(\lambda_1, \ldots, \lambda_d)$ and $\lambda_1, \ldots, \lambda_d$ are the eigenvalues of S. Denoting by $v_1, \ldots v_d$ an orthogonal basis of eigenvectors of S, the diagonalization of S an also be written as

$$S = \sum_{i=1}^{d} \lambda_i v_i v_i^T, \tag{6.26}$$

where the (rank-one and SDP) matrix $v_i^T v_i$ is nothing but the matrix of the orthogonal projector on $\mathbb{R}v_i$. Of course $S \succeq 0$ is equivalent to the nonnegativity of its eigenvalues $\lambda_1, \ldots, \lambda_d$. A remarkable property of the SDP cone is that it is self-polar.

Lemma 6.4. *Let $S \in \mathcal{S}_d$. Then $S \succeq 0$. If and only if*

$$S \cdot F = \text{tr}(SF) \geq 0 \quad \text{for every } F \succeq 0. \tag{6.27}$$

Proof. Let $h \in \mathbb{R}^d$. Then

$$Sh \cdot h = h^T S h = \text{tr}(h^T S h) = \text{tr}(Sh h^T) = S \cdot h h^T$$

(where we have used the fact that $\text{tr}(MN) = \text{tr}(NM)$ whenever M and N are $m \times l$ and $l \times m$ matrices, respectively). In particular, if S satisfies (6.27) choosing $F = hh^T$ (which is obviously SDP) gives that $Sh \cdot h \geq 0$ so that $S \succeq 0$. Conversely, assume that $S \succeq 0$ and let $F \succeq 0$, writing the diagonalization of F as $F = \sum_{i=1}^{d} \alpha_i h_i h_i^T$ with $\alpha_i \geq 0$, we obtain

$$S \cdot F = \sum_{i=1}^{d} \alpha_i \text{tr}(Sh_i h_i^T) = \sum_{i=1}^{d} \alpha_i Sh_i \cdot h_i \geq 0.$$
$$\square$$

Now, given $S_0, S_1, \ldots, S_m \in \mathcal{S}_d(\mathbb{R})^{m+1}$ and $c \in \mathbb{R}^m$, let us consider the (primal SDP) problem

$$\inf_{x \in \mathbb{R}^m} \left\{ c \cdot x : S_0 + \sum_{i=1}^{m} x_i S_i \succeq 0 \right\} \tag{6.28}$$

which we write in Fenchel–Rockafellar form as

$$\inf_{x \in \mathbb{R}^m} \{G(x) + H(\Lambda(x))\}$$

with $G(x) = c \cdot x$, $\Lambda \in \mathcal{L}(\mathbb{R}^m, \mathcal{S}_d(\mathbb{R}))$ is defined by $\Lambda(x) := \sum_{i=1}^m x_i S_i$ and H corresponds to the SDP constraint:

$$H(S) = \begin{cases} 0 & \text{if } S_0 + S \succeq 0, \\ +\infty & \text{otherwise.} \end{cases}$$

Let us compute the adjoint of Λ:

$$\Lambda^*(S) \cdot x = \sum_{i=1}^m x_i S_i \cdot S$$

so that $\Lambda^*(S) = (S_1 \cdot S, \dots, S_m \cdot S)$. We thus have

$$G^*(\Lambda^*(S)) = \begin{cases} 0 & \text{if } S_i \cdot S = c_i \text{ for } i = 1, \dots, m, \\ +\infty & \text{otherwise,} \end{cases}$$

and

$$H^*(-S) = \sup\{-S \cdot F : F \succeq -S_0\} = \sup\{S \cdot \Sigma : \Sigma \preceq S_0\}.$$

Using Lemma 6.4 this gives

$$H^*(-S) = \begin{cases} S_0 \cdot S & \text{if } S \succeq 0, \\ +\infty & \text{otherwise.} \end{cases}$$

The Fenchel–Rockafellar dual of (6.28) thus reads

$$\sup\{-S_0 \cdot S : S_i \cdot S = c_i, i = 1, \dots, m, \ S \succeq 0\}, \tag{6.29}$$

which is called the dual SDP problem of (6.28). As usual, one easily checks the weak duality inequality

$$\inf(6.28) \geq \sup(6.29). \tag{6.30}$$

As a consequence of the Fenchel–Rockafellar theorem, we have the following duality theorem for SDP programming.

Theorem 6.6. *If the value of (6.28) is finite and there exists $x \in \mathbb{R}^m$ such that $S_0 + \sum_{i=1}^m x_i S_i \in \mathcal{S}_d^{++}(\mathbb{R})$ (i.e., $S_0 + \sum_{i=1}^m x_i S_i$ is positive definite), then the infimum of (6.28) coincides with*

$$\max\{-S_0 \cdot S : S_i \cdot S = c_i, i = 1, \dots, m, \ S \succeq 0\}.$$

Note that the qualification conditions in Theorem 6.6 not only requires that there exists an admissible point[a] for (6.28) but also that the admissible set has non-empty interior. Once again, duality enables us to find optimality conditions:

Theorem 6.7. *Under the assumptions of Theorem 6.6, x solves (6.6) if and only if it is admissible for (6.6) and there exists $S \succeq 0$ such that*

$$S_i \cdot S = c_i, \quad i = 1, \ldots, m \tag{6.31}$$

and the complementarity condition

$$\left(S_0 + \sum_{i=1}^{m} x_i S_i \right) S = 0 \ (\textit{as a matrix}). \tag{6.32}$$

Proof. If $S \succeq 0$ and (6.31) and (6.32) hold, then

$$c \cdot x = \sum_{i=1}^{m} x_i S_i \cdot S = -S_0 \cdot S \leq \sup(6.29),$$

so that x solves (6.28) thanks to (6.30). Conversely, assume that x solves (6.6), let S be a solution of (6.29) then $S \succeq 0$ and (6.31) holds. Moreover, since the values of (6.28) and (6.29) coincide we have

$$\sum_{i=1}^{m} c_i x_i = -S_0 \cdot S.$$

By (6.31), we thus get

$$\left(\sum_{i=1}^{m} c_i S_i \right) \cdot S = -S_0 \cdot S, \quad \text{i.e.,} \quad \text{tr}\left(\left(S_0 + \sum_{i=1}^{m} x_i S_i \right) S \right) = 0.$$

Since both S and $S_0 + \sum_{i=1}^{m} x_i S_i$ are SDP we claim that this implies (6.32). To see this, diagonalize S: $S = \sum_{i=1}^{d} \lambda_i v_i v_i^T$ with $\lambda_i \geq 0$ and set $S(x) = S_0 + \sum_{i=1}^{m} x_i S_i$ then

$$0 = \text{tr}(S(x)S) = \sum_{i=1}^{d} \lambda_i (S(x)v_i) \cdot v_i,$$

each term being non-negative this gives $\lambda_i (S(x)v_i) \cdot v_i = 0$ for every $i = 1, \ldots, d$. We claim that $\lambda_i S(x)v_i = 0$. Indeed, let us now diagonalize the

---

[a]For instance if $S_0 = -\,\text{id}$ and $S_i = 0$ for $i = 1, \ldots, m$ then the admissible set for (6.28) is empty and the primal value is $+\infty$.

SDP matrix $S(x)$ as $S(x) = \sum_{j=1}^{d} \alpha_j w_j w_j^T$ with $\alpha_j \geq 0$ and $(w_1, \ldots, w_d)$ an orthogonal basis, then we have

$$0 = \lambda_i(S(x)v_i) \cdot v_i = \lambda_i \sum_{j=1}^{d} \alpha_j (w_j \cdot v_i)^2,$$

each term in this sum being non-negative, we get $\lambda_i \alpha_j (w_j \cdot v_i)^2 = 0$ so that

$$\lambda_i S(x)v_i = \lambda_i \sum_{j=1}^{d} \alpha_j (w_j \cdot v_i)w_j = 0.$$

In particular $\lambda_i S(x)v_i v_i^T = 0$ for every $i = 1, \ldots, d$ and then

$$S(x)S = \sum_{i=1}^{d} \lambda_i S(x)v_i v_i^T = 0.$$

$\square$

A simple illustration for eigenvalue computation is as follows.

Exercise 6.3. *Let $A \in S_d(\mathbb{R})$ and $\lambda_{\max}(A)$ be its largest eigenvalue. Show that*

$$\lambda_{\max}(A) = \min_{x \in \mathbb{R}}\{x \,:\, x\,\mathrm{id} - A \succeq 0\} = \max\{A \cdot X : \mathrm{tr}(X) = 1, \ X \succeq 0\}.$$

6.3.4. *Link with KKT and Lagrangian duality*

Given a normed space E, $m \in \mathbb{N}^*$ and $(f_0, f_1, \ldots, f_m) \in \Gamma_0(E)^{m+1}$, we consider the constrained convex problem

$$\inf\{f_0(x) : x \in E, \ f_i(x) \leq 0, \ \forall i \in \{1, \ldots, m\}\}. \tag{6.33}$$

We make the following assumption of Slater type:

$$\exists x_0 \in E : f_0(x_0) < +\infty \quad \text{and} \quad f_i(x_0) < 0, \forall i \in \{1, \ldots, m\}. \tag{6.34}$$

The perturbation function for (6.33) consists in changing the right-hand side of the constraints, which amounts to considering for every $(x, y) \in E \times \mathbb{R}^m$:

$$\Phi(x, y) := \begin{cases} f_0(x) & \text{if } f_i(x) \leq y_i \ \forall i \in \{1, \ldots, m\}, \\ +\infty & \text{otherwise.} \end{cases} \tag{6.35}$$

Note that $\Phi \in \Gamma_0(E \times \mathbb{R}^m)$, indeed, it can be expressed as

$$\Phi(x, y) = f_0(x) + \sum_{i=1}^{m} \chi_{\mathrm{epi}(f_i)}(y_i, x)$$

hence it is convex, lsc and it is proper thanks to (6.34). Moreover, thanks to (6.34), $y \mapsto \Phi(x_0, y)$ is constant equal to $f(x_0)$ in a neighborhood of 0. Thus, applying Theorem 6.2, we deduce that whenever

$$\inf(6.33) = \inf_{x \in E} \Phi(x, 0) \in \mathbb{R}, \tag{6.36}$$

then

$$\inf(6.33) = \max_{q \in \mathbb{R}^m} -\Phi^*(0, q). \tag{6.37}$$

Let $q \in \mathbb{R}^m$, and let us compute $\Phi^*(0, q)$:

$$\Phi^*(0, q) = \sup\{q \cdot y - f_0(x) : x \in E, \ y \in \mathbb{R}^m, \ f_i(x) \le y_i, \ i = 1, \dots, m\}. \tag{6.38}$$

If one of the components of q, q_i, is positive, taking $x = x_0$, $y_j = 0$ for $j \ne i$ and letting y_i tend to $+\infty$, we get $\Phi^*(0, q) = +\infty$. If, on the contrary, $q \in \mathbb{R}^m_-$, in (6.38) the term $\sum_{i=1}^m q_i y_i$ is maximized when $y_i = f_i(x)$ hence $\Phi^*(0, q)$ can be written as the value of a maximization in x only:

$$\Phi^*(0, q) := \sup_{x \in E} \left[\sum_{i=1}^m q_i f_i(x) - f_0(x) \right].$$

Now it is convenient to set $\lambda = -q$ and rewrite the previous computation as

$$-\Phi^*(0, -\lambda) = \begin{cases} \inf_E \left\{ \sum_{i=1}^m \lambda_i f_i + f_0 \right\} & \text{if } \lambda \ge 0, \\ +\infty & \text{otherwise.} \end{cases} \tag{6.39}$$

In this form, we recognize the Lagrangian:

$$\mathscr{L}(x, \lambda) := f_0(x) + \sum_{i=1}^m \lambda_i f_i(x), \quad x \in E, \quad \lambda = (\lambda_1, \dots, \lambda_m) \in \mathbb{R}^m. \tag{6.40}$$

Note that the primal problem (6.33) can be rewritten in Lagrangian inf sup form:

$$\inf_{x \in E} \sup_{\lambda \in \mathbb{R}^m_+} \mathscr{L}(x, \lambda) \tag{6.41}$$

since obviously

$$\sup_{\lambda \in \mathbb{R}^m_+} \mathscr{L}(x, \lambda) = \begin{cases} f_0(x) & \text{if } f_i(x) \leq 0 \; \forall i \in \{1, \ldots, m\}, \\ +\infty & \text{otherwise.} \end{cases}$$

The duality relation (6.37) together with (6.39) then express that:

- the infimum in x and the supremum in λ can be switched;
- the dual of (6.33) with respect to the perturbation (6.35) is the problem obtained from (6.41) when switching the infimum in x and the supremum in λ:

$$\sup_{\lambda \in \mathbb{R}^m_+} \inf_{x \in E} \mathscr{L}(x, \lambda); \tag{6.42}$$

- the dual has a solution.

Let us summarize these findings as follows.

Theorem 6.8 (Lagrangian duality in convex programming). *Let $(f_0, f_1, \ldots, f_m) \in \Gamma_0(E)^{m+1}$ and define the corresponding Lagrangian by (6.40) assume that (6.34) holds and that $\inf(6.33)$ is finite. Then we have*

$$\inf_{x \in E} \sup_{\lambda \in \mathbb{R}^m_+} \mathscr{L}(x, \lambda) = \max_{\lambda \in \mathbb{R}^m_+} \inf_{x \in E} \mathscr{L}(x, \lambda). \tag{6.43}$$

The extremality relation resulting from the duality between (6.33) and (6.42) leads to the following KKT type result.

Theorem 6.9. *Under the same assumptions as in Theorem 6.8, assume that $\lambda \in \mathbb{R}^m_+$ is a solution of the dual (6.42). Then x solves the primal (6.33) if and only if it satisfies the KKT conditions:*

- *primal feasibility: $f_i(x) \leq 0$, $i = 1, \ldots, m$;*
- *Lagrangian minimization:*

$$f_0(x) + \sum_{i=1}^m \lambda_i f_i(x) \leq f_0(z) + \sum_{i=1}^m \lambda_i f_i(z), \quad \forall z \in E;$$

- *complementarity slackness: $\lambda_i f_i(x) = 0$, $i = 1, \ldots, m$.*

Proof. By (6.43), x solves (6.33) if and only if it is primal feasible and

$$f_0(x) = \inf_{z \in E} \mathscr{L}(z, \lambda).$$

But since $\lambda \geq 0$, primal feasibility also gives

$$\mathscr{L}(x, \lambda) \leq f_0(x) \leq \mathscr{L}(z, \lambda), \quad \forall z \in E.$$

This implies that x is an (unconstrained) minimizer of $\mathscr{L}(\cdot, \lambda)$ but also that $f_0(x) \leq f_0(x) + \sum_{i=1}^{m} \lambda_i f_i(x)$ so that $\sum_{i=1}^{m} \lambda_i f_i(x) = 0$ which implies complementarity slackness. Finally, the three conditions obviously imply optimality for (6.33). $\qquad \square$

The previous theorem allows for another interpretation of the solutions of the dual: (up to a minus sign) they are multipliers associated with the constraints $f_i(x) \leq 0$, keeping in mind that dual solutions form the subdifferential of the value function. This enables us to interpret (as economists often do) vectors of KKT multipliers as the marginal prices of the inequality constraints.

Exercise 6.4. *Under the same assumptions as in Theorem 6.8, let* $\lambda \in \mathbb{R}_+^m$ *and* x *be feasible for (6.33) show equivalence between the following assertions:*

- *x solves (6.33) and λ solves (6.42);*
- *x minimizes $\mathscr{L}(., \lambda)$ over E and $\lambda_i f_i(x) = 0$ for $i = 1, \ldots, m$;*
- *$0 \in \partial(f_0 + \sum_{i=1}^{m} \lambda_i f_i)(x)$ and $\lambda_i f_i(x) = 0$ for $i = 1, \ldots, m$;*
- *(x, λ) is a saddle-point of the Lagrangian, i.e.,*

$$\mathscr{L}(z, \lambda) \geq \mathscr{L}(x, \lambda) \geq \mathscr{L}(x, \mu), \quad \forall (z, \mu) \in E \times \mathbb{R}_+^m.$$

The Fenchel–Rockafellar duality also has a min–max/saddle point interpretation (see Exercise 6.22).

6.4. On the optimal transport problem

6.4.1. *Kantorovich duality*

An illuminating application (in infinite dimensions) of convex duality for linear programming concerns the optimal transport problem of Monge and Kantorovich. We refer to the textbooks of Santambrogio [89] and Villani [98], [97] for a detailed presentation of this rich topic and its numerous applications. Initially formulated by Monge in 1781, the problem consists in finding the cheapest way to *transport* a probability measure onto another. Given a cost function and two probability measures representing a distribution of sources and a distribution of targets, respectively, we look for a transport *plan* between the sources and the targets which minimizes the total transport cost.

We are denoting by X and Y the source and target spaces and assume for simplicity that both are compact metric spaces. The source distribution μ is a Borel probability measure (which we denote $\mu \in \mathcal{P}(X)$) on X and the target distribution $\nu \in \mathcal{P}(Y)$ is a Borel probability measure on Y. The transport cost is denoted c and assumed to be continuous $c \in C(X \times Y, \mathbb{R})$ and $c(x, y)$ represents the cost of transporting one unit of mass from x to y. The unknown is a plan or coupling $\gamma \in \mathcal{P}(X \times Y)$ between μ and ν where $\gamma(A \times B)$ represents the amount of mass sent from A to B. Mass conservation imposes the constraints that γ should have marginals μ and ν, i.e.,

$$\pi_{1\#}\gamma = \mu, \quad \pi_{2\#}\gamma = \nu,$$

where $\pi_{1\#}\gamma$ and $\pi_{2\#}\gamma$ are defined by

$$\pi_{1\#}\gamma(A) := \gamma(A \times Y), \quad \pi_{2\#}\gamma(B) := \gamma(X \times B) \tag{6.44}$$

for every Borel subsets A and B of X and Y, respectively. Another way to define the marginals of γ is by test-functions[b] that is for every $(\varphi, \psi) \in C(X) \times C(Y)$:

$$\int_X \varphi(x)\mathrm{d}\pi_{1\#}\gamma(x) = \int_{X \times Y} \varphi(x)\mathrm{d}\gamma(x, y), \quad \text{and}$$

$$\int_Y \psi(y)\mathrm{d}\pi_{2\#}\gamma(y) = \int_{X \times Y} \psi(y)\mathrm{d}\gamma(x, y).$$

We denote by $\Pi(\mu, \nu)$ the set of transport plans/couplings between μ and ν:

$$\Pi(\mu, \nu) := \{\gamma \in \mathcal{P}(X \times Y) : \pi_{1\#}\gamma = \mu, \ \pi_{2\#}\gamma = \nu\}.$$

Note that $\Pi(\mu, \nu)$ is non-empty (it contains the independent coupling $\mu \otimes \nu$) and convex. The Monge–Kantorovich optimal transport problem then reads

$$\inf_{\gamma \in \Pi(\mu, \nu)} \int_{X \times Y} c(x, y)\mathrm{d}\gamma(x, y). \tag{6.45}$$

This is a linear programming problem (*a priori* in infinite dimensions unless X and Y are finite).

---

[b] A representation theorem due to Riesz, see, for instance [88], allows to identify the topological dual of $C(X)$, equipped with the uniform norm, with the space of Borel measures, $\mathcal{M}(X)$ on X through $\varphi \in C(X) \mapsto p_\eta(f) := \int_X \varphi\mathrm{d}\eta$ with $\eta \in \mathcal{M}(X)$.

It turns out, as was discovered by Kantorovich in the 1940s, that (6.45) is the dual of the following problem:

$$\sup\left\{\int_X \varphi d\mu + \int_Y \psi d\nu \ : \ (\varphi, \psi) \in C(X) \times C(Y), \ \varphi \oplus \psi \leq c\right\}, \quad (6.46)$$

where

$$(\varphi \oplus \psi)(x, y) = \varphi(x) + \psi(y), \quad \forall (x, y) \in X \times Y.$$

Indeed, rewrite (6.46) as

$$- \inf_{(\varphi,\psi)\in C(X)\times C(Y)} G(\varphi, \psi) + H(\Lambda(\varphi, \psi)),$$

where $\Lambda : C(X) \times C(Y) \to C(X \times Y)$ is defined by

$$\Lambda(\varphi, \psi) = \varphi \oplus \psi,$$

so that $\Lambda \in \mathcal{L}_c(C(X) \times C(Y), C(X \times Y))$, G is the continuous linear form

$$G(\varphi, \psi) = -\int_X \varphi d\mu - \int_Y \psi d\nu$$

and $H \in \Gamma_0(C(X \times Y))$ is defined for every $\theta \in C(X \times Y)$ by

$$H(\theta) = \begin{cases} 0 & \text{if } \theta \leq c, \\ +\infty & \text{otherwise.} \end{cases}$$

The Fenchel–Rockafellar dual of (6.46) then reads (pay attention to the minus sign which leads to a minimization in the dual)

$$- \sup_{\gamma \in \mathcal{M}(X \times Y)} \{-G^*(-\Lambda^*(\gamma)) - H^*(\gamma)\}$$

$$= \inf_{\gamma \in \mathcal{M}(X \times Y)} \{G^*(-\Lambda^*(\gamma)) + H^*(\gamma)\}.$$

Now for $\gamma \in \mathcal{M}(X \times Y) = (C(X \times Y))^*$ and $(\varphi, \psi) \in C(X) \times C(Y)$, we have

$$\Lambda^*(\gamma)(\varphi, \psi) = \int_{X \times Y} \varphi \oplus \psi d\gamma$$

$$= \int_X \varphi d\pi_{1\#}\gamma + \int_Y \psi d\pi_{2\#}\gamma$$

$$= (\pi_{1\#}\gamma, \pi_{2\#}\gamma)(\varphi, \psi)$$

so that $\Lambda^*(\gamma) = (\pi_{1\#}\gamma, \pi_{2\#}\gamma)$. A direct computation gives

$$G^*(-\Lambda^*(\gamma)) = \begin{cases} 0 & \text{if } (\pi_{1\#}\gamma, \pi_{2\#}\gamma) = (\mu, \nu), \\ +\infty & \text{otherwise} \end{cases}$$

likewise

$$H^*(\gamma) = \sup_{\theta \leq c} \int_{X \times Y} \theta \,\mathrm{d}\gamma = \begin{cases} \int_{X \times Y} \theta \,\mathrm{d}\gamma & \text{if } \gamma \geq 0, \\ +\infty & \text{otherwise.} \end{cases}$$

Now observe that G is everywhere continuous and H is continuous at $\Lambda(\varphi_0, \psi_0)$ for $\varphi_0 = \min_{X \times Y} c - 1$ and $\psi_0 = 0$. We thus deduce from the Fenchel–Rockafellar theorem, the so-called Kantorovich duality formula

$$\min_{\gamma \in \Pi(\mu, \nu)} \int_{X \times Y} c \,\mathrm{d}\gamma = \sup \left\{ \int_X \varphi \,\mathrm{d}\mu + \int_Y \psi \,\mathrm{d}\nu : (\varphi, \psi) \in C(X) \times C(Y), \right.$$

$$\left. \varphi \oplus \psi \leq c \right\}. \tag{6.47}$$

In particular (6.45) admits a solution (which could have easily been deduced from the Banach–Alaoglu–Bourbaki theorem, see [23]). As for the existence of a solution to (6.46), it follows neither from the Fenchel–Rockafellar theorem nor from Theorem 6.1 (there is no coercivity in (6.46) and $C(X) \times C(Y)$ is not reflexive, see Exercise 1.20). The existence of a solution to (6.46) requires specific arguments and the use of Ascoli's theorem.

Theorem 6.10. *Problem (6.46) admits at least a solution (φ, ψ) (such a solution is called a pair of Kantorovich potentials).*

Proof. Let us denote by K the linear function $K = -G$, i.e.,

$$K(\varphi, \psi) = \int_X \varphi \,\mathrm{d}\mu + \int_Y \psi \,\mathrm{d}\nu$$

and observe that if λ is a constant then, since μ and ν have the same total mass 1

$$K(\varphi + \lambda, \psi - \lambda) = K(\varphi, \psi). \tag{6.48}$$

Note also that the constraint $\varphi \oplus \psi \leq c$ can be equivalently formulated as

$$\psi(y) \leq \varphi^c(y), \quad \forall y \in Y$$

where

$$\varphi^c(y) := \min_{x \in X}\{c(x,y) - \varphi(x)\}, \ \forall y \in Y. \tag{6.49}$$

The function φ^c is called the c-concave transform of φ. By construction, $\varphi^c \oplus \varphi \le c$ and, since $\psi \le \varphi^c$ and ν is non-negative, we have

$$K(\varphi, \psi) \le K(\varphi, \varphi^c).$$

But we can use this c-*concavification* trick once more, defining the c-concave envelope of φ as

$$\varphi^{c\check{c}}(x) := \min_{y \in Y}\{c(x,y) - \varphi^c(y)\}, \quad \forall x \in X, \tag{6.50}$$

we have $\varphi^{c\check{c}} \oplus \varphi^c \le c$ and since $\varphi(x) \le c(x,y) - \varphi^c(y)$ taking the minimum in x gives $\varphi \le \varphi^{c\check{c}}$ hence

$$K(\varphi, \psi) \le K(\varphi, \varphi^c) \le K(\varphi^{c\check{c}}, \varphi^c). \tag{6.51}$$

Now the key compactness argument is that one has a control on the modulus of continuity of φ^c. Indeed, denoting by dist_Y the distance on Y (which makes it compact) and setting

$$\omega(\varepsilon) := \sup\{|c(x,y) - c(x,y')|, x \in X, \ (y,y') \in Y^2, \ \mathrm{dist}_Y(y,y') \le \varepsilon\},$$

$$\varepsilon \ge 0,$$

we have thanks to the uniform continuity of c (Theorem 1.3)

$$\omega(\varepsilon) \to 0 \quad \text{as } \varepsilon \to 0^+$$

and for every $(y,y') \in Y^2$ since $c(x,y') \le c(x,y) + \omega(\mathrm{dist}_Y(y,y'))$ for every $x \in X$, we have

$$\varphi^c(y') \le \varphi^c(y) + \omega(\mathrm{dist}_Y(y,y')).$$

Hence φ^c admits ω as modulus of continuity:

$$|\varphi^c(y') - \varphi^c(y)| \le \omega(\mathrm{dist}_Y(y,y')), \quad \forall (y,y') \in Y^2. \tag{6.52}$$

In other words, the set $\{\varphi^c, \varphi \in C(X)\}$ is uniformly equicontinuous in $C(Y)$. In a similar way, the set $\{\varphi^{c\check{c}}, \varphi \in C(X)\}$ is uniformly equicontinuous in $C(X)$. Now let (φ_n, ψ_n) be a maximizing sequence for (6.46), thanks to (6.51), $(\varphi_n^{c\,\check{c}}, \varphi_n^c)$ is also a maximizing sequence for (6.46). Let us finally define

$$\widetilde{\varphi}_n := (\varphi_n^c - \min_Y \varphi_n^c)^{\check{c}} = \varphi_n^{c\,\check{c}} + \min_Y \varphi_n^c, \quad \widetilde{\psi}_n := \varphi_n^c - \min_Y \varphi_n^c.$$

By (6.48), $(\widetilde{\varphi}_n, \widetilde{\psi}_n)$ is yet another maximizing sequence for (6.46), it is easy to check that it is uniformly bounded and both $(\widetilde{\varphi}_n)$ and $(\widetilde{\psi}_n)$ are uniformly equicontinuous. Thanks to Ascoli's (Theorem 1.20), $(\widetilde{\varphi}_n, \widetilde{\psi}_n)$ admits a subsequence which converges uniformly to some (φ, ψ); obviously $\varphi \oplus \psi \leq c$ and (φ, ψ) solve (6.46). $\qquad\square$

6.4.2. *Characterization of optimal plans*

The Kantorovich duality enables one to characterize optimal transport plans through generalized convex analysis concepts. Let us first introduce some definitions.

Definition 6.1. Let $\varphi : X \to \mathbb{R} \cup \{-\infty\}$. Then φ is called c-concave if there exists a non-empty subset $\mathcal{A}$ of $Y \times \mathbb{R}$ such that

$$\varphi(x) = \inf_{(y,\lambda) \in \mathcal{A}} \{c(x,y) - \lambda\}, \quad \forall x \in X. \tag{6.53}$$

Lemma 6.5. *Let $\varphi : X \to \mathbb{R} \cup \{-\infty\}$ be c-concave. Then,*

- *either $\varphi \equiv -\infty$,*
- *or $\varphi \in C(X)$.*

If φ is not identically $-\infty$, then $\varphi = \varphi^{c\check{c}}$ (where φ^c and $\varphi^{c\check{c}}$ are defined by (6.49) and (6.50), respectively).

Proof. Let us write φ in the form (6.53) and define

$$\alpha := \sup\{\lambda \in \mathbb{R} : \exists y \in Y \text{ such that } (y,\lambda) \in \mathcal{A}\}.$$

If $\alpha = +\infty$, then obviously $\varphi \equiv -\infty$. Assume then that $\alpha < +\infty$, then φ is bounded, and it is immediate to check as in the proof of Theorem 6.10 that φ is continuous (using the uniform continuity of c again). We have already observed that $\varphi \leq \varphi^{c\check{c}}$. To show the converse inequality when φ is of the form (6.53), let $(y,\lambda) \in \mathcal{A}$ then $\varphi^c(y) \geq \lambda$ so that for every $x \in X$,

$\varphi^{c\check{c}}(x) \le c(x, y) - \lambda$, taking the infimum with respect to $(y, \lambda) \in \mathcal{A}$ we get $\varphi^{c\check{c}} \le \varphi$. $\qquad\qquad\qquad\qquad\qquad\qquad\qquad\qquad\qquad\qquad$ □

Definition 6.2 (c-concave analysis concepts). Given $\varphi \in C(X)$, the c-superdifferential of φ-at $x \in X$ denoted by $\partial^c\varphi(x)$ is the (possibly empty) set

$$\partial^c\varphi(x) := \{y \in Y : \varphi(x) + \varphi^c(y) = c(x, y)\}$$
$$= \{y \in Y : \varphi(x') \le \varphi(x) + c(x', y) - c(x, y), \ \forall x' \in X\}.$$

The c-superdifferential of φ, simply denoted $\partial^c\varphi$, is the graph of $x \mapsto \partial^c\varphi(x)$, i.e.,

$$\partial^c\varphi := \{(x, y) \in X \times Y : y \in \partial^c\varphi(x)\}.$$

A non-empty subset S of $X \times Y$ is called c-cyclically monotone if for every $N \in \mathbb{N}$ and every $(x_i, y_i)_{i=0,\ldots,N} \in S^{N+1}$ setting $x_{N+1} = x_0$ one has

$$\sum_{i=0}^{N} c(x_{i+1}, y_i) \ge \sum_{i=0}^{N} c(x_i, y_i). \qquad\qquad (6.54)$$

It is immediate to check by standard continuity/compactness arguments that if $\varphi \in C(X)$ is c-concave then for every $x \in X$, $\partial^c\varphi(x) \ne \emptyset$ (use (6.53), pick a minimizing sequence (y_n, λ_n) and extract a convergent subsequence with limit (y, λ), checking that $y \in \partial^c\varphi(x)$ is then straightforward). In this case, $\partial^c\varphi$ is closed since it is the set where the continuous function $\varphi \oplus \varphi^c - c$ achieves its maximum on the compact set $X \times Y$.

Definition 6.3. Let $\gamma \in \mathcal{P}(X \times Y)$, the support of γ, denoted $\mathrm{supp}(\gamma)$, is the smallest closed subset of $X \times Y$ which has mass 1 for γ (equivalently its complement is the largest open set which is γ-negligible).

A first consequence of the Kantorovich duality formula (6.47) and of the attainment result from Theorem 6.10 is the following proposition.

Proposition 6.4. *Let $\gamma \in \Pi(\mu, \nu)$. Then γ solves the Monge–Kantorovich problem (6.45) if and only if there exists $\varphi \in C(X)$ which is c-concave and such that $\mathrm{supp}(\gamma) \subset \partial^c\varphi$.*

Proof. Let $\gamma \in \Pi(\mu, \nu)$ solve (6.45), and φ, ψ be a solution of the Kantorovich dual (6.47). We may assume that φ is c-concave by the

c-concavification trick, then we have

$$\int_{X\times Y} c\,d\gamma = \int_X \varphi\,d\mu + \int_Y \varphi^c\,d\nu = \int_{X\times Y}(\varphi\oplus\varphi^c)\,d\gamma$$

but since $\varphi\oplus\varphi^c \leq c$ the previous equality implies that $\varphi\oplus\varphi^c = c$ γ a.e. so that $\gamma(\partial^c\varphi) = 1$ since $\partial^c\varphi$ is closed this gives $\mathrm{supp}(\gamma) \subset \partial^c\varphi$. Conversely, if there exists $\varphi \in C(X)$ which is c-concave and such that $\mathrm{supp}(\gamma) \subset \partial^c\varphi$ then $\varphi\oplus\varphi^{\tilde{c}} = c$ γ-a.e. so that

$$\int_{X\times Y} c\,d\gamma = \int_X \varphi\,d\mu + \int_Y \varphi^c\,d\nu \leq \max(6.47) = \min(6.45),$$

which shows the optimality of γ. $\qquad\square$

A more intrinsic characterization of optimality is given by the following result which has been established in various contexts and levels of generality, by several authors, see Smith and Knott [92], Brezis [24, 25], Rochet [83], Gangbo and McCann [55]. This characterization has its roots in a well-known theorem proved by Rockafellar in 1966 which characterizes subgradients of convex functions [84] which we will recall below.

Theorem 6.11 (c-cyclical monotonicity of the optimal support).
Let S be a non-empty subset of $X \times Y$. Then the following statement are equivalent

1) *there exists a c-concave and finite function φ such that $S \subset \partial^c\varphi$;*
2) *S is c-cyclically monotone.*

As a consequence, $\gamma \in \Pi(\mu,\nu)$ is optimal for (6.45) if and only if $\mathrm{supp}(\gamma)$ is c-cyclically monotone.

Proof. If $S \subset \partial^c\varphi$ for some finite c-concave (hence continuous) potential φ, if $(x_i,y_i)_{i=0,\dots,N} \in S^{N+1}$ and $x_{N+1} = x_0$ then since $y_i \in \partial^c\varphi(x_i)$ we have

$$\varphi(x_{i+1}) - \varphi(x_i) \leq c(x_{i+1},y_i) - c(x_i,y_i), \quad i = 0,\dots,N.$$

Summing over i and using the fact that $x_{N+1} = x_0$ we get

$$\sum_{i=0}^{N}(c(x_{i+1},y_i) - c(x_i,y_i)) \geq 0,$$

so that S is c-cyclically monotone.

Conversely, assume that S is c-cyclically monotone and let us fix $(x_0, y_0) \in S$. Define then for every $x \in X$:

$$\varphi(x) := \inf \Big[\sum_{i=0}^{N-1} (c(x_{i+1}, y_i) - c(x_i, y_i)) : N \geq 1,$$

$$(x_i, y_i)_{i=1,\ldots,N-1} \in S^{N-1}, \; x_N = x \Big].$$

By construction, φ is c-concave and since S is c-cyclically monotone, one has $\varphi(x_0) = 0$. Hence φ is everywhere finite and continuous by Lemma 6.5. Now let $(x, y) \in S$ and $x' \in X$, let $N \in \mathbb{N}$ and $(x_1, y_1), \ldots, (x_{N-1}, y_{N-1}) \in S^{N-1}$ let $(x_N, y_N) = (x, y)$ and $x_{N+1} = x'$ we thus have

$$\varphi(x') \leq \sum_{i=0}^{N-1} (c(x_{i+1}, y_i) - c(x_i, y_i)) + c(x', y) - c(x, y).$$

Taking the infimum with respect to $(x_1, y_1), \ldots, (x_{N-1}, y_{N-1}) \in S^{N-1}$ we get

$$\varphi(x') \leq \varphi(x) + c(x', y) - c(x, y)$$

for every $x' \in X$ so that $(x, y) \in \partial^c \varphi$. $\qquad\square$

As already mentioned, the proof above is directly inspired by the seminal paper of Rockafellar [84] where the following theorem is established.

Theorem 6.12 (A cyclical characterization of subsets of subdifferentials). *Let E be a normed space and let S be a non-empty subset of $E \times E^*$. Then the following statements are equivalent:*

1. *There exists $f \in \Gamma_0(E)$ such that $S \subset \partial f := \{(x, p) \in E \times E^* : p \in \partial f(x)\}$.*
2. *For every $N \in \mathbb{N}$ every $(x_0, p_0), \ldots, (x_N, p_N) \in S^{N+1}$ setting $x_{N+1} = x_0$ one has:*

$$\sum_{i=0}^{N} p_i(x_i - x_{i+1}) \geq 0.$$

6.4.3. *Monge solutions*

The previous abstract characterizations of optimal transport plans can under certain additional conditions lead to the conclusion that optimal plans are in fact very sparse in the sense that they are supported by the graph of a certain map, this is the concept of Monge solution.

Definition 6.4. Let T be a Borel map: $X \to Y$. The pushforward of μ through μ (also known as the image of μ by T) is the Borel probability measure on Y denoted by $T_{\#}\mu$ and defined equivalently by:

$$T_{\#}\mu(B) = \mu(T^{-1}(B)), \quad \forall B \subset Y \text{ Borel}$$

or by

$$\int_Y \psi dT_{\#}\mu = \int_X \psi \circ T d\mu, \quad \forall \psi \in C(Y).$$

One says that T transports μ to ν if $T_{\#}\mu = \nu$, in such a case the measure $\gamma_T := (\mathrm{id}, T)_{\#}\mu \in \mathcal{P}(X \times Y)$ defined by

$$\int_{X \times Y} \eta(x,y) d\gamma_T(x,y) = \int_X \eta(x, T(x)) d\mu(x), \quad \forall \eta \in C(X \times Y)$$

belongs to $\Pi(\mu, \nu)$. An optimal $\gamma \in \Pi(\mu, \nu)$ for (6.45) is said to be a Monge solution if it is of the form $\gamma = \gamma_T$ for some T transporting μ to ν. In this case, T solves the (nonlinear) Monge problem

$$\inf_{T : T_{\#}\mu = \nu} \int_X c(x, T(x)) d\mu(x). \tag{6.55}$$

Note that there might exist no transport map between μ and ν, for instance, if μ is a Dirac mass at some $x \in X$ then $T_{\#}\mu$ is the Dirac mass at $T(x)$ hence it is impossible to transport μ to ν by a map unless ν is itself a Dirac mass.

The unidimensional case and the monotone transport

Let us consider the unidimensional case, where:

- $X = Y = [a, b]$ is a compact interval of the real line;
- μ is nonatomic, i.e., $\mu(\{x\}) = 0$ for every $x \in [a, b]$, ν is an arbitrary Borel probability measure on $[a, b]$;
- $c \in C([a, b] \times [a, b])$ is strictly submodular, i.e., for every x_1 and x_2 in $[a, b]$ if $x_1 < x_2$ then

$$y \in [a, b] \mapsto c(x_2, y) - c(x_1, y) \text{ is decreasing.}$$

There are many examples of strictly submodular costs, we let the reader check the following:

Exercise 6.5 (Examples of submodular costs). *Show that:*

1) $c(x, y) = -xy$ *is strictly submodular;*
2) *if C is strictly convex, then $c(x,y) := C(x - y)$ is strictly submodular;*
3) *if g is strictly concave, then $c(x,y) := g(x + y)$ is strictly submodular;*
4) *if c is strictly submodular, then so are $(x,y) \mapsto c(x,y) + f(x) + g(y)$ and $(x,y) \mapsto c(u(x), v(y))$ where u and v are increasing and continuous;*
5) *if $c \in C^2([a,b] \times [a,b])$ and $\partial_{xy}^2 c < 0$ on $[a,b] \times [a,b]$, then c is strictly submodular.*

Intuitively, for transport problems, a supermodular cost induces the so-called positive sorting, i.e., the fact that large values of x should be matched with large values of y. This suggests some monotonicity of optimal plans which is intimately related to the important notion of monotone transport which is the object of the following exercise.

Exercise 6.6 (Monotone transport). *Let $\mu \in \mathcal{P}([a,b])$ be non-atomic. Define the cumulative distribution functions of μ and ν by*

$$F_\mu(t) := \mu([a, t]), \quad F_\nu(t) := \nu([a, t]), \ \forall t \in [a, b].$$

1) *Show that a monotone function $F : [a,b] \to \mathbb{R}$ has at most countably many points of discontinuity and that this discontinuity set is μ-negligible.*
2) *Show that F_μ and F_ν are non-decreasing, right-continuous, usc and that F_μ is continuous.*
3) *Show that for every $\alpha \in [0,1]$ one has*

$$\mu(\{x \in [a,b] : F_\mu(x) \leq \alpha\}) = \alpha.$$

4) *Define the monotone transport, T^* between μ and ν as follows. For every $x \in [a,b]$, set*

$$T^*(x) := \inf\{t \in [a,b] : F_\nu(t) \geq F_\mu(x)\}.$$

Show that T^* is non-decreasing and that $F_\nu(t) \geq F_\mu(x) \iff t \geq T^*(x)$.
5) *Show that $T^*_\# \mu = \nu$. (Hint: use the previous questions to deduce that $T^*_\# \mu$ and ν agree on all intervals of the form $[a,t]$ and then conclude by approximation of continuous functions by step functions.)*

6) *Show that if T is non-decreasing and $T_{\#}\mu = \nu$ then $\mu(\{T = T^*\}) = 1$ (Hint: show that $F_\nu \circ T \geq F_\mu$, deduce that $T \geq T^*$ and use the fact that $0 = \int_a^b (T - T^*)d\mu$.)*

Having the construction of the monotone transport from the exercise above in mind, we have the following theorem.

Theorem 6.13. *Let μ and ν be Borel probability measures on the compact interval $[a, b]$ with μ nonatomic, let $c \in C([a, b] \times [a, b])$ be strictly submodular and consider the optimal transport problem*

$$\inf_{\gamma \in \Pi(\mu, \nu)} \int_{[a,b] \times [a,b]} c(x, y) d\gamma(x, y). \tag{6.56}$$

Then (6.56) has a unique solution γ. Moreover γ is of Monge type and is given by $\gamma = \gamma_{T^*}$ where T^* is the monotone transport between μ and ν as defined in Exercise 6.6.

Proof. Step 1: monotonicity

Let γ be a solution of (6.56) and let us set $S := \operatorname{supp}(\gamma)$. It follows from Theorem 6.11 that if $x_1 < x_2$ and $((x_1, y_1), (x_2, y_2)) \in S^2$ then

$$c(x_1, y_1) + c(x_2, y_2) \leq c(x_1, y_2) + c(x_2, y_1).$$

Together with the strict submodularity of c this is easily seen to imply that $y_1 \leq y_2$. It is not difficult to check that

$$\operatorname{supp}(\mu) = \{x \in [a, b] : \exists y \in [a, b] \text{ s.t. } (x, y) \in S\}$$

so that whenever $x \in \operatorname{supp}(\mu)$ the corresponding slice S_x is non-empty, i.e.,

$$S_x := \{y \in [a, b] \text{ s.t. } (x, y) \in S\} \neq \emptyset.$$

Since S_x is compact and non-empty we can define for every $x \in \operatorname{supp}(\mu)$:

$$t^+(x) := \max S_x, \ t^-(x) := \min S_x.$$

Of course, $t^- \leq t^+$ and, due to the monotonicity of the set S, whenever $x' < x$ one has $t^+(x') \leq t^-(x)$. Let us define the set

$$N := \{x \in \operatorname{supp}(\mu) : t^+(x) > t^-(x)\}$$

and observe that, again thanks to the monotonicity of S, if $(x', x) \in N \times N$ with $x' \neq x$ the non-empty open intervals $(t^-(x), t^+(x))$ and $(t^-(x'), t^+(x'))$

are disjoint. Hence, for each $x \in N$, we may choose $q_x \in \mathbb{Q} \cap (t^-(x), t^+(x))$, and the map $x \in N \mapsto q_x \in \mathbb{Q}$ is one to one so that N is at most countable. Since μ is nonatomic, this implies that $\mu(N) = 0$.

Setting $\alpha := \min \operatorname{supp}(\mu)$, $\beta := \max \operatorname{supp}(\mu)$, we can extend $t^\pm$ by the non-decreasing functions defined on the whole of $[a, b]$ (and still denoted $t^\pm$) by

$$t^\pm(x) := \begin{cases} a & \text{if } x \in [a, \alpha), \\ \sup\{t^\pm(x'), x' \in \operatorname{supp}(\mu),\ x' \leq x\} & \text{if } x \in [\alpha, \beta], \\ b & \text{if } x \in (\beta, b]. \end{cases}$$

Step 2: γ is of Monge type

Whenever $x \in \operatorname{supp}(\mu) \setminus N$ we have $t^+(x) = t^-(x)$, hence S_x is reduced to the singleton $\{t^+(x)\} = \{t^-(x)\}$ for μ-almost every x. We deduce from

$$1 = \gamma((\operatorname{supp}(\mu) \setminus N) \times [a, b]))$$

that $(x, y) = (x, t^\pm(x))$ γ-almost everywhere, i.e., the graphs of t^+ and t^- have full measure for γ. Let A and B be two Borel subsets of $[a, b]$. Since the first marginal of γ is μ, we have

$$\begin{aligned} \gamma(A \times B) &= \gamma((A \cap (\operatorname{supp}(\mu) \setminus N)) \times B) \\ &= \gamma(\{x \in A \cap (\operatorname{supp}(\mu) \setminus N) : t^\pm(x) \in B\}) \\ &= \mu(A \cap (\operatorname{supp}(\mu) \setminus N) \cap t^{\pm^{-1}}(B)) \\ &= \mu(A \cap t^{\pm^{-1}}(B)) = \gamma_{t^\pm}(A \times B) \end{aligned}$$

a standard monotone class argument (see, for instance, [88, Chapter 8]) implies that $\gamma = \gamma_{t^+} = \gamma_{t^-}$; γ is therefore of Monge type and $t^\pm_\# \mu = \nu$.

Step 3: conclusion

Let us now show that $t^+ = T^*$ μ-a.e which will imply that $\gamma = \gamma_{t^+} = \gamma_{T^*}$ and end the proof. Firstly, since $t^+_\# \mu = \nu$ we have for every $x \in [a, b]$,

$$\mu(\{x' \in [a, b] : t^+(x') \leq t^+(x)\}) = F_\nu(t^+(x))$$

but since t^+ is non-decreasing, the left-hand side of the previous identity is larger than $F_\mu(x)$ hence

$$F_\mu(x) \leq F_\nu(t^+(x))$$

which, by the very definition of $T^*(x)$, shows that $t^+(x) \geq T^*(x)$ for every $x \in [a, b]$. Finally, since $t^+_\# \mu = T^*_\# \mu = \nu$, we have $\int_{[a,b]} (t^+(x) - T^*(x)) d\mu(x) = 0$, implying $t^+ = T^*$ μ-a.e. and completing the proof. $\quad\square$

The quadratic case, Brenier's theorem

Now let us consider the optimal transport problem with a quadratic cost in several dimensions. This problem was solved by Brenier [22] whose results totally renewed the interest for the Monge–Kantorovich problem and stimulated a huge stream of research in the last three decades. For the sake of simplicity, we shall consider the compactly supported case, i.e., we fix some closed ball $B = \overline{B}(0, r)$ of $\mathbb{R}^d$ and consider marginals μ and ν in $\mathcal{P}(B)$ with

$$\mu \text{ absolutely continuous wrt the } d\text{-dimensional Lebesgue measure.}$$
(6.57)

Denoting by $|.|$ the Euclidean distance, we consider the optimal transport problem

$$\inf_{\gamma \in \Pi(\mu, \nu)} \frac{1}{2} \int_{B \times B} |x - y|^2 \mathrm{d}\gamma(x, y).$$
(6.58)

Expanding the square and using the constraint $\gamma \in \Pi(\mu, \nu)$, it is immediate to check that (6.58) is equivalent to

$$\sup_{\gamma \in \Pi(\mu, \nu)} \int_{B \times B} x \cdot y \mathrm{d}\gamma(x, y).$$
(6.59)

We know from Kantorovich duality that the supremum in (6.59) is attained by some $\gamma \in \Pi(\mu, \nu)$ and coincides with (pay attention to the change of convention here since in (6.59) we maximize the integral)

$$\inf_{(u,v) \in C(B)^2} \left[\int_B u \mathrm{d}\mu + \int_B v \mathrm{d}\nu : u(x) + v(y) \geq x \cdot y \right].$$
(6.60)

Theorem 6.10 ensures that (6.60) admits a pair of optimal potentials u, v such that

$$u(x) = \max_{y \in B} \{x \cdot y - v(y)\}, \quad \forall x \in B$$
(6.61)

so that u is convex and Lipschitz (since B is bounded). Moreover, γ and (u, v) are related through the complementary slackness condition:

$$u(x) + v(y) = x \cdot y, \quad \forall (x, y) \in \text{supp}(\gamma)$$

which, with (6.61), is easily seen to imply that

$$(x, y) \in \text{supp}(\gamma) \Rightarrow y \in \partial u(x)$$

(one may extend u by $+\infty$ outside of B if necessary). It now follows from Rademacher's theorem (Theorem 5.3) that u is differentiable at every $x \in B \setminus N$ where N is Lebesgue-negligible so that (6.57) implies that $\partial u(x) = \{\nabla u(x)\}$ for every x except perhaps on a negligible set. Arguing as in the unidimensional case, one can then check that γ is of Monge type:

$$\gamma = \gamma_{\nabla u}.$$

In particular $\nabla u_{\#}\mu = \nu$, the map ∇u therefore transports μ to ν, it is called the Brenier's map between μ and ν and may be viewed as a generalization of the monotone transport in several dimensions. Brenier's theorem precisely says that the quadratic optimal transport problem has Monge solutions and that the optimal maps are gradient of convex functions. Brenier's theorem was extended by Robert McCann (more general assumptions on μ taking advantage of the fact that convex functions are in some sense more often differentiable than just Lipschitz ones), see McCann [74]. The same kind of arguments as above enable one to conclude that optimal plans necessarily are of Monge type for more general cost functions (those who satisfy the so-called twist condition), see McCann and Gangbo [55], Caffarelli [29], Gangbo [54], Levin [69], Carlier [31] or have a look at Exercise 6.28.

In case the Brenier's map is a diffeomorphism and the measures μ and ν have densities f and g with respect to the Lebesgue measure, the transport property $\nabla u_{\#}\mu = \nu$ can be expressed thanks to the change of variables theorem by the Monge–Ampère equation:

$$f = \det(D^2 u)g(\nabla u).$$

A deep and important regularity theory for Brenier's maps and solutions of Monge–Ampère equations was developed by Luis Caffarelli in the 1990s and extended more recently by Alessio Figalli and Guido de Philippis. An excellent reference on the subject is the book [48] of Fields medallist Alessio Figalli.

6.4.4.   *The discrete case*

We now pay attention to the discrete case where the source and target measures have a finite support which we denote respectively $S := \{x_i,\ i = 1,\ldots,I\}$, $T := \{y_j,\ i = 1,\ldots,J\}$. The fixed marginals therefore are of the form

$$\mu := \sum_{i=1}^{I} \mu_i, \quad \nu := \sum_{J=1}^{J} \nu_j,$$

where the weight vectors $\boldsymbol{\mu} := (\mu_1, \ldots, \mu_I)$ and $\boldsymbol{\nu} := (\nu_1, \ldots, \nu_J)$ belong to $(0, +\infty)^I$ and $(0, +\infty)^J$. The weights $\boldsymbol{\mu}$ and $\boldsymbol{\nu}$ also satisfy the compatibility condition

$$\sum_{i=1}^{I} \mu_i = \sum_{j=1}^{J} \nu_j$$

that is the total amount of mass available at the sources coincides with the total demand of the targets (and it is not necessarily normalized to 1 here). The cost of transporting one unit of mass from x_i to y_j being denoted by $c_{ij} = c(x_i, y_j)$, the cost matrix C is the $I \times J$ matrix with entries c_{ij}. The corresponding transport problem then reads:

$$\inf_{\gamma \in T(\boldsymbol{\mu}, \boldsymbol{\nu})} \operatorname{tr}(C^T \gamma) := \sum_{i=1}^{I} \sum_{j=1}^{J} c_{ij} \gamma_{ij}, \qquad (6.62)$$

where $T(\boldsymbol{\mu}, \boldsymbol{\nu})$ is the transport polytope consisting of all $I \times J$ matrices γ with non-negative entries γ_{ij} which have prescribed row and sum columns given by the marginals, i.e.,

$$\sum_{j=1}^{J} \gamma_{ij} = \mu_i, \ \forall i = 1, \ldots, I, \quad \sum_{i=1}^{I} \gamma_{ij} = \nu_j, \ \forall j = 1, \ldots, J. \qquad (6.63)$$

Note that (6.62) is a linear programming problem such as the ones we studied in Section 6.3.2. Since (6.62) has a compact feasible set, it obviously admits a solution, hence (by Theorem 6.5 for instance or as a very special case of Theorem 6.10) so does its dual which takes the discrete Kantorovich form:

$$\sup \left[\sum_{i=1}^{I} \varphi_i \mu_i + \sum_{j=1}^{J} \psi_j \nu_j \ : \ \varphi_i + \psi_j \le c_{ij}, \ \forall (i,j) \right]. \qquad (6.64)$$

Being a linear programming problem with a compact feasible set $T(\boldsymbol{\mu}, \boldsymbol{\nu})$, we know from Proposition 3.4 that the minimum in (6.62) is achieved at at least one extreme point of the transport polytope $T(\boldsymbol{\mu}, \boldsymbol{\nu})$. These extreme points are well-known and characterized by a certain acyclicity condition.

Definition 6.5. Let $\gamma \in \mathbb{R}_+^{I \times J}$ be a matrix with nonnegative entries, one says that γ contains a cycle if there are pairwise distinct row indices $i_1, \ldots, i_k$, pairwise distinct column indices $j_1, \ldots, j_k$ such that

$$\gamma_{i_l j_l} > 0, \ \forall l = 1, \ldots, k, \quad \gamma_{i_l j_{l+1}} > 0, \ \forall l = 1, \ldots, k-1 \quad \text{and} \quad \gamma_{i_k j_1} > 0.$$

The matrix γ is called acyclic if it contains no cycle.

Theorem 6.14. *Let* $\gamma \in T(\boldsymbol{\mu}, \boldsymbol{\nu})$. *Then* γ *belongs to the set of extreme points of the transport polytope* $\mathrm{Ext}(T(\boldsymbol{\mu}, \boldsymbol{\nu})$ *if and only if* γ *is acyclic.*

Proof. Let us assume that γ contains a cycle that is a collection of pairwise distinct row and column indices $i_1, \ldots, i_k, j_1, \ldots, j_k$ such that $\gamma_{i_l j_l} \gamma_{i_l j_{l+1}} > 0$ for $l = 1, \ldots, k$ (where we have closed the cycle by setting $j_{k+1} = j_1$). Define then the matrix H with entries H_{ij} by

$$
H_{ij} = \begin{cases} 1 & \text{if } (i,j) = i_l, j_l \text{ for some } l = 1, \ldots, k, \\ -1 & \text{if } (i,j) = i_l, j_{l+1} \text{ for some } l = 1, \ldots, k, \\ 0 & \text{otherwise.} \end{cases}
$$

Note that by construction H has zero sum on every row and every column and that $\gamma_{ij} > 0$ whenever $H_{ij} \neq 0$, hence for $\varepsilon > 0$ and ε smaller than the smallest non-zero entry of γ, the matrix $\gamma - \varepsilon H$ has non-negative entries, it therefore belongs to $T(\boldsymbol{\mu}, \boldsymbol{\nu})$ as well as $\gamma + \varepsilon H$ and since

$$
\gamma = \frac{1}{2}(\gamma - \varepsilon H) + \frac{1}{2}(\gamma + \varepsilon H)
$$

we deduce that γ does not belong to $\mathrm{Ext}(T(\boldsymbol{\mu}, \boldsymbol{\nu}))$. Extreme points therefore are acyclic.

Now assume that γ is not extreme hence can be written for some $\lambda \in (0,1)$ and some $(\pi, \theta) \in T(\boldsymbol{\mu}, \boldsymbol{\nu})$ with $\pi \neq \theta$ as $\gamma = (1 - \lambda)\pi + \lambda\theta$. Of course, whenever $\gamma_{ij} = 0$ one necessarily have $\pi_{ij} = \theta_{ij} = 0$. Let us set $E_+ := \{(i,j) \in \{1, \ldots, I\} \times \{1, \ldots, J\} : \pi_{ij} > \theta_{ij}\}$, $E_- := \{(i,j) \in \{1, \ldots, I\} \times \{1, \ldots, J\} : \pi_{ij} < \theta_{ij}\}$, then $\gamma_{ij} > 0$ for every $(i,j) \in E_+ \cup E_-$. Since π and θ share the same row and column sums whenever $(i,j) \in E_\pm$ there exists $j' \neq j$ such that $(i,j') \in E_\mp$ (likewise there exists $i' \neq i$ such that $(i',j) \in E_\mp$). Let us pick $(i_1, j_1) \in E_+$, then there is some $j_2 \neq j_1$ such that $(i_1, j_2) \in E_-$ and then also some $i_2 \neq i_1$ such that $(i_2, j_2) \in E_+$ and some $j_3 \neq j_2$ such that $(i_2, j_3) \in E_-$. If $j_3 = j_1$, the matrix γ contains a cycle, if $j_3 \neq j_1$, we can continue this excursion and find some $i_3 \neq i_2$ such that $(i_3, j_3) \in E_+$. If $i_3 = i_1$ then $(i_3, j_3), (i_1, j_2), (i_2, j_2), (i_2, j_3)$ is a cycle. If $i_3 \neq i_1$ (and $i_3 \neq i_2$ by construction) we continue the excursion to another position in the matrix on the i_3th row. This scheme necessarily has to stop after finitely many steps, more precisely when we reach a row of column that has been previously visited. This necessarily implies that there is a cycle in the matrix. $\square$

The following exercise introduces a convenient graph theoretic frame-work to acyclicity and shows that extreme points of $T(\boldsymbol{\mu}, \boldsymbol{\nu})$ are sparse in the sense that they have at most $I + J - 1$ non-zero entries.

Exercise 6.7. *A graph* $G = (N, E)$ *consists of a non-empty finite set* N *whose elements are called the nodes of the graph* G *and* $E \subset N \times N$ *is a (possibly empty) collection of edges which are pairs of unordered distinct nodes (unordered here means that we do not distinguish the edge* (x, y) *from* (y, x), *i.e., that the set* E *is symmetric). If* x *and* y *are distinct nodes and* $(x, y) \in E$ *one says that the nodes* x *and* y *are connected by the edge* $e = (x, y)$ *or that* x *and* y *are neighbors. The degree of* $x \in N$ *is the number of neighbors of* x *(it is* 0 *if* x *is not connected to any other node). A cycle in the graph* $G = (N, E)$ *is a collection of distinct nodes* $(x_1, \ldots, x_k)$ *such that* x_l *is a neighbor of* x_{l+1} *for* $l = 1, \ldots, k$ *where we have set* $x_{k+1} = x_1$. *Finally, the graph* G *is called acyclic if it has no cycle.*

1) *Let* $G = (N, E)$ *be an acyclic graph show that there is an* $x \in N$ *with degree* 0 *or* 1 *(Hint: assume on the contrary each node have at least two neighbors, starting from* $x_1 \in N$, *pick* x_2 *a neighbor of* $x_1, \ldots$ *stop once you reach an already visited node and exhibit a cycle from this construction.)*

2) *Let* $G = (N, E)$ *be an acyclic graph and denote by* n *the cardinality of* N. *Show that* E *has cardinality less than* $n - 1$ *(argue by induction on* n, *for the inductive step, use the previous question and remove a node with* 0 *or* 1 *degree).*

3) *Let* $\gamma \in T(\boldsymbol{\mu}, \boldsymbol{\nu})$ *and define the graph* $G_\gamma = (N_\gamma, E_\gamma)$ *by* $N_\gamma := S \cup T$ *where* N_+ *and* N_- *are disjoint sets with respective cardinality* I *and* J *(representing sources and targets, respectively) and* E_γ *consists of all edges connecting a source* i *and a target* j *for which* $\gamma_{ij} > 0$. *Show that* G_γ *is an acyclic graph if and only if* γ *is an acyclic matrix.*

4) *Show that (6.62) has a solution with at most* $I + J - 1$ *non-zero entries.*

The assignment problem

The assignment problem consists in assigning N different tasks to N different persons so as to minimize a certain total disutility, denoting by c_{ij} the disutility of person i for the task j, this reads as

$$\inf_{\sigma \in \mathfrak{S}_N} \sum_{i=1}^{N} c_{i\sigma(i)}, \tag{6.65}$$

where $\mathfrak{S}_N$ denotes the group of permutations of $\{1, \ldots, N\}$ (and an assignment is a $\sigma \in \mathfrak{S}_N$ which prescribes the task $\sigma(i)$ for person i). The assignment problem is a discrete optimization problem over the finite set $\mathfrak{S}_N$. However, enumerating the $N!$ elements of $\mathfrak{S}_N$ is not a feasible option when N is large (curse of dimensionality). Note also that (6.65) consists in looking for a Monge solution to the LP transport problem

$$\inf_{\gamma \in \mathrm{BS}_N} \mathrm{tr}(C^T \gamma) := \sum_{i=1}^{N} \sum_{j=1}^{N} c_{ij} \gamma_{ij}, \tag{6.66}$$

where BS_N is the set of bistochastic matrices, i.e., the set of $N \times N$ matrices with non-negative entries γ_{ij} which have row and columns sums equal to 1 which is a special instance of (6.62) with $I = J = N$ and $\mu = \nu = (1, \ldots, 1)$. Of course, each $\sigma \in \mathfrak{S}_N$ defines the permutation matrix P^σ by:

$$P_{ij}^\sigma := \begin{cases} 1 & \text{if } j = \sigma(i), \\ 0 & \text{otherwise} \end{cases}$$

and the set of permutation matrices

$$\mathrm{Perm}_N := \{P^\sigma, \ \sigma \in \mathfrak{S}_N\}$$

is clearly included in BS_N; in fact, permutation matrices are exactly those bistochastic matrices which only have $\{0, 1\}$ entries (and therefore exactly one 1 on each line and each column). Since

$$\mathrm{tr}(C^T P^\sigma) = \sum_{i=1}^{N} \sum_{j=1}^{N} c_{ij} P_{ij}^\sigma = \sum_{i=1}^{N} c_{i\sigma(i)},$$

the inequality $\min(6.65) \geq \min(6.66)$ is straightforward. The converse inequality follows from the Birkhoff–von Neumann theorem which states that permutation matrices are the extreme points of the (convex compact) set of bistochastic matrices:

Theorem 6.15 (Birkhoff–von Neumann). $\mathrm{Ext}(\mathrm{BS}_N) = \mathrm{Perm}_N$.

Proof. Let $\sigma \in \mathfrak{S}_N$, $\lambda \in (0, 1)$ and P, Q in BS_N such that $P^\sigma = (1-\lambda)P + \lambda Q$, if $j = \sigma(i)$ we have $1 = (1 - \lambda)P_{ij} + \lambda Q_{ij}$ hence $P_{ij} = Q_{ij} = 1$ since both P_{ij} and Q_{ij} are less than 1. Similarly if $j \neq \sigma(i)$ then $P_{ij} = Q_{ij} = 0$. Hence, we have $P = Q$ which proves that $P^\sigma \in \mathrm{Ext}(\mathrm{BS}_N)$.

Now assume that $\gamma \in \mathrm{Ext}(\mathrm{BS}_N)$ and let us show that $\gamma \in \mathrm{Perm}_N$, i.e., $\gamma \in \{0, 1\}^{N \times N}$. Assume on the contrary that $\gamma_{i_1 j_1} \in (0, 1)$ for some i_1, j_1

then since γ is bistochastic there exists $j_2 \neq j_1$ such that $\gamma_{i_1 j_2} \in (0, 1)$. Since the sum of the entries of the j_2th column of γ sum to 1, there is an $i_2 \neq i_1$ such that $\gamma_{i_2, j_2} \in (0, 1)$, we then generate, as in the proof of Theorem 6.14, an excursion in the matrix which visits positions with entries in $(0, 1)$ and stops once one reaches one previously visited row or column, this implies that γ contains a cycle which contradicts its extremality by Theorem 6.14.

$\square$

Combining the Birkhoff-von Neumann theorem and Proposition 3.4, we deduce the following corollary.

Corollary 6.1. *The LP problem (6.66) admits at least one solution which is a permutation matrix P^σ; in particular σ solves the assignment problem (6.65) and the value of (6.65) coincides with the value of the LP problem (6.66).*

It is very tempting to try to generalize Theorems 6.14 and 6.15 to continuous cases, with the hope of proving that the extreme measures with prescribed marginals often are of Monge type (this unfortunately needs not be the case). For recent results on this fascinating (and much more subtle than one would expect) question, we refer to the article of Ahmad, Kim and McCann [3] and the references therein (in particular Lindenstrauss [70] and Hestir and Williams [62]).

Entropic regularization

We have seen in Exercise 6.7 that the discrete transport problem (6.62) admits sparse solutions with less than $I + J - 1$ non-zero entries, but if I and J are large, discovering where these entries are is a very expensive task in practice. An alternative approach is to force positivity by adding a suitable entropy term which rules out zero coefficient. This is a classic idea in LP, in particular a very elegant and precise convergence analysis of this entropic penalty method (and its dual counterpart) has been performed by Cominetti and San-Martin [36]. In the context of discrete optimal transport, this idea has been *remise au goût du jour* more recently, independently by Cuturi [38] and Galichon and Salanié [53]. Entropic regularization is at the heart of fast solvers thanks to the famous Sinkhorn scaling algorithm (aka IPFP: Iterative Proportional Fitting Procedure), the recent book of Cuturi and Peyré [82] is an excellent reference on this very active area, also see [14] for various applications. For the sake of simplicity, we will here restrict ourselves to the discrete case, but the entropic regularization of continuous optimal transport problems is tightly related to large deviations theory and the so-called Schrödinger bridge problem which goes back to [90], see in particular Léonard [68].

Given a regularization parameter $\varepsilon > 0$ the entropic regularization of the discrete transport problem (6.62) is

$$\inf_{\gamma \in T(\boldsymbol{\mu}, \boldsymbol{\nu})} J^{\varepsilon}(\gamma) := \operatorname{tr}(C^T \gamma) + \varepsilon \operatorname{Ent}(\gamma), \tag{6.67}$$

where

$$\operatorname{Ent}(\gamma) := \sum_{i=1}^{I} \sum_{j=1}^{J} \Phi(\gamma_{ij}) \quad \text{with} \quad \Phi(s) := \begin{cases} s \log(s) & \text{if } s > 0, \\ 0 & \text{if } s = 0 \end{cases}$$

is the entropy of γ. It is easy to check that (6.67) admits a unique (by strict convexity of Ent) solution which we denote by γ^{ε}. Before characterizing γ^{ε}, let us observe that the fact that $\Phi'(0^+) = -\infty$ implies that γ^{ε} only has positive entries:

Lemma 6.6. *Let γ^{ε} be the solution of (6.67). Then for every $(i,j) \in \{1,\dots,I\} \times \{1,\dots,J\}$ one has $\gamma_{ij}^{\varepsilon} > 0$.*

Proof. Assume on the contrary that $N_0 := \{(i,j) \in \{1,\dots,I\} \times \{1,\dots,J\} : \gamma_{ij}^{\varepsilon} = 0\} \neq \emptyset$. We will derive a contradiction by finding a competitor of γ^{ε} with a strictly smaller value of the cost J^{ε}. Let us denote by $m := \sum_{i=1}^{I} \mu_i = \sum_{j=1}^{J} \nu_j$ and $\theta_{ij} := \frac{1}{m} \mu_i \nu_j$ so that θ belong to $T(\boldsymbol{\mu}, \boldsymbol{\nu})$ and has positive entries, for $t \in (0,1)$ set $\theta(t) := \gamma^{\varepsilon} + t(\theta - \gamma^{\varepsilon})$, then

$$J^{\varepsilon}(\theta(t)) - J^{\varepsilon}(\gamma^{\varepsilon}) = t \operatorname{tr}(C^T(\theta - \gamma^{\varepsilon})) + \varepsilon \sum_{(i,j) \notin N_0} (\Phi(\theta_{ij}(t)) - \Phi(\gamma_{ij}^{\varepsilon}))$$

$$+ \varepsilon \sum_{(i,j) \in N_0} t\theta_{ij} \log(t\theta_{ij}).$$

Note that there is an $M > 1$ such that for every $t \in (0,1)$ and every $(i,j) \notin N_0$, $[\gamma_{ij}^{\varepsilon}, \theta_{ij}(t)] \subset [1/M, M]$, the mean value theorem thus yields

$$\Phi(\theta_{ij}(t)) - \Phi(\gamma_{ij}^{\varepsilon}) \leq t|\theta_{ij} - \gamma_{ij}^{\varepsilon}| \sup_{\alpha \in [1/M, M]} |1 + \log(\alpha)|$$

so that

$$\varepsilon \sum_{(i,j) \notin N_0} (\Phi(\theta_{ij}(t)) - \Phi(\gamma_{ij}^{\varepsilon})) = O(t)$$

hence

$$J^\varepsilon(\theta(t)) - J^\varepsilon(\gamma^\varepsilon) \le O(t) + \varepsilon \sum_{(i,j)\in N_0} t\theta_{ij}\log(t\theta_{ij})$$

so that $J^\varepsilon(\theta(t)) < J^\varepsilon(\gamma^\varepsilon)$ for $t > 0$ small enough, contradicting the optimality of γ^ε. $\square$

Thanks to Lemma 6.6 we know that no non-negativity constraint is binding for the minimizer γ^ε of (6.67). In other words, only the affine marginal constraints (6.63) have to be taken into account. Using the results from Section 4.3.1 and the fact that the marginal constraints are affine, we deduce that the optimality conditions characterizing γ^ε are given by the existence of Lagrange multipliers $u_1,\ldots,u_I$ and $v_1,\ldots,v_J$ such that for every (i,j) one has

$$c_{ij} + \varepsilon(1 + \log(\gamma^\varepsilon_{ij})) = u_i + v_j, \quad \text{i.e.,} \quad \gamma^\varepsilon_{ij} = \exp\left(\frac{u_i + v_j - \varepsilon - c_{ij}}{\varepsilon}\right)$$

which, setting $\varphi_i = u_i - \varepsilon = \varepsilon\log(a_i)$ and $\psi_j = v_j - \varepsilon = \varepsilon\log(b_j)$, we can rewrite as

$$\gamma^\varepsilon_{ij} = a_i b_j e^{-\frac{c_{ij}}{\varepsilon}} = e^{\frac{\varphi_i + \psi_j - c_{ij}}{\varepsilon}}.$$

Now the fact that the marginals of γ^ε are μ and ν gives the $I+J$ following conditions for the $I+J$ unknown coefficients a_i, b_j

$$\mu_i = a_i \sum_{j=1}^J b_j e^{-\frac{c_{ij}}{\varepsilon}} = e^{\frac{\varphi_i}{\varepsilon}} \sum_{j=1}^J e^{\frac{\psi_j - c_{ij}}{\varepsilon}}, \quad i = 1,\ldots,I, \tag{6.68}$$

and

$$\nu_j = b_j \sum_{j=1}^I a_i e^{-\frac{c_{ij}}{\varepsilon}} = e^{\frac{\psi_j}{\varepsilon}} \sum_{i=1}^I e^{\frac{\varphi_i - c_{ij}}{\varepsilon}}, \quad j = 1,\ldots,J. \tag{6.69}$$

Note that the system of nonlinear equations (6.68)–(6.69) is nothing but the system of optimality conditions for the (smooth and unconstrained) concave maximization problem

$$\sup_{(\varphi_i)_i,(\psi_j)_j} \sum_{i=1}^I \varphi_i\mu_i + \sum_{j=1}^J \psi_j\nu_j - \varepsilon\sum_{i=1}^I\sum_{j=1}^J e^{\frac{\varphi_i+\psi_j-c_{ij}}{\varepsilon}}, \tag{6.70}$$

which is the dual of (6.67) and is a regularization of the dual (6.64) where the constraint has been replaced by an exponential penalty term. Note that, as for the Kantorovich dual, the functional in (6.70) is unchanged when adding a constant to the φ_i's and substracting it to the ψ_j's, this is the only source of non-uniqueness for the Lagrange multipliers and can be easily fixed by some suitable normalization. In other words, the system (6.68)–(6.69) has a unique solution up to this invariance with respect to constants.

Exercise 6.8. *Let γ^ε be the solution of (6.67), show that when $\varepsilon \to 0^+$, γ^ε converges to the solution of*

$$\inf_{\gamma \in T_{\mathrm{opt}(\mu,\nu)}} \mathrm{Ent}(\gamma),$$

where $T_{\mathrm{opt}(\mu,\nu)}$ is the set if solutions of (6.62).

6.5. Exercises

Exercise 6.9. *For $x \in \mathbb{R}$, set*

$$f(x) := \begin{cases} x\log(x) - x & \text{if } x > 0, \\ 0 & \text{if } x = 0, \\ +\infty & \text{otherwise,} \end{cases} \qquad g(x) := \begin{cases} 0 & \text{if } x \leq 0, \\ +\infty & \text{otherwise.} \end{cases}$$

Show that f and g belong to $\Gamma_0(\mathbb{R})$ and compute f^* and g^*. Solve

$$\inf_{x \in \mathbb{R}} \{f(x) + g(x)\},$$

compute the value of the Fenchel-Rockafellar dual

$$\sup\{-f^*(-p) - g^*(p)\}$$

and show that this supremum is not attained. Comment on this.

Exercise 6.10 (An infinite duality gap). *Let E be a Hilbert space and A and B be two non-empty closed convex subsets of E such that*

$$0 \notin (A - B), \quad 0 \in \overline{A - B}. \tag{6.71}$$

Define $f = \chi_A$, and $g = \chi_B$.

1) *Show that (6.71) holds in the particular case where $E = \mathbb{R}^2$, $A := \mathbb{R}_+ \times \{0\}$ and $B := \{(x,y) \in \mathbb{R}_+^2 : xy \geq 1\}$.*

2) *Show that the value of the primal problem*

$$\inf_{x \in E} \{f(x) + g(x)\} \qquad (6.72)$$

is $+\infty$.

3) *Show that the value of the dual problem*

$$\sup_{p \in E} \{-f^*(p) - g^*(-p)\} \qquad (6.73)$$

is $(\chi_A \square \chi_{-B})^{**}(0)$.

4) *Thanks to (6.71), choose* $x_n \in E^N$ *such that* $x_n \to 0$ *and* $x_n = a_n - b_n$ *with* $a_n \in A$ *and* $b_n \in B$ *for every n, use the lower semicontinuity of* $(\chi_A \square \chi_{-B})^{**}$ *to show that*

$$(\chi_A \square \chi_{-B})^{**}(0) = 0.$$

5) *Deduce that there is an infinite duality gap*

$$\inf(6.72) = +\infty > 0 = \sup(6.73)$$

and conclude.

Exercise 6.11. *Let E be a Hilbert space,* $(F_1, \ldots, F_N) \in \Gamma_0(E)^N$ *set*

$$F := (F_1 \square \ldots \square F_N)$$

and consider the two problems

$$\inf \left[\sum_{i=1}^{N} F_i(x_i) : (x_1, \ldots, x_N) \in E^N, \ \sum_{i=1}^{N} x_i = 0 \right] \qquad (6.74)$$

and

$$\sup_{p \in E} \left\{ -\sum_{i=1}^{N} F_i^*(p) \right\}. \qquad (6.75)$$

1) *Check that* $F(0) = \inf(6.74)$ *and* $F^{**}(0) = \sup(6.75)$.
2) *Show that if* $\partial F(0) \neq \emptyset$ *then* $\inf(6.74) = \max(6.75)$ *and that the set of solutions of (6.75) is* $\partial F(0)$.
3) *Show that if* $F(0) \in \mathbb{R}$ *and there exists* $h_1, \ldots, h_N$ *such that* $\sum_{i=1}^{N} h_i = 0$, F_1 *is finite and continuous at* h_1 *and* $F_i(h_i) < +\infty$ *for every* $i = 2, \ldots, N$ *then* $\inf(6.74) = \max(6.75)$.

4) *Interpret the solutions of (6.75) as multipliers associated with the constraint in (6.74).*

Exercise 6.12. *Let f_0 and f_1 be in $\Gamma_0(\mathbb{R}^d)$ such that $\mathrm{dom}(f_0) \cap \mathrm{dom}(f_1) \neq \emptyset$, f_0 strictly convex, coercive and $f_1 \geq 0$. For $\lambda \in \mathbb{R}_+$ consider*

$$v(\lambda) := \inf_{x \in \mathbb{R}^d} \{f_0(x) + \lambda f_1(x)\}. \tag{6.76}$$

1) *Show that (6.76) admits a unique solution x_λ and that $\lambda \in \mathbb{R}_+^* \mapsto x_\lambda$ is continuous. Is $\lambda \mapsto x_\lambda$ continuous at $\lambda = 0$?*
2) *Show that $\lambda \in \mathbb{R}_+ \mapsto f_1(x_\lambda)$ is non-increasing and $\lambda \in \mathbb{R}_+ \mapsto f_0(x_\lambda)$ is non-decreasing.*
3) *Show that v is non-decreasing, concave on $\mathbb{R}_+$ and continuous on $\mathbb{R}_+^*$.*
4) *Show that v is of class C^1 on $\mathbb{R}_+^*$ and that $v'(\lambda) = f_1(x_\lambda)$ for every $\lambda \in \mathbb{R}_+^*$.*

Exercise 6.13. *Let f_0 and f_1 be as in Exercise 6.12 and for $\alpha \in \mathbb{R}_+$, consider*

$$w(\alpha) := \inf_{y \in \mathbb{R}^d} \{f_0(y) : f_1(y) \leq \alpha\} \tag{6.77}$$

and define

$$\alpha_0 := \inf\{\alpha \in \mathbb{R} : \mathrm{dom}(f_0) \cap \{f_1 \leq \alpha\} \neq \emptyset\}.$$

1) *Show that $w(\alpha) = +\infty$ for $\alpha < \alpha_0$ and that $w(\alpha) < +\infty$ for $\alpha > \alpha_0$. Is $w(\alpha_0)$ necessarily finite? In the sequel, we will always take $\alpha > \alpha_0$.*
2) *Show that, for every $\alpha > \alpha_0$, (6.77) admits a unique solution y_α.*
3) *Show that w is convex, non-increasing and continuous on $(\alpha_0, +\infty)$.*
4) *Show that for any $\alpha > \alpha_0$, one has $w(\alpha) = \sup_{\lambda \geq 0}\{v(\lambda) - \alpha\lambda\}$ (where $v(\lambda)$ is as in Exercise 6.12) and that the previous supremum is attained.*
5) *Show that for any $\alpha > \alpha_0$, there exists a multiplier $\lambda \in \mathbb{R}_+$ such that $y_\alpha = x_\lambda$ (where x_λ is as in Exercise 6.12).*
6) *Show that for any $\alpha > \alpha_0$, $y_\alpha = x_\lambda$ if and only if λ maximizes $\lambda \in \mathbb{R}_+ \mapsto v(\lambda) - \alpha\lambda$, we denote by $\Lambda(\alpha)$ the set of such multipliers.*
7) *Show that for every $\lambda \in \Lambda(\alpha)$ one has $-\lambda \in \partial w(\alpha)$.*
8) *Show that if $\beta \geq \alpha$, $\mu \in \Lambda(\beta)$, $\lambda \in \Lambda(\alpha)$ then $\lambda \geq \mu$ (in other words, the stronger the constraint the higher the multipliers).*

Exercise 6.14. *Given $x_0 \in \mathbb{R}^d$ and $\mu > 0$, consider*

$$\inf_{x \in \mathbb{R}^d} \mu \|x\|_1 + \|x - x_0\|_\infty \qquad (6.78)$$

(where $\|x\|_1 := \sum_{i=1}^d |x_i|$, $\|y\|_\infty := \max(|y_1|, \ldots, |y_d|)$). Show that (6.78) admits solutions and use convex duality to show that x solves (6.78) if and only if there exists $p \in \mathbb{R}^d$ such that

$$\|p\|_\infty \le 1, \quad \mu \|p\|_1 \le 1, \quad p \cdot x = \|x\|_1, \quad \mu p \cdot (x_0 - x) = \|x - x_0\|_\infty.$$

Exercise 6.15. *Given $y \in \mathbb{R}^d$, consider*

$$\inf_{x \in \mathbb{R}^d} \max(x_1, \ldots, x_d) + \frac{1}{2} \|x - y\|_2^2 \qquad (6.79)$$

(where $\|z\|_2^2 := \sum_{i=1}^d z_i^2$).

1) Show that

$$\max(x_1, \ldots, x_d) = \sigma_\Delta(x) \text{ where } \Delta := \left\{ p \in \mathbb{R}_+^d : \sum_{i=1}^d p_i = 1 \right\}.$$

2) Show that (6.79) admits a unique solution x and that x is characterized by the conditions

$$y - x \in \Delta, (y - x) \cdot x = \max(x_1, \ldots, x_d).$$

3) Show that x is of the form $x_i := \min(\alpha, y_i)$ where α is the unique root of

$$\sum_{i=1}^d (y_i - \alpha)_+ = 1.$$

Exercise 6.16 (Toland duality). *Let E be a normed space, f and g be in $\Gamma_0(E)$ and consider the non-convex problem*

$$\inf_{x \in \text{dom}(f)} \{ f(x) - g(x) \} \qquad (6.80)$$

and its Toland dual

$$\inf_{p \in \text{dom}(g^*)} \{ g^*(p) - f^*(p) \}. \qquad (6.81)$$

1) Show that $\inf(6.80) = \inf(6.81)$. (Hint: show that any minorant of $f - g$ is a minorant of $g^* - f^*$.)

2) *Assume that* inf(6.80) $\in \mathbb{R}$ *show that if* x *solves (6.80) and* $p \in \partial g(x)$
   then p *solves (6.81).*

Exercise 6.17. *Let* k *and* p *be positive integers* $A \in \mathcal{S}_k^{++}(\mathbb{R})$, $B \in$
$\mathcal{M}_{k \times p}(\mathbb{R})$ $C \in \mathcal{S}_p(\mathbb{R})$ *and let* $S \in \mathcal{S}_{k+p}(\mathbb{R})$ *be given in block form as*

$$S := \begin{pmatrix} A & B \\ B^T & C \end{pmatrix}.$$

Show that $S \succeq 0$ *if and only if*

$$C \succeq B^T A^{-1} B$$

(the matrix $C - B^T A^{-1} B$ *is called the Schur complement of* A *in* S*). (Hint:*
compute $(x^T y^T) S \begin{pmatrix} x \\ y \end{pmatrix}$ *and first minimize the result with respect to* $x \in \mathbb{R}^k$
and then express the minimum as a quadratic function of $y \in \mathbb{R}^p$ *only.)*

Exercise 6.18. *Let* d *and* k *be positive integers,* $c \in \mathbb{R}^d$, $\alpha \in \mathbb{R}$, $M \in$
$\mathcal{M}_{k,d}(\mathbb{R})$ *for* $x \in \mathbb{R}^d$ *define* $S(x) \in \mathcal{S}_{k+1}(\mathbb{R})$ *by*

$$S(x) := \begin{pmatrix} I_k & Mx \\ x^T M^T & b^T x + \alpha \end{pmatrix}.$$

Show that $b^T x + \alpha \geq \|Mx\|^2 := x^T M^T M x$ *if and only if* $S(x) \succeq 0$*. Then,*
given $c \in \mathbb{R}^d$*, reformulate the quadratic program*

$$\inf_{x \in \mathbb{R}^d} \{ c^T x : b^T x + \alpha \geq \|Mx\|^2 \}$$

as an SDP program.

Exercise 6.19. *Consider*

$$\inf_{x \in \mathbb{R}, \, x^2 \leq 0} x$$

and write it as

$$\inf_{x \in \mathbb{R}} \sup_{\lambda \in \mathbb{R}_+} \mathscr{L}(x, \lambda) \quad \text{with} \quad \mathscr{L}(x, \lambda) := x + \lambda x^2.$$

Compute for $\lambda \geq 0$ *the quantity* $q(\lambda) := \inf_{x \in \mathbb{R}} \mathscr{L}(x, \lambda)$*. Does* q *admit a*
maximizer on $\mathbb{R}_+$*? Conclude.*

Exercise 6.20 (Uniqueness holds for *most* LP problems). *Let C be a non-empty convex compact subset of $\mathbb{R}^d$, given $p \in \mathbb{R}^d$ consider*

$$\sigma_C(p) := \sup_{x \in C} p \cdot x. \tag{6.82}$$

1) *Show that x solves (6.82) if and only if $x \in \partial \sigma_C(p)$.*
2) *Show that σ_C is convex and Lipschitz.*
3) *Show that there exists a subset R of full measure in $\mathbb{R}^d$ such that for every $p \in R$, problem (6.82) has a unique solution. (Hint: use Rademacher's theorem.)*

Exercise 6.21 (Hopf–Lax formulas). *Let $L : \mathbb{R}^d \to \mathbb{R}$ and $u_0 : \mathbb{R}^d \to \mathbb{R}$ be convex functions: Show that for every $t > 0$ and $x \in \mathbb{R}^d$ the following (Hopf-Lax formula) holds and that the sup in the right-hand side is attained*

$$\inf_{y \in \mathbb{R}^d} \left\{ tL\left(\frac{x-y}{t}\right) + u_0(y) \right\} = \sup_{q \in \mathbb{R}^d} \left\{ -u_0^*(q) + q \cdot x - tL^*(q) \right\}.$$

Exercise 6.22. *Let E and F be two normed spaces, $\Lambda \in \mathcal{L}_c(E, F)$, $J \in \Gamma_0(E \times F)$ and consider the problem (6.12) as well as its dual (6.14) and assume that the value of (6.12) is finite. Define the Lagrangian of problem (6.12) as*

$$\mathcal{L}(x, z, \lambda) := J(x, z) + \lambda(\Lambda x - z), \quad \forall (x, z, \lambda) \in E \times F \times F^*.$$

1) *Show that (6.12) is equivalent to*

$$\inf_{(x,z) \in E \times F} \sup_{\lambda \in F^*} \mathcal{L}(x, z, \lambda).$$

2) *Show that (6.14) is equivalent (as in Section 6.3.4, it is convenient to change q into $\lambda := -q$) to*

$$\sup_{\lambda \in F^*} \inf_{(x,z) \in E \times F} \mathcal{L}(x, z, \lambda).$$

3) *Show that if the qualification condition (6.15) holds then*

$$\inf_{(x,z) \in E \times F} \sup_{\lambda \in F^*} \mathcal{L}(x, z, \lambda) = \max_{\lambda \in F^*} \inf_{(x,z) \in E \times F} \mathcal{L}(x, z, \lambda).$$

4) *If (6.15) holds show that $x \in E$ solves (6.12) and $q \in E^*$ solve (6.12) if and only if $(x, \Lambda(x), \lambda) = (x, \Lambda(x), -q)$ is a saddle-point of the Lagrangian $\mathcal{L}$, i.e.,*

$$\mathcal{L}(y, z, \lambda) \geq \mathcal{L}(x, \Lambda(x), \lambda) \geq \mathcal{L}(x, \Lambda(x), \mu)$$

for every $(y, z, \mu) \in E \times F \times F^*$.

Exercise 6.23 (Augmented Lagrangian). *Adopt the same assumptions and notations as in Exercise 6.22 including the qualification condition (6.15) and also that E and F are Hilbert space (with F^* identified with F). Given $r > 0$, the augmented Lagrangian for (6.12) is*

$$\mathscr{L}_r(x, z, \lambda) := \mathscr{L}(x, z, \lambda) + \frac{r}{2}\|\Lambda(x) - z\|^2, \forall (x, z, \lambda) \in E \times F^2.$$

Prove that x solves (6.12) and q solves (6.14) if and only if $(x, \Lambda x, \lambda)$ with $\lambda = -q$ is a saddle-point of $\mathscr{L}_r$. What is in your opinion the advantage of looking for saddle-points of $\mathscr{L}_r$ rather than for saddle-points of $\mathscr{L}$?

Exercise 6.24. *Let A and B be two non-empty sets and $f: A \times B \to \mathbb{R}$.*

1) *Show that*

$$\inf_{a \in A} \sup_{b \in B} f(a, b) \geq \sup_{b \in B} \inf_{a \in A} f(a, b).$$

2) *Give examples where the previous inequality is strict.*
3) *Prove that there is an equality when A and B are convex compact subsets of $\mathbb{R}^d$ and $\mathbb{R}^m$ and f is continuous and bilinear. (Hint: rewrite the inf sup problem in Fenchel–Rockafellar form using χ_A and σ_B.) Give a game-theoretic interpretation of this result in terms of zero-sum games.*

Exercise 6.25. *Let X and Y be compact metric spaces, $c \in C(X, \times Y)$, given $\psi \in C(Y)$, recall that $\psi^{\check{c}}$ is defined by*

$$\psi^{\check{c}}(x) := \min_{y \in Y}\{c(x, y) - \psi(y)\}, \quad \forall x \in X.$$

Assume that $x \in X$ is such that there is a unique $y \in Y$ such that $\psi^{\check{c}}(x) = c(x, y) - \psi(y)$ and let $h \in C(Y)$. Show that $t \in \mathbb{R} \mapsto (\psi + th)^{\check{c}}(x)$ is differentiable at $t = 0$ and that

$$\frac{d}{dt}((\psi + th)^{\check{c}}(x))(0) = -h(y).$$

Exercise 6.26. *Let $X = [0, 1]$, $Y = [-1, 1]$ and μ and ν be the uniform probability measures on X and Y, respectively. Consider the mass transport problem*

$$\min_{\gamma \in \Pi(\mu, \nu)} \int_{X \times Y} (x^2 - y^2)^2 \, d\gamma(x, y). \tag{6.83}$$

1) *Solve (6.83).*
2) *Show that (6.83) does not admit any Monge solution.*

Exercise 6.27 (Kantorovich–Rubinstein formula). *Let (x, d) be a compact metric space, $(\mu, \nu) \in \mathcal{P}(X)^2$. Consider the optimal transport problem with the distance as cost:*

$$\min_{\gamma \in \Pi(\mu, \nu)} \int_{X \times X} d(x, y) d\gamma(x, y). \tag{6.84}$$

1) *Show that* min*(6.84) coincides with*

$$\max \left[\int_X u \, d(\mu - \nu)(x) : \ u : X \to \mathbb{R}, \ 1\text{-}Lipschitz \right]. \tag{6.85}$$

2) *Consider $X = Y = [0, 1]$ with the usual distance on $\mathbb{R}$, μ uniform on $[0, \frac{1}{2}]$ and ν uniform on $[\frac{1}{2}, 1]$. Show that any $\gamma \in \Pi(\mu, \nu)$ solves (6.84).*

Exercise 6.28. *Let $r > 0$, $B = \overline{B}(0, r)$ be a closed ball in $\mathbb{R}^d$, let μ and ν be Borel probability measures on B with μ absolutely continuous with respect to the d-dimensional Lebesgue measure. Let $C \in C^1(\mathbb{R}^d)$ be strictly convex. The aim of this exercise is to show that the solutions of the corresponding Monge–Kantorovich problem*

$$\inf_{\gamma \in \Pi(\mu, \nu)} \int_{B \times B} C(x - y) d\gamma(x, y) \tag{6.86}$$

are of Monge type (a result which was first proved by McCann and Gangbo [55]). We fix γ a solution of (6.86).

1) *Show that ∇C is one to one.*
2) *Show that the Kantorovich dual of (6.86) has a solution (φ, ψ) such that φ is Lipschitz and differentiable μ almost everywhere.*
3) *Show that if $(x, y) \in \mathrm{supp}(\gamma)$ with $|x| < r$ and φ is differentiable at x then $\nabla \varphi(x) = \nabla C(x - y)$ and that $y = x - \nabla C^*(\nabla \varphi(x))$.*
4) *Deduce that γ is of Monge type and that (6.86) has a unique solution.*

Chapter 7

Iterative Methods for Convex Minimization

Iterative methods are at the core of numerical methods for convex optimization. As such, they play an increasing role in applications, in particular in large-scale problems arising in data sciences (we shall review some of these in Chapter 8). The aim of this chapter is to give an overview of some of these iterative methods\to find approximate solutions of (constrained or unconstrained) convex minimization problems. Of course, the choice of a particular method depends very much on the structure of the problem at stake, whether there are constraints or not, whether the cost function is smooth or not...

We shall insist here on global convergence: converge of the iterates to a global minimizer or, at least, convergence of values to the minimal cost (with a speed of convergence if possible). With this respect, in the smooth case, Nesterov accelerated gradient method plays a distinguished role.

This is just an introduction to this extremely wide and rapidly developing subject of great practical importance. We refer the interested reader wishing to have a more complete knowledge of the subject to the textbooks by Bonnans *et al.* [17], Lemaréchal and Hiriart-Urruty [64] and the references therein. Regarding proximal methods and more generally monotone operator methods for convex optimization, we highly recommend the book of Bauschke and Combettes [11].

7.1. On Newton's method

Consider a C^1 function $F : \mathbb{R}^d \to \mathbb{R}^d$. We are interested in solving

$$F(x) = 0, \tag{7.1}$$

which is a set of d (nonlinear in general) equations with d unknowns. Newton's method consists at each step in finding a root of the linear approximation of F instead of the nonlinear map F itself. More precisely, starting from an initialization $x_0 \in \mathbb{R}^d$, Newton's method is as follows. If x_k is known at step k, x_{k+1} is obtained by solving

$$L_k(x) = 0 \quad \text{with} \quad L_k(x) := F(x_k) + F'(x_k)(x - x_k). \tag{7.2}$$

If $F'(x_k)$ is singular then it might be the case that $F(x_k)$ is not in the range of $F'(x_k)$ so that (7.2) has no solution but if $F'(x_k)$ is invertible then the Newton's algorithm amounts to

$$x_{k+1} = x_k - F'(x_k)^{-1} F(x_k). \tag{7.3}$$

For Newton iterates (7.3) to be well-defined, we see that $F'(x_k)$ should remain invertible and in practice one has to solve the linear system $F'(x_k)(x_{k+1} - x_k) = -F(x_k)$ at each step which may be quite costly. Even if F' is invertible at each point and $(F')^{-1}$ remains bounded, there are no guarantees that Newton iterates converge to a solution of (7.1) if one starts with an initialization which is far from a solution, Newton iterates may have cycles or diverge quite dramatically (even in dimension 1, see Exercises 7.1 and 7.2). However, if one starts really close to a solution of (7.1), the convergence is ensured and actually very fast, it is quadratic (in the sense of (7.6)).

Theorem 7.1. *Assume that $\overline{x} \in \mathbb{R}^d$ is a solution of (7.1) and that for some $r > 0$ there holds*

$$F'(x) \text{ is invertible } \forall x \in \overline{B}(\overline{x}, r) \quad \text{and} \quad M := \sup_{x \in \overline{B}(\overline{x}, r)} \|F'(x)^{-1}\| < +\infty \tag{7.4}$$

and that for some $C > 0$

$$F' \text{ is } C\text{-Lipschitz on } \overline{B}(\overline{x}, r). \tag{7.5}$$

If $r' \leq \min(r, \frac{2}{MC})$ and $x_0 \in \overline{B}(\overline{x}, r')$, the iterates of Newton's method starting from x_0 are well-defined, remain in $\overline{B}(\overline{x}, r')$ and converge to $\overline{x}$. Moreover, the convergence is quadratic in the sense that

$$|x_{k+1} - \overline{x}| \leq \frac{MC}{2}|x_k - \overline{x}|^2, \quad \forall k \in \mathbb{N}. \tag{7.6}$$

Proof. Assume that $x_k \in \overline{B}(\overline{x}, r')$ so that thanks to (7.4), $F'(x_k)$ is invertible and x_{k+1} is well defined, since $F(\overline{x}) = 0$ we may rewrite (7.3) as

$$x_{k+1} - \overline{x} = x_k - \overline{x} - F'(x_k)^{-1}(F(x_k) - F(\overline{x}))$$
$$= -F'(x_k)^{-1}(F(x_k) - F(\overline{x}) - F'(x_k)(x_k - \overline{x}))$$

which, thanks to (7.4), gives

$$|x_{k+1} - \overline{x}| \leq M|F(\overline{x}) - F(x_k) - F'(x_k)(\overline{x} - x_k)|. \qquad (7.7)$$

Now observe that

$$F(\overline{x}) - F(x_k) - F'(x_k)(\overline{x} - x_k) = \int_0^1 (F'(x_k + t(\overline{x} - x_k)) - F'(x_k))(\overline{x} - x_k)\mathrm{d}t$$

and that, thanks to (7.5), for every $t \in [0, 1]$ one has

$$|(F'(x_k + t(\overline{x} - x_k)) - F'(x_k))(\overline{x} - x_k)| \leq Ct|x_k - \overline{x}|^2,$$

integrating and replacing in (7.7) gives

$$|x_{k+1} - \overline{x}| \leq \frac{MC}{2}|x_k - \overline{x}|^2 \leq \frac{MCr'^2}{2} \leq r',$$

which shows (7.6) as well as $x_{k+1} \in \overline{B}(\overline{x}, r')$. This also implies the convergence of x_k to $\overline{x}$ since $|x_{k+1} - \overline{x}| \leq \frac{MCr'}{2}|x_k - \overline{x}|$ and $r' \leq \frac{2}{MC}$. $\qquad \square$

Exercise 7.1. *Let us define the function $F: \mathbb{R} \to \mathbb{R}$ by*

$$F(x) := \begin{cases} 1 + x & \text{if } x \geq 1, \\ 1 + x + (x-1)^3 & \text{if } x \in [0, 1], \\ -1 + x + (x+1)^3 & \text{if } x \in [-1, 0], \\ -1 + x & \text{if } x < -1. \end{cases}$$

1) *Show that F is odd, increasing, C^1 and that F' is Lipschitz.*
2) *Compute the iterates of Newton's method starting from $x_0 = 1$ and $x_0 = -5$, respectively, and conclude.*

Exercise 7.2. *We are now considering the Newton method for $F(x) := \arctan(x)$ for every $x \in \mathbb{R}$.*

1) *Show that the equation $\alpha \arctan(\alpha) = 3$ has a unique positive root which we denote α^*.*

2) *Show that if $|x_0| \geq \alpha^*$ then the iterates x_k of Newton's method are well defined, alternate sign and that $|x_{k+1}| \geq 2|x_k|$ for every k. Conclude.*

Of course, Newton's method may be used for $F = \nabla f$ where $f \in C^2(\mathbb{R}^d, \mathbb{R})$ is a function we want to minimize (or at least to find some critical point of). In this case the iterates of Newton's method (provided the Hessian of f remains invertible at x_k of course) reads

$$x_{k+1} = x_k - [D^2 f(x_k)]^{-1}(\nabla f(x_k)).$$

Exercise 7.3. *Let $f \in C^2(\mathbb{R}^d, \mathbb{R})$ with $D^2 f(x)$ positive definite for every x. Show that Newton's iterates are well defined and that Newton's method is equivalent to a succession of linear quadratic minimization problems:*

$$x_{k+1} = \arg\min Q_k,$$

with

$$Q_k(x) := f(x_k) + \nabla f(x_k)(x - x_k) + \frac{1}{2} D^2 f(x_k)(x - x_k) \cdot (x - x_k).$$

Newton's method is a powerful tool to solve nonlinear equations and Theorem 7.1 is just a basic result on this vast topic. For more on Newton's method and other general iterative methods to solve nonlinear equations, the textbook of Ortega and Rheinboldt [80] is a very complete classical reference. In the context of convex minimization, Newton's method is at the heart of modern interior point methods which use so-called self-concordant barriers, see the textbook of Boyd and Vandenberghe [19].

7.2. The gradient method

In this section, we consider gradient methods for the unconstrained smooth and convex minimization problem

$$\inf_{x \in E} f(x), \tag{7.8}$$

where E is a Hilbert space and $f \colon E \to \mathbb{R}$ is convex and differentiable so that solving (7.8) is equivalent to find a point where ∇f is 0. The smoothness assumption we will typically make on f in this section is that $f \in C^{1,1}(E)$, which means that f is C^1 and ∇f is Lipschitz, that is, for some $M > 0$, we have

$$\|\nabla f(x) - \nabla f(y)\| \leq M\|x - y\|, \quad \forall (x, y) \in E \times E. \tag{7.9}$$

Another assumption we will often (but not always) make in this section is that f is strongly convex, i.e., that there is an $\alpha > 0$ such that

$$f(y) \geq f(x) + \nabla f(x) \cdot (y - x) + \frac{\alpha}{2}\|x - y\|^2, \quad \forall (x, y) \in E \times E. \quad (7.10)$$

Note that (7.10) implies both that f is coercive and strictly convex so it admits a minimizer (Theorem 6.1) which is unique by strict convexity, we shall denote by x^* this minimizer which is also the unique critical point of f. Before going further, we advise the reader to make the following exercises to become more familiar with the assumptions above.

Exercise 7.4. *Let $f \in C^1(E)$. Show that the following conditions are equivalent:*

1) *f satisfies (7.10);*
2) *$f - \frac{\alpha}{2}\|.\|^2$ is convex;*
3) *for every $(x, y) \in E^2$*

$$(\nabla f(x) - \nabla f(y)) \cdot (x - y) \geq \alpha \|x - y\|^2. \quad (7.11)$$

Finally, show that, if f satisfies (7.9) and (7.10), then $M \geq \alpha$.

Exercise 7.5. *Let $f \in C^1(E)$ (not necessarily convex). Show that f satisfies (7.9), then*

$$|f(y) - f(x) - \nabla f(x)(y - x)| \leq \frac{M}{2}\|x - y\|^2, \quad \forall (x, y) \in E^2. \quad (7.12)$$

Assume now that f satisfies (7.12), observing that for every $(x, h, v) \in E^3$, one has

$$
\begin{aligned}
(\nabla f(x + h) &- \nabla f(x)) \cdot v \\
&= f\left(x + \frac{h - v}{2}\right) - f(x + h) + \nabla f(x + h) \cdot \frac{v + h}{2} \\
&\quad + f(x + h) - f\left(x + \frac{v + h}{2}\right) + \nabla f(x + h) \cdot \frac{v - h}{2} \\
&\quad + f\left(x + \frac{v + h}{2}\right) - f(x) - \nabla f(x) \cdot \frac{v + h}{2} \\
&\quad + f(x) - f\left(x + \frac{h - v}{2}\right) - \nabla f(x) \cdot \frac{v - h}{2},
\end{aligned}
$$

show that ∇f is M-Lipschitz.

Exercise 7.6. *Show that if f is strongly convex and $C^{1,1}$ then so is f^*.*

7.2.1. *Convergence of iterates*

The gradient method with fixed step ρ consists in generating iteratively a sequence x_n starting from an initialization x_0 by

$$x_{n+1} = x_n - \rho \nabla f(x_n). \tag{7.13}$$

The advantage of this method is that it is explicit (provided the gradient of f can be computed easily of course). Its linear convergence[a] is guaranteed by the following result.

Theorem 7.2. *Let f satisfy (7.9) and (7.10). Then, any sequence x_n generated by the gradient method with fixed step $\rho > 0$ (7.13) converges to the solution of (7.8), x^*, provided $\rho \in \left(0, \frac{2\alpha}{M^2}\right)$.*

Proof. Set

$$T_\rho(x) := x - \rho \nabla f(x), \quad \forall x \in E.$$

Let $(x, y) \in E^2$, using (7.9) and (7.10) (in the form (7.11)) we get

$$\|T_\rho(x) - T_\rho(y)\|^2 = \|x - y\|^2 + \rho^2 \|\nabla f(x) - \nabla f(y)\|^2$$
$$- 2\rho(\nabla f(x) - \nabla f(y)) \cdot (x - y)$$
$$\leq \|x - y\|^2 (1 + \rho^2 M^2 - 2\alpha\rho)$$

so that T_ρ is a contraction as soon as

$$\rho^2 M^2 - 2\alpha\rho < 0, \tag{7.14}$$

i.e., $\rho \in \left(0, \frac{2\alpha}{M^2}\right)$. By virtue of the Banach's fixed-point theorem (Theorem 1.4), (x_n) therefore converges (linearly) to the unique fixed point of T_ρ which is x^*. □

Of course, if one chooses ρ too large, the gradient method with fixed step ρ may not converge (just take a quadratic function f) but choosing it too small may result in a very slow convergence, a reasonable compromise is given by $\rho = \frac{\alpha}{M^2}$ (which minimizes the quantity in (7.14)).

---

[a]That is, $\|x_{n+1} - x^*\| \leq \beta \|x_n - x^*\|$ for some $\beta \in (0, 1)$.

Let us now consider a constrained variant of (7.8)

$$\inf_{x \in C} f(x), \tag{7.15}$$

where C is a non-empty closed convex subset of E and f satisfies (7.9) and (7.10) as before. By the same arguments as in the unconstrained case, there exists a unique solution x^* which, by Lemma 6.2, is characterized by

$$x^* \in C, \quad \nabla f(x^*) \cdot (x - x^*) \geq 0, \quad \forall x \in C,$$

which, by the projection theorem (Theorem 1.24), is equivalent to the fact that, given any step $\rho > 0$ one has

$$x^* = \text{proj}_C(x^* - \rho \nabla f(x^*)), \tag{7.16}$$

where proj_C denotes the projection onto C. The projected gradient method with step ρ consists, starting from an arbitrary initial condition x_0, in iterating the map $\text{proj}_C \circ (\text{id} - \rho \nabla f)$:

$$x_{n+1} = \text{proj}_C(x_n - \rho \nabla f(x_n)). \tag{7.17}$$

Its convergence is ensured by the following theorem.

Theorem 7.3. *Let f satisfy (7.9) and (7.10). Then any sequence x_n generated by the projected gradient method with fixed step ρ (7.17) converges to the solution of (7.15), x^*, provided $\rho \in \left(0, \frac{2\alpha}{M^2}\right)$.*

Proof. Defining T_ρ as in the proof of Theorem 7.3 we observe that $\text{proj}_C \circ T_\rho$ is a contraction whenever T_ρ is since proj_C is 1-Lipschitz (see Proposition 1.15). The proof then again follows from Banach's fixed-point theorem (Theorem 1.4). $\qquad \square$

7.2.2. *Convergence of values*

We now consider the convergence of the values of f along the iterates of the gradient method with fixed step. We shall assume that f is convex with an M-Lipschitz gradient, i.e., satisfies (7.9) and admits a minimizer x^*, but we will not assume that it is strongly convex. Starting from $x_0 \in E$, we consider the sequence generated by the gradient method with fixed step $\frac{1}{M}$:

$$x_{n+1} := x_n - \frac{1}{M} \nabla f(x_n). \tag{7.18}$$

Let us first observe that the choice of the step $\frac{1}{M}$ ensures that $f(x_n)$ is non-increasing. Indeed, for every $x \in E$, we have

$$f(x - M^{-1}\nabla f(x)) = f(x) - \frac{1}{M} \int_0^1 \nabla f(x - tM^{-1}\nabla f(x)) \cdot \nabla f(x) dt$$

$$= f(x) - \frac{1}{M}\|\nabla f(x)\|^2$$

$$- \frac{1}{M} \int_0^1 (\nabla f(x - tM^{-1}\nabla f(x)) - \nabla f(x)) \cdot \nabla f(x) dt$$

$$\leq f(x) - \frac{1}{M}\|\nabla f(x)\|^2 + \frac{1}{2M}\|\nabla f(x)\|^2,$$

where we have used the Cauchy–Schwarz inequality and the fact that ∇f is M-Lipschitz in the last line. We thus have

$$f(x - M^{-1}\nabla f(x)) \leq f(x) - \frac{1}{2M}\|\nabla f(x)\|^2. \qquad (7.19)$$

Lemma 7.1. *Let f be convex and with an M-Lipschitz gradient on E. Then for every $(x, y) \in E^2$, one has*

$$f(y) - f(x) \leq \nabla f(y) \cdot (y - x) - \frac{1}{2M}\|\nabla f(y) - \nabla f(x)\|^2, \qquad (7.20)$$

and

$$(\nabla f(x) - \nabla f(y)) \cdot (x - y) \geq \frac{1}{M}\|\nabla f(y) - \nabla f(x)\|^2. \qquad (7.21)$$

Proof. Since f is convex and differentiable, we have

$$f(y) - f(z) \leq \nabla f(y) \cdot (y - z), \quad \forall (y, z) \in E^2 \qquad (7.22)$$

but since ∇f is M-Lipschitz, we also have

$$f(z) - f(x) \leq \nabla f(x) \cdot (z - x) + \frac{M}{2}\|x - z\|^2, \quad \forall (x, z) \in E^2 \qquad (7.23)$$

so that

$$f(y) - f(x) \leq \inf_{z \in E} \left[\nabla f(y) \cdot (y - z) + \nabla f(x) \cdot (z - x) + \frac{M}{2}\|x - z\|^2 \right].$$

The minimizer of the right-hand side is explicit and given by

$$z = x + \frac{1}{M}(\nabla f(y) - \nabla f(x)).$$

We let the reader check that replacing exactly gives inequality (7.20). Reversing the role of x and y in (7.20) yields

$$f(x) - f(y) \leq \nabla f(x) \cdot (x - y) - \frac{1}{2M} \|\nabla f(y) - \nabla f(x)\|^2.$$

Summing with (7.20) we get

$$0 \leq (\nabla f(y) - \nabla f(x)) \cdot (y - x) - \frac{1}{M} \|\nabla f(y) - \nabla f(x)\|^2$$

and thus deduce (7.21). □

We can now deduce an $O(\frac{1}{n})$ convergence result for the values of $f(x_n)$ to the minimal value of f.

Theorem 7.4. *Let f be convex, with an M-Lipschitz gradient on E and admitting a minimum at x^*. Then defining the sequence $(x_n)_n$ starting from x_0 by (7.18) we have*

$$f(x_n) - f(x^*) \leq \frac{2M\|x_0 - x^*\|^2}{n}, \quad \forall n \geq 1. \tag{7.24}$$

Proof. We already know that $f(x_n)$ is non-increasing because of (7.19):

$$f(x_{n+1}) - f(x_n) \leq -\frac{1}{2M} \|\nabla f(x_n)\|^2 \tag{7.25}$$

and since $f(x_n) \geq f(x^*)$, $f(x_n)$ converges in a monotone way. Now we use (7.21) and the fact that $\nabla f(x^*) = 0$ to get

$$\nabla f(x_n) \cdot (x_n - x^*) \geq \frac{1}{M} \|\nabla f(x_n)\|^2$$

and thus obtain

$$\|x_{n+1} - x^*\|^2 = \|x_n - M^{-1}\nabla f(x_n) - x^*\|^2$$

$$= \|x_n - x^*\|^2 + \frac{1}{M^2} \|\nabla f(x_n)\|^2 - \frac{2}{M} \nabla f(x_n) \cdot (x_n - x^*)$$

$$\leq \|x_n - x^*\|^2 + \frac{1}{M^2} \|\nabla f(x_n)\|^2 - \frac{2}{M^2} \|\nabla f(x_n)\|^2$$

$$= \|x_n - x^*\|^2 - \frac{1}{M^2} \|\nabla f(x_n)\|^2$$

so that $\|x_n - x^*\|$ is non-increasing and thus converges.

Set $\theta_k := f(x_k) - f(x^*)$, we already know that θ_k is non-increasing and non-negative. By convexity of f, the Cauchy–Schwarz inequality and using that $\|x_k - x^*\| \leq \|x_0 - x^*\|$ we find

$$\theta_k \leq \nabla f(x_k) \cdot (x_k - x^*) \leq \|\nabla f(x_k)\| \ \|x_0 - x^*\|$$

so that

$$\|\nabla f(x_k)\|^2 \geq \frac{\theta_k^2}{\|x_0 - x^*\|^2}$$

hence, using (7.25) yields

$$\theta_{k+1} - \theta_k \leq -\frac{1}{2M}\|\nabla f(x_k)\|^2 \leq -\alpha\theta_k^2, \quad \text{with} \quad \alpha := \frac{1}{2M\|x_0 - x^*\|^2}.$$

Dividing the inequality $\alpha\theta_k^2 \leq \theta_k - \theta_{k+1}$ by $\theta_k\theta_{k+1}$ (which we assume to be positive) and using that $\theta_{k+1} \leq \theta_k$, we get

$$\alpha \leq \alpha\frac{\theta_k}{\theta_{k+1}} \leq \frac{1}{\theta_{k+1}} - \frac{1}{\theta_k}.$$

Summing these inequalities for $k = 0, \ldots, n-1$ gives

$$n\alpha \leq \frac{1}{\theta_n} - \frac{1}{\theta_0} \leq \frac{1}{\theta_n},$$

i.e.,

$$\theta_n \leq \frac{1}{n\alpha} = \frac{2M\|x_0 - x^*\|^2}{n}. \qquad \square$$

A continuous-time analog of Theorem 7.4 is given by the following exercise.

Exercise 7.7. *Let f be as in Theorem 7.4, let $x_0 \in E$ and let $x : \mathbb{R}_+ \to E$ be the gradient flows of f starting from x_0, i.e.,*

$$\dot{x}(t) = -\nabla f(x(t)), \quad t \geq 0, \quad x(0) = x_0.$$

Show (by simply adapting the arguments in the proof above to the continuous-time setting) that there is a constant C such that

$$f(x(t)) - f(x^*) \leq \frac{C\|x_0 - x^*\|^2}{t}, \quad \forall t \geq 0.$$

7.2.3. *Nesterov acceleration*

One of the most fascinating and important results in modern optimization is the accelerated gradient method found by Nesterov in 1983 [79]. Nesterov's acceleration enables us to reach an $O\left(\frac{1}{n^2}\right)$ bound for the values of the objective function along the iterates of a sequence generated by a smart and totally explicit gradient method. The $O\left(\frac{1}{n^2}\right)$ rate in Nesterov's method is actually optimal among first-order methods for the minimization of convex (but not necessarily strongly convex) functions with a Lipschitz gradient, see [78].

The framework is exactly the same as in Section 7.2.2, our aim is to minimize a convex function f with an M-Lipschitz gradient over a Hilbert space E. Nesterov's method, consists, starting from $x_0 \in E$ and setting $y_0 = x_0$ in constructing two sequences (x_k, y_k) by the recursion:

$$y_{k+1} = x_k - \frac{1}{M}\nabla f(x_k), \quad x_{k+1} = y_{k+1} + \frac{\lambda_k - 1}{\lambda_{k+1}}(y_{k+1} - y_k), \quad (7.26)$$

where $\lambda_0 = 0$ and

$$\lambda_{k+1} = \frac{1 + \sqrt{1 + 4\lambda_k^2}}{2}, \quad \forall k \geq 0. \quad (7.27)$$

For further use, let us mention some properties of the sequence λ_k:

- by induction on k, it is easy to see that for every k, one has

$$\lambda_k \geq \frac{k}{2}; \quad (7.28)$$

- taking the square of the relation $2\lambda_{k+1} - 1 = \sqrt{1 + 4\lambda_k^2}$ gives

$$\lambda_{k+1}^2 - \lambda_{k+1} = \lambda_k^2. \quad (7.29)$$

We are now ready to prove Nesterov's quadratic rate, the proof below is directly inspired from the arguments of Beck and Teboulle [13].

Theorem 7.5. *Let f be convex, with an M-Lipschitz gradient on E and admitting a minimum at x^*. Then defining the sequence $(x_n, y_n)_n$ starting from (x_0, y_0) by Nesterov's iterates (7.26)–(7.27) we have*

$$f(y_n) - f(x^*) \leq \frac{2M\|x_0 - x^*\|^2}{(n-1)^2}, \quad \forall n \geq 2. \quad (7.30)$$

Proof. Given x and y in E summing the inequality $f(x) - f(y) \leq \nabla f(x) \cdot (x - y)$ with (7.19), we get

$$f(x - M^{-1}\nabla f(x)) - f(y) \leq -\frac{1}{2M}\|\nabla f(x)\|^2 + \nabla f(x) \cdot (x - y). \quad (7.31)$$

Applying (7.31) when $x = x_k$ and $y = y_k$ and using (7.26) gives

$$f(y_{k+1}) - f(y_k) \leq -\frac{M}{2}\|x_k - y_{k+1}\|^2 + M(x_k - y_{k+1}) \cdot (x_k - y_k). \quad (7.32)$$

Applying (7.31) when $x = x_k$ and $y = x^*$ then gives

$$f(y_{k+1}) - f(x^*) \leq -\frac{M}{2}\|x_k - y_{k+1}\|^2 + M(x_k - y_{k+1}) \cdot (x_k - x^*). \quad (7.33)$$

To shorten notations, let us set $\theta_k := f(y_k) - f(x^*)$. For $k \geq 1$, $\lambda_k \geq 1$, multiplying (7.32) by $(\lambda_k - 1)$ and adding (7.33) then gives

$$\lambda_k\theta_{k+1} - (\lambda_k - 1)\theta_k \leq -\frac{M\lambda_k}{2}\|x_k - y_{k+1}\|^2 + M(x_k - y_{k+1}) \cdot z_k, \quad (7.34)$$

where

$$z_k := \lambda_k x_k - (\lambda_k - 1)y_k - x^*. \quad (7.35)$$

Here comes the trick which explains the somehow mysterious recursive formula (7.27). As observed in (7.29), (7.27) implies that $\lambda_k(\lambda_k - 1) = \lambda_{k-1}^2$, hence multiplying (7.34) by λ_k gives

$$\lambda_k^2\theta_{k+1} - \lambda_{k-1}^2\theta_k \leq -\frac{M}{2}(\|\lambda_k(y_{k+1} - x_k)\|^2 + 2\lambda_k(y_{k+1} - x_k) \cdot z_k). \quad (7.36)$$

Note that the left-hand side of this inequality is the term of a telescopic sum. Our next goal is to write the right-hand side as the term of a telescopic sum as well. To do so, first observe that the identity $\|a\|^2 + 2a \cdot b = \|a+b\|^2 - \|b\|^2$ applied to $a = \lambda_k(y_{k+1} - x_k)$ and $b = z_k$ gives

$$\lambda_k^2\theta_{k+1} - \lambda_{k-1}^2\theta_k \leq -\frac{M}{2}(\|\lambda_k(y_{k+1} - x_k) + z_k\|^2 - \|z_k\|^2). \quad (7.37)$$

Next, we observe that

$$\lambda_k(y_{k+1} - x_k) + z_k = \lambda_k y_{k+1} - (\lambda_k - 1)y_k - x^*$$

and that, by (7.26),

$$\begin{aligned}
z_{k+1} &= \lambda_{k+1} x_{k+1} - (\lambda_{k+1} - 1)y_{k+1} - x^* \\
&= \lambda_{k+1}\left(y_{k+1} + \frac{\lambda_k - 1}{\lambda_{k+1}}(y_{k+1} - y_k) \right) - (\lambda_{k+1} - 1)y_{k+1} - x^* \\
&= \lambda_k y_{k+1} - (\lambda_k - 1)y_k - x^*
\end{aligned}$$

so that

$$z_{k+1} = \lambda_k(y_{k+1} - x_k) + z_k.$$

Hence, as promised, (7.37) rewrites as an inequality between terms of two telescopic sums:

$$\lambda_k^2 \theta_{k+1} - \lambda_{k-1}^2 \theta_k \leq \frac{M}{2}(\|z_k\|^2 - \|z_{k+1}\|^2). \tag{7.38}$$

Summing for $k = 1, \ldots, n-1$ and using the fact that $\lambda_0^2 \theta_1 = 0$ we therefore obtain

$$\lambda_{n-1}^2 \theta_n \leq \frac{M}{2}\|z_1\|^2 = \frac{M}{2}\|x_1 - x^*\|^2$$

but we know from (7.28) that $\lambda_{n-1}^2 \geq \frac{1}{4}(n-1)^2$, since $x_1 = x_0$ this yields the desired inequality

$$\theta_n = f(y_n) - f(x^*) \leq \frac{2M\|x_0 - x^*\|^2}{(n-1)^2}. \qquad \square$$

7.3. The proximal point method

Now we are interested in minimizing a possibly non-smooth convex function $f \in \Gamma_0(E)$ on the Hilbert space E. Our aim is to present and prove the convergence of the proximal point algorithm, introduced by Martinet [72, 73] and further analyzed by Rockafellar [85, 86]. The proximal point algorithm with step $\rho > 0$, consists, starting from an initial condition x_0, in constructing a sequence by successive minimization of the form

$$\text{given } x_n, \ x_{n+1} \text{ minimizes } y \mapsto \frac{1}{2\rho}\|y - x_n\|^2 + f(y). \tag{7.39}$$

Defining the function

$$\Phi_{\rho,f,x}(y) := \frac{1}{2\rho}\|y-x\|^2 + f(y), \quad \forall(x,y) \in E^2,$$

$\Phi_{\rho,f,x} \in \Gamma_0(E)$, it is strictly convex and coercive (because f has an affine continuous minorant) hence it admits a unique minimizer, in particular the scheme (7.39) is well-defined. Moreover, the minimizer y of $\Phi_{\rho,f,x}$ is characterized by the condition

$$0 \in \partial\Phi_{\rho,f,x}(y).$$

Since $\Phi_{\rho,f,x}$ is the sum of a differentiable quadratic term and f, one has (see Exercise 3.24):

$$\partial\Phi_{\rho,f,x}(y) = \frac{1}{\rho}(y-x) + \partial f(y).$$

We thus have an equivalence between:

- y is the minimizer of $\Phi_{\rho,f,x}$;
- $x \in y + \rho\partial f(y)$ which we shall write as

$$y \in (\mathrm{id} + \rho\partial f)^{-1}(x).$$

Following the fundamental contributions of Moreau [75], the minimizer of $\Phi_{\rho,f,x}$ as a function of x is called the proximal mapping of ρf and denoted $\mathrm{prox}_{\rho f}$:

$$\mathrm{prox}_{\rho f}(x) = (\mathrm{id} + \rho\partial f)^{-1}(x). \tag{7.40}$$

It is characterized by the minimality property

$$y = \mathrm{prox}_{\rho f}(x) \iff \frac{1}{2}\|y-x\|^2 + \rho f(y) \le \frac{1}{2}\|z-x\|^2 + \rho f(z), \quad \forall z \in E.$$

The iterations of the proximal point algorithm can be rewritten more concisely as

$$x_{n+1} = \mathrm{prox}_{\rho f}(x_n). \tag{7.41}$$

Before studying the properties and the convergence of these iterations, it may be useful to point out an analogy with gradient flows in continuous time. Imagine just for a moment that f is $C^{1,1}$ and consider the *gradient flow* dynamics

$$\dot{x}(t) = -\nabla f(x(t)), \quad t \geq 0, \quad x(0) = x_0. \tag{7.42}$$

When ∇f is Lipschitz, it follows from the Cauchy–Lipschitz theorem that this ODE has a unique solution starting from x_0. Since

$$\frac{d}{dt} f(x(t)) = \nabla f(x(t)) \cdot \dot{x}(t) = -\|\nabla f(x(t))\|^2,$$

f is non-increasing along gradient flows trajectories and it is natural to expect that, under some strong convexity assumptions, for large time, trajectories of (7.42) converge to minimizers of f (see Exercise 7.9). It is therefore very tempting to use time discretizations of the gradient flow (7.42) as iterative schemes for minimizing f. A first possible discretization is the *explicit* Euler scheme. Given a time step ρ, the explicit Euler scheme iteratively computes x_n by

$$\frac{x_{n+1} - x_n}{\rho} = -\nabla f(x_n),$$

which is nothing but the gradient method with step ρ. Another scheme is the *implicit* Euler scheme where the iterates are obtained by iteratively solving

$$\frac{x_{n+1} - x_n}{\rho} = -\nabla f(x_{n+1}), \quad \text{i.e.,} \quad x_{n+1} + \rho \nabla f(x_{n+1}) = x_n,$$

which when f is convex (or more generally $f + \frac{1}{2\rho}\|\cdot\|^2$ is convex) amounts to solve

$$\inf_y \frac{1}{2\rho} \|y - x_n\|^2 + f(y),$$

i.e., $x_{n+1} = \mathrm{prox}_{\rho f}(x_n)$. The proximal point algorithm is the discretization of the implicit Euler scheme for the gradient flow of f. As we shall see, no smoothness of f is required for its convergence, the price to pay is that each step is itself a (strongly convex) minimization problem instead of an explicit update as in the case of the gradient method.

Let us now consider a (useful in some applications) explicit example.

Example 7.1. For $x \in \mathbb{R}^d$ set

$$|x|_1 := \sum_{i=1}^d |x_i|,$$

it is neither smooth nor strongly convex but it is a *simple* convex function in the sense that its proximal mapping is explicit. Indeed, given $x \in \mathbb{R}^d$, $y = \text{prox}_{\rho|.|_1}$ is characterized by the inclusion

$$\frac{x - y}{\rho} \in \partial |\cdot|_1(y),$$

i.e., recalling the expression for the subdifferential of the l^1-norm from Example 3.3, for every $i = 1, \ldots, d$

$$x_i - y_i = \rho \text{ if } y_i > 0, x_i - y_i \in [-\rho, \rho] \text{ if } y_i{=}0 \text{ and } x_i - y_i = -\rho \text{ if } y_i < 0.$$

This gives an explicit expression for y_i

$$y_i = \begin{cases} x_i - \rho & \text{if } x_i \geq \rho, \\ x_i + \rho & \text{if } x_i \leq -\rho, \\ 0 & \text{if } |x_i| \leq \rho, \end{cases}$$

which can be expressed as $y_i = \text{shrink}_\rho(x_i)$ where shrink_ρ denotes the so-called *shrinkage* function

$$\text{shrink}_\rho(x) := (|x| - \rho)_+ \text{sign}(x) \quad \forall x \in \mathbb{R}.$$

The explicit expression of the proximal mapping thus reads

$$\text{prox}_{\rho|.|_1}(x) = (\text{shrink}_\rho(x_1), \ldots, \text{shrink}_\rho(x_d)). \tag{7.43}$$

Let us now go back to the analysis of the proximal point method (7.41). Let us start with some basic properties of the proximal operator.[b]

Proposition 7.1. *Let $f \in \Gamma_0(E)$ and $\rho > 0$. Then*

1) *$x = \text{prox}_{\rho f}(x)$ if and only if x is a minimizer of f;*
2) *$\text{prox}_{\rho f}$ is firmly non-expansive, i.e., for every $(x_1, x_2) \in E^2$ one has*

$$(\text{prox}_{\rho f}(x_1) - \text{prox}_{\rho f}(x_2)) \cdot (x_1 - x_2)$$
$$\geq \|\text{prox}_{\rho f}(x_1) - \text{prox}_{\rho f}(x_2)\|^2, \tag{7.44}$$

which implies that $\text{prox}_{\rho f}$ is both monotone and 1-Lipschitz.

---

[b]Note the similarity with the properties of projections onto a closed convex set C (see Proposition 1.15), this is of course not a coincidence since $\text{proj}_C = \text{prox}_{\chi_C}$.

Proof. By (7.40), $x = \text{prox}_{\rho f}(x)$ if and only if $0 \in \partial f(x)$ which means that x is a minimizer of f. Let us now prove non-expansivity, set $y_i := \text{prox}_{\rho f}(x_i)$ for $i = 1, 2$, then $x_i - y_i \in \rho \partial f(y_i)$ so that

$$\rho(f(y_1) - f(y_2)) \geq (x_2 - y_2) \cdot (y_1 - y_2),$$
$$\rho(f(y_2) - f(y_1)) \geq (x_1 - y_1) \cdot (y_2 - y_1).$$

Summing these two inequalities yields

$$0 \geq \|y_1 - y_2\|^2 - (y_1 - y_2) \cdot (x_1 - x_2),$$

which gives (7.44). In particular, $(y_1 - y_2) \cdot (x_1 - x_2) \geq 0$ (monotonicity of $\text{prox}_{\rho f}$) and by the Cauchy–Schwarz inequality we also have $\|y_1 - y_2\| \leq \|x_1 - x_2\|$ so that $\text{prox}_{\rho f}$ is 1-Lipschitz. $\square$

The fact that $\text{prox}_{\rho f}$ is 1-Lipschitz alone is not enough to guarantee that its iterates converge (think of iterates of a rotation) but, as we will see, (7.44) is a much stronger property (also see Exercise 7.15). More specific arguments are needed to show convergence and good estimates of the proximal point algorithm among which the following lemma.

Lemma 7.2 (Opial's lemma). *If (x_n) weakly converges to x (in the Hilbert space E) and $y \in E \setminus \{x\}$ then*

$$\liminf_n \|x_n - x\| < \liminf_n \|x_n - y\|.$$

Proof. This is a direct consequence of

$$\|x_n - y\|^2 = \|x_n - x\|^2 + \|x - y\|^2 + 2(x_n - x) \cdot (x - y)$$

together with the fact that the weak convergence of x_n to x gives $(x_n - x) \cdot (x - y) \to 0$ as $n \to \infty$. $\square$

An obvious property of iterates of the proximal point algorithm (7.41) is that by construction

$$\frac{1}{2}\|x_{k+1} - x_k\|^2 \leq \rho(f(x_k) - f(x_{k+1})) \tag{7.45}$$

but summing these inequalities, thanks to the fact that the right-hand side is a telescopic sum gives

$$\frac{1}{2}\sum_{k=0}^{n} \|x_{k+1} - x_k\|^2 \leq \rho(f(x_0) - f(x_{n+1})). \tag{7.46}$$

So that, if f is bounded from below (which is in particular the case if it has a minimizer), we have for free that $f(x_n)$ is non-increasing and bounded from below thus converges, and the right-hand side of (7.46) is bounded so that $\sum_k \|x_{k+1} - x_k\|^2$ is convergent, in particular $x_{n+1} - x_n$ converges to 0. A finer estimate on the decay of $f(x_n)$ is as follows.

Proposition 7.2. *Let (x_n) be generated by the proximal point algorithm (7.41). Then for every $x \in E$*

$$(f(x_n) - f(x)) \leq \frac{1}{2\rho n} \left(\|x_0 - x\|^2 - \|x_n - x\|^2 - \sum_{k=1}^{n} k\|x_k - x_{k-1}\|^2 \right).$$
(7.47)

In particular, $f(x_n)$ converges to $\inf_E f$ and if $\inf_E f$ is finite then $f(x_n) - \inf_E f = O\left(\frac{1}{n}\right)$.

Proof. Let $x \in E$, since by construction $x_{k-1} - x_k \in \rho \partial f(x_k)$ we have

$$2\rho(f(x) - f(x_k)) \geq 2(x_{k-1} - x_k) \cdot (x - x_k)$$
$$= \|x_{k-1} - x_k\|^2 + \|x_k - x\|^2 - \|x_{k-1} - x\|^2.$$

Summing for $k = 1, \ldots, n$ gives

$$2\rho n f(x) \geq 2\rho \sum_{k=1}^{n} f(x_k) + \sum_{k=1}^{n} \|x_k - x_{k-1}\|^2 + \|x_n - x\|^2 - \|x_0 - x\|^2.$$
(7.48)

Multiplying

$$2\rho f(x_k) + \|x_k - x_{k-1}\|^2 \leq 2\rho f(x_{k-1})$$

by $(k-1)$ and rearranging terms also gives

$$2\rho f(x_k) \geq (k-1)\|x_k - x_{k-1}\|^2 + 2\rho(k f(x_k) - (k-1)f(x_{k-1})),$$

summing again for $k = 1, \ldots, n$, we obtain

$$2\rho \sum_{k=1}^{n} f(x_k) \geq \sum_{k=1}^{n} (k-1)\|x_k - x_{k-1}\|^2 + 2\rho n f(x_n).$$
(7.49)

Combining (7.49) with (7.48) thus yields

$$2\rho n(f(x) - f(x_n)) \geq \sum_{k=1}^{n} k\|x_k - x_{k-1}\|^2 + \|x_n - x\|^2 - \|x_0 - x\|^2$$

so that (7.47) holds. □

As for convergence of iterates themselves, it is the object of the following result.

Theorem 7.6. *If f admits a minimizer, then any sequence (x_n) generated by the proximal point algorithm (7.41) weakly converges to a minimizer of f.*

Proof. Let y be a minimizer of f then $\text{prox}_{\rho f}(y) = y$ by Proposition 7.1, using the fact that $\text{prox}_{\rho f}$ is 1-Lipschitz thus gives

$$\|x_{n+1} - y\| = \|\text{prox}_{\rho f}(x_n) - \text{prox}_{\rho f}(y)\| \leq \|x_n - y\| \qquad (7.50)$$

so that the non-increasing sequence of distances $\|x_n - y\|$ converges (and this holds for any minimizer y). In particular (x_n) is bounded hence admits weakly convergent subsequences, let y be a weak cluster point of (x_n), thanks to Proposition 7.2 and the weak lower semicontinuity of f we have

$$f(y) \leq \lim_n f(x_n) = \inf_E f.$$

Weak cluster points of (x_n) therefore are minimizers of f (equivalently fixed points of $\text{prox}_{\rho f}$). To show that the bounded sequence (x_n) weakly converges, it is enough to prove that it has a unique weak cluster point (see Exercise 1.21). Let y and z be two such weak cluster points that is along suitable subsequences

$$x_{\varphi(n)} \rightharpoonup y, \quad x_{\psi(n)} \rightharpoonup z \text{ as } n \to +\infty$$

since both y and z are minimizers of f, the non-increasing sequences $\|x_n - y\|$ and $\|x_n - z\|$ converge, but if $y \neq z$, we deduce from Lemma 7.2 that

$$\lim_n \|x_n - y\| = \liminf_n \|x_{\varphi(n)} - y\|$$

$$< \liminf_n \|x_{\varphi(n)} - z\| = \lim_n \|x_n - z\| = \liminf_n \|x_{\psi(n)} - z\|$$

$$< \liminf_n \|x_{\psi(n)} - y\| = \lim_n \|x_n - y\|,$$

which yields the desired contradiction and the weak convergence of the full sequence (x_n). □

7.4. Splitting methods

Now we consider the minimization of the sum of two[c] convex functions f and g. Of course, if both f and g are smooth (which rules out the presence of constraints), one can use gradient methods. But, if one of them is not smooth, this is not an option. Moreover, the proximal point algorithm for $f + g$ might be tedious and it may be the case that both prox_f and prox_g are easy to compute whereas prox_{f+g} is not. This is one of the many situations where splitting strategies are particularly adapted.

7.4.1. *Forward–Backward splitting*

Let us start with the possibility of combining (explicit or forward) gradient steps with (implicit or backward) proximal steps. Consider indeed the problem

$$\inf_{x \in E} \{f(x) + g(x)\}, \tag{7.51}$$

where E is a Hilbert space, $g \in \Gamma_0(E)$ and f is convex and differentiable. To fix ideas, let us first assume that f is strongly convex and has a Lipschitz gradient, i.e., satisfies (7.9) and (7.10). In this case, (7.51) has a unique solution, which we may of course compute by the proximal-point algorithm for $f + g$. But this might not be a smart strategy: on the one hand, prox_{f+g} might be complicated to compute (even if both prox_f and prox_g have closed forms) and on the other hand, it does not take advantage of the nice properties of f. A more efficient way to proceed is to combine the proximal method for the possibly non-smooth term g and gradient iterates for the nice term f. This leads to the following scheme:

$$x_{n+1} = \mathrm{prox}_{\rho g}(x_n - \rho \nabla f(x_n)). \tag{7.52}$$

Since $\mathrm{prox}_{\rho g}$ is 1-Lipschitz and $\mathrm{id} - \rho \nabla f$ is a contraction provided $\rho \in (0, \frac{2\alpha}{M^2})$ (Theorem 7.2), in this case, it directly follows from Banach's fixed-point theorem (Theorem 1.4) that sequences generated by (7.52)) converge (at least linearly) to the only solution of

$$x = \mathrm{prox}_{\rho g}(x - \rho \nabla f(x)),$$

---

[c]The minimization of the sum of more than two convex functions can actually be brought down to this case thanks to a certain lifting trick, see Exercise 7.20.

which is equivalent to $x - \rho\nabla f(x) \in x + \rho\partial g(x)$, i.e., $0 \in \partial g(x) + \nabla f(x)$ which is the optimality condition for (7.51). Note that if $g = \chi_C$, the scheme (7.52) is nothing but the projected gradient method.

The scheme (7.52) is called ISTA[d] by Beck and Teboulle [13]. Now, if f is only convex (i.e., without assuming (7.10)) but has an M-Lipschitz gradient and if (7.51) admits solutions, for a step choice $\rho = M^{-1}$, Beck and Teboulle [13] proved that the sequence generated by ISTA satisfies the error estimate: $f(x_n) + g(x_n) - \inf_E\{f + g\} = O(1/n)$. More importantly, Beck and Teboulle proposed an acceleration method called FISTA (fast iterative shrinkage thresholding algorithm) in the spirit of Nesterov's method to achieve: $f(x_n) + g(x_n) - \inf_E\{f + g\} = O(1/n^2)$. It goes without saying that the results of Beck and Teboulle [13] had a tremendous impact on numerical convex optimization in the last decade.

7.4.2. Douglas–Rachford method

We now consider the minimization of $f + g$ as in (7.51) but in the case where f and g are no better than $\Gamma_0(E)$. We also assume that there exists at least a point $\overline{x} \in E$ such that

$$0 \in \partial f(\overline{x}) + \partial g(\overline{x}). \tag{7.53}$$

Note that (7.53) is a sufficient condition for $\overline{x}$ to be a minimizer of $f + g$ since $\partial f(\overline{x}) + \partial g(\overline{x}) \subset \partial(f+g)(\overline{x})$ (but the inclusion might be strict and we saw in Theorem 3.11 that some qualification condition ensures that these two sets coincide). If (7.53) holds, there is a $\overline{p} \in E$ such that $\overline{p} \in \partial f(\overline{x})$ and $-\overline{p} \in \partial g(\overline{x})$ so that $f(\overline{x}) + g(\overline{x}) = -f^*(\overline{p}) - g^*(-\overline{p})$ hence $\overline{p}$ solves the dual of (7.51):

$$\sup_{p \in E}\{-f^*(p) - g^*(-p)\}. \tag{7.54}$$

The optimal primal–dual pair $(\overline{x}, \overline{p})$ therefore satisfies

$$\overline{x} + \overline{p} \in \overline{x} + \partial f(\overline{x}), \quad \overline{x} - \overline{p} \in \overline{x} + \partial g(\overline{x}),$$

which in proximal terms rewrites as

$$\overline{x} = \text{prox}_f(\overline{x} + \overline{p}) = \text{prox}_g(\overline{x} - \overline{p}). \tag{7.55}$$

---

[d]ISTA means iterative shrinkage thresholding algorithm. This terminology comes from the fact that when g is the l^1-norm, $\text{prox}_{\rho g}$ is given by the explicit shrinkage formula (7.43).

Setting

$$\overline{v} := \overline{x} - \overline{p}$$

since $\overline{x} + \overline{p} = 2\overline{x} - \overline{v} = 2\text{prox}_g(\overline{v}) - \overline{v}$ we thus have $\overline{x} = \text{prox}_g(\overline{v}) = \text{prox}_f(2\overline{x} - \overline{v})$ and then

$$0 = \text{prox}_f(2\text{prox}_g(\overline{v}) - \overline{v}) - \text{prox}_g(\overline{v}).$$

So that

$$\overline{v} = S\overline{v}, \tag{7.56}$$

where S is defined by

$$S := \text{prox}_f \circ (2\text{prox}_g - \text{id}) + \text{id} - \text{prox}_g. \tag{7.57}$$

The backward–forward splitting scheme which consists in iterating the operator S:

$$v_{k+1} = S(v_k) \tag{7.58}$$

was initially proposed by Douglas and Rachford [42] and was analyzed by Lions and Mercier [71], also see Eckstein and Bertsekas [43] and Combettes [35]. Note that (7.58) can be rewritten as

$$x_k = \text{prox}_g(v_k), \quad v_{k+1} = \text{prox}_f(2x_k - v_k) + v_k - x_k. \tag{7.59}$$

Douglas–Rachford algorithm is therefore particularly adapted to the case where prox_f and prox_g are easy to compute (or even better have a closed form as in Example 7.1). Note also that Douglas–Rachford generalizes the proximal point algorithm in the sense that the latter corresponds to (7.59) when $g = 0$. As observed first by Lions and Mercier, the convergence of the Douglas–Rachford algorithm is based on properties of firmly non-expansive operators,

Definition 7.1. A map $T : E \to E$ is called firmly non-expansive if

$$(T(x) - T(y)) \cdot (x - y) \geq \|T(x) - T(y)\|^2, \quad \forall (x, y) \in E \times E.$$

Lemma 7.3. *Let T be a firmly non-expansive map $E \to E$. Then for every $(x, y) \in E^2$ one has*

$$\|x - y\|^2 \geq \|T(x) - T(y)\|^2 + \|(\text{id} - T)(x) - (\text{id} - T)(y)\|^2 \tag{7.60}$$

and $\text{id} - T$ is firmly non-expansive.

Proof. We directly compute:

$$\|x - y\|^2 = \|T(x) - T(y)\|^2 + \|(\mathrm{id}\,-T)(x) - (\mathrm{id}\,-T)(y)\|^2$$
$$+ 2(T(x) - T(y)) \cdot (x - y) - 2\|T(x) - T(y)\|^2$$

since T is firmly non-expansive, the term on the second line is non-negative which exactly gives (7.60). As for the firm nonexpansivity of id $-T$, we have

$$\|(\mathrm{id}\,-T)(x) - (\mathrm{id}\,-T)(y)\|^2 = \|x - y\|^2 + \|T(x) - T(y)\|^2$$
$$- 2(x - y) \cdot (T(x) - T(y))$$
$$\leq \|x - y\|^2 - (x - y) \cdot (T(x) - T(y))$$
$$= ((\mathrm{id}\,-T)(x) - (\mathrm{id}\,-T)(y)) \cdot (x - y),$$

where we have used the firm nonexpansivity of T at the second line. $\qquad\square$

We know from Proposition 7.1 that prox_f and prox_g are firmly non-expansive. Another class of firmly non-expansive maps is given by the following lemma due to Lions and Mercier [71].

Lemma 7.4 (Lions and Mercier). *If T_1 and T_2 are two firmly non-expansive maps $E \to E$, then*

$$T := T_1 \circ (2T_2 - \mathrm{id}) + (\mathrm{id}\,-T_2)$$

is firmly non-expansive. In particular, if f and g are in $\Gamma_0(E)$, the operator S defined by (7.57) is firmly nonexpansive.

Proof. For $(x, y) \in E^2$, one has

$$\|T(x) - T(y)\|^2$$
$$=: \|T_1(2T_2(x) - x) - T_1(2T_2(y) - y)\|^2$$
$$+ \|(\mathrm{id}\,-T_2)(x) - (\mathrm{id}\,-T_2)(y)\|^2$$
$$+ 2((\mathrm{id}\,-T_2)(x) - (\mathrm{id}\,-T_2)(y)) \cdot (T_1(2T_2(x) - x) - T_1(2T_2(y) - y)).$$

Since T_1 is firmly nonexpansive

$$\|T_1(2T_2(x) - x) - T_1(2T_2(y) - y)\|^2$$
$$\leq (T_1(2T_2(x) - x) - T_1(2T_2(y) - y)) \cdot (2T_2(x) - x - (2T_2(y) - y)),$$

hence

$$\|T(x) - T(y)\|^2 \le \|(\mathrm{id} - T_2)(x) - (\mathrm{id} - T_2)(y)\|^2$$
$$+ (T_1(2T_2(x) - x) - T_1(2T_2(y) - y)) \cdot (x - y)$$
$$= ((\mathrm{id} - T_2)(x) - (\mathrm{id} - T_2)(y)) \cdot (x - y)$$
$$- (T_2(x) - T_2(y)) \cdot ((\mathrm{id} - T_2)(x) - (\mathrm{id} - T_2)(y))$$
$$+ (T_1(2T_2(x) - x) - T_1(2T_2(y) - y)) \cdot (x - y).$$

Since T_2 is firmly nonexpansive, $(T_2(x) - T_2(y)) \cdot ((\mathrm{id} - T_2)(x) - (\mathrm{id} - T_2)(y)) \ge 0$, and then

$$\|T(x) - T(y)\|^2 \le ((\mathrm{id} - T_2)(x) - (\mathrm{id} - T_2)(y)) \cdot (x - y)$$
$$+ (T_1(2T_2(x) - x) - T_1(2T_2(y) - y)) \cdot (x - y)$$
$$= (T(x) - T(y)) \cdot (x - y). \qquad \square$$

The following result is a (very) special case of a famous characterization of maximal monotone operators due to Minty.

Lemma 7.5. *Let E be a Hilbert space and $A : E \to E$ be continuous and monotone, i.e.,*

$$(A(u) - A(v)) \cdot (u - v) \ge 0, \quad \forall (u, v) \in E^2.$$

Let $x \in E$ and $p \in E$. Then the following statements are equivalent:

$$p = A(x) \tag{7.61}$$

and

$$(A(y) - p) \cdot (y - x) \ge 0, \quad \forall y \in E. \tag{7.62}$$

Proof. The fact that (7.61) implies (7.62) obviously follows from the monotonicity of A. Assume now that p satisfies (7.62), take $h \in E$, $\lambda > 0$ setting $y := x + \lambda h$ in (7.62) and dividing by λ gives $(A(x + \lambda h) - p) \cdot h \ge 0$, since A is continuous letting $\lambda \to 0^+$ gives $(A(x) - p) \cdot h \ge 0$ but since h is arbitrary this finally implies that $A(x) = p$. $\qquad \square$

We are now ready to prove convergence of the Douglas–Rachford algorithm:

Theorem 7.7. *Let f and g be in $\Gamma_0(E)$, assume that there exists $\bar{x} \in E$ such that $0 \in \partial f(\bar{x}) + \partial g(\bar{x})$. Let $v_0 \in E$ and define the sequence v_n by the iterates of the Douglas–Rachford algorithm (7.57)–(7.58). Then v_n*

converges weakly to some v_∞ which is a fixed point of S defined by (7.57) and $x_\infty := \text{prox}_g(v_\infty)$ satisfies $0 \in \partial f(x_\infty) + \partial g(x_\infty)$.

Proof. To shorten notations let us set $Q := \text{id} - S$ so that $v_k - v_{k+1} = Q(v_k)$. We know from (7.56) that S admits at least one fixed point, let $\overline{v}$ be such a fixed-point $S(\overline{v}) = \overline{v}$, i.e., $Q(\overline{v}) = 0$. Since S is firmly nonexpansive by Lemma 7.4, it is 1-Lipschitz hence

$$\|v_{k+1} - \overline{v}\| = \|S(v_k) - S(\overline{v})\| \le \|v_k - \overline{v}\|$$

so that $\|v_{k+1} - \overline{v}\|$ converges monotonically (and this holds for any fixed point $\overline{v}$ of S), in particular $(v_k)_k$ is bounded. It now follows from (7.60) in Lemma 7.3 that

$$\|v_k - \overline{v}\|^2 \ge \|S(v_k) - S(\overline{v})\|^2 + \|Q(v_k) - Q(\overline{v})\|^2 = \|v_{k+1} - \overline{v}\|^2 + \|Q(v_k)\|^2$$

but since $\|v_k - \overline{v}\|^2$ converges, this implies

$$Q(v_k) = v_k - v_{k+1} \to 0 \text{ strongly as } k \to \infty. \tag{7.63}$$

Since $(v_k)_k$ is bounded, it admits weak cluster points, let $\widetilde{v}$ be such a cluster point, i.e., $v_{k_n} \rightharpoonup \widetilde{v}$ along a suitable subsequence k_n. Let us now show that $S(\widetilde{v}) = \widetilde{v}$, i.e., $Q(\widetilde{v}) = 0$. Since Q is firmly non-expansive by Lemma 7.3 it is continuous and monotone so that to show that $Q(\widetilde{v}) = 0$, thanks to Lemma 7.5 it is enough to show that

$$Q(v) \cdot (v - \widetilde{v}) \ge 0, \quad \forall v \in E \tag{7.64}$$

knowing, by monotonicity of Q that

$$(Q(v) - Q(v_{k_n})) \cdot (v - v_{k_n}) \ge 0$$

and that $Q(v_{k_n})$ converges strongly to 0 and v_{k_n} converges weakly to $\widetilde{v}$, passing to the limit $n \to \infty$ in the above inequality enables us to conclude that (7.64) holds hence $S(\widetilde{v}) = \widetilde{v}$.

To sum up, we know that (v_k) admits weak cluster points, that all such weak cluster points are fixed points of S and the distance between v_k and any fixed point of S converges. Using Opial's lemma and arguing as in the proof of Theorem 7.6, we can conclude that the whole sequence (v_k) weakly converges to some v_∞ which is a fixed point of S. Setting $x_\infty := \text{prox}_g(v_\infty)$ and recalling the definition of S (7.57), we have

$$v_\infty = \text{prox}_f(2x_\infty - v_\infty) + v_\infty - x_\infty$$

so that $x_\infty = \text{prox}_g(v_\infty) = \text{prox}_f(2x_\infty - v_\infty)$, that is

$$v_\infty \in x_\infty + \partial g(x_\infty), \quad 2x_\infty - v_\infty \in x_\infty + \partial f(x_\infty)$$

so that

$$v_\infty - x_\infty \in \partial g(x_\infty), \quad x_\infty - v_\infty \in \partial f(x_\infty),$$

hence $0 \in \partial f(x_\infty) + \partial g(x_\infty)$. □

7.4.3. Link with augmented Lagrangian methods

In this final section, our aim is to show the link between the Douglas–Rachford algorithm and an augmented Lagrangian algorithm also known as alternating direction method of multipliers (ADMM). For more on these methods, we refer the reader to the book of Fortin and Glowinski [51], Gabay and Mercier [52] and Bertsekas and Eckstein [43].

Given f and g in $\Gamma_0(E)$, we again consider the minimization of $f + g$ as in (7.51) and rewrite it as

$$\inf\{f(y) + g(x), \ (x,y) \in E \times E, \ x = y\}. \tag{7.65}$$

We introduce the augmented Lagrangian[e] for this problem:

$$\mathscr{L}_1(x, y, \lambda) := f(y) + g(x) + \lambda \cdot (y - x) + \frac{1}{2}\|y - x\|^2, \quad \forall (x, y, \lambda) \in E^3.$$

Using the augmented Lagrangian $\mathscr{L}_1$, we can rewrite (7.65) in inf–sup form as

$$\inf_{(x,y) \in E \times E} \sup_{\lambda \in E} \mathscr{L}_1(x, y, \lambda). \tag{7.66}$$

The ADMM augmented Lagrangian method consists, starting from (x_0, y_0, λ_0) in iterating the following steps, given (x_k, y_k, λ_k):

• **Step 1:** Find y_{k+1} by solving

$$\inf_{y \in E} \mathscr{L}_1(x_k, y, \lambda_k).$$

---

[e]Here, for notational simplicity, we augment the usual Lagrangian with the quadratic term $\frac{1}{2}\|y - x\|^2$, but we could as well put a positive parameter in front of the quadratic term, or equivalently multiply f and g by some positive ρ.

- **Step 2:** Find x_{k+1} by solving

$$\inf_{x \in E} \mathscr{L}_1(x, y_{k+1}, \lambda_k).$$

- **Step 3:** Update the multiplier by a gradient ascent of $\mathscr{L}_1$, i.e., set

$$\lambda_{k+1} := \lambda_k + (y_{k+1} - x_{k+1}). \tag{7.67}$$

Observing that minimizing $\mathscr{L}_1(x_k, y, \lambda_k)$ with respect to y is equivalent to

$$\inf_{y \in E} \left\{ f(y) + \frac{1}{2} \| y - (x_k - \lambda_k) \|^2 \right\}.$$

Step 1 of ADMM can be rewritten in proximal terms as

$$y_{k+1} = \text{prox}_f(x_k - \lambda_k) \tag{7.68}$$

likewise, **Step 2** reads

$$x_{k+1} = \text{prox}_g(y_{k+1} + \lambda_k). \tag{7.69}$$

It turns out that ADMM is nothing but the Douglas–Rachford algorithm (slightly in disguise). Indeed, recall that Douglas–Rachford iterates are given by

$$v_{k+1} = v_k - x_k + \text{prox}_f(2x_k - v_k), \quad \text{with} \quad x_k := \text{prox}_g(v_k)$$

introducing $y_{k+1} := \text{prox}_f(2x_k - v_k)$ this rewrites as

$$y_{k+1} := \text{prox}_f(2x_k - v_k), \quad v_{k+1} = v_k + y_{k+1} - x_k, \quad x_{k+1} := \text{prox}_g(v_{k+1}).$$

Setting $\lambda_k := v_k - x_k$, we have $2x_k - v_k = x_k - \lambda_k$ and $v_{k+1} = y_{k+1} + \lambda_k$ so that the updates for y_{k+1} and x_{k+1} above reformulate as

$$y_{k+1} = \text{prox}_f(x_k - \lambda_k), \quad x_{k+1} = \text{prox}_g(y_{k+1} + \lambda_k),$$

which is nothing but (7.68)–(7.69). As for λ_{k+1} we have

$$\lambda_{k+1} = v_{k+1} - x_{k+1} = \lambda_k + y_{k+1} - x_{k+1},$$

which is exactly (7.67).

7.5.   Exercises

Exercise 7.8. *Let $f \in C^1(\mathbb{R}^d \times \mathbb{R}^m, \mathbb{R})$ be strongly convex (i.e., satisfy (7.10) for some $\alpha > 0$). We are interested in the unconstrained minimization*

$$\inf_{(x,y) \in \mathbb{R}^d \times \mathbb{R}^m} f(x,y).$$

The coordinate descent method simply consists in minimizing f alternatively with respect to one variable at a time. That is, given (x_0, y_0) one generates inductively the sequence (x_k, y_k) by

$$x_{k+1} = \operatorname{argmin}_{\mathbb{R}^d} f(\cdot, y_k), \ y_{k+1} = \operatorname{argmin}_{\mathbb{R}^m} f(x_{k+1}, \cdot).$$

1) *Explain why the coordinate scheme above is well defined and why f admits a unique minimizer (which we now denote (x^*, y^*)).*
2) *Show that*

$$f(x_k, y_k) - f(x_{k+1}, y_{k+1}) \geq \frac{\alpha}{2}(|x_{k+1} - x_k|^2 + |y_{k+1} - y_k|^2).$$

3) *Show that $x_{k+1} - x_k$ and $y_{k+1} - y_k$ converge to 0 and that (x_k, y_k) converges to (x^*, y^*).*

Exercise 7.9. *Let E be a Hilbert space and $f \in C^1(E)$ be convex and satisfy (7.9) and (7.10) and let x^* be the point where f achieves its minimum. The gradient flow of f is the ODE*

$$\dot{x}(t) = -\nabla f(x(t)), \quad \forall t \geq 0.$$

Show that for any initial condition $x(0)$ the previous ODE has a unique solution and that this trajectory converges exponentially fast to x^* in the sense that

$$\|x(t) - x^*\| \leq e^{-\alpha t} \|x(0) - x^*\|.$$

(Hint: differentiate $\frac{1}{2}\|x(t) - x^*\|^2$ and use (7.10).)

Exercise 7.10. *Let E be a Hilbert space and $f : E \to \mathbb{R}$ (not necessarily convex) satisfy (7.9).*

1) *Show that for every $(x,y) \in E^2$,*

$$|f(y) - f(x) - \nabla f(x)(y - x)| \leq \frac{M}{2}\|x - y\|^2.$$

2) *Show that for every* $\rho \in \left(0, \frac{2}{M}\right)$, *one has*

$$f(x - \rho \nabla f(x)) \leq f(x), \quad \forall x \in E.$$

3) *What does it imply for the gradient method for a (possibly nonconvex)* $C^{1,1}$ *function?*

Exercise 7.11. *In this exercise, we consider a second-order ODE called the heavy ball dynamics (with damping). Having classical mechanics in mind, it corresponds to the motion of a heavy particle subject to a potential force (which is the gradient of a certain convex function) and friction. The following results follow from the elegant analysis of Alvarez [4]. Let* $f \in C^2(\mathbb{R}^d)$ *be bounded from below, convex, with* ∇f *Lipschitz (on the whole of* $\mathbb{R}^d$*). For* $\lambda > 0$ *consider the heavy ball second-order ODE :*

$$\ddot{x}(t) + \lambda \dot{x}(t) + \nabla f(x(t)) = 0, \quad t \geq 0, \quad x(0) = x_0, \quad \dot{x}(0) = v_0, \quad (7.70)$$

where $(x_0, v_0) \in \mathbb{R}^d \times \mathbb{R}^d$ *is given.*

1) *Show that (7.70) has a unique solution which we now denote* $x(\cdot)$.
2) *The energy of the particle is defined by*

$$E(x, v) := f(x) + \frac{1}{2}|v|^2,$$

show that

$$e(t) := E(x(t), \dot{x}(t)) = E(x_0, v_0) - \lambda \int_0^t |\dot{x}(s)|^2 \, ds.$$

3) *Show that* e *is non-increasing and converges as* $t \to +\infty$, *that* $\dot{x}$ *is bounded and that* $\int_0^{+\infty} |\dot{x}|^2 \, dt < +\infty$.
4) *Let* $y \in \mathbb{R}^d$ *and set* $\varphi(t) := \frac{1}{2}|x(t) - y|^2$. *Show that*

$$\ddot{\varphi} + \lambda \dot{\varphi} = -\nabla f(x)(x - y) + |\dot{x}|^2$$

and deduce that

$$\ddot{\varphi} + \lambda \dot{\varphi} \leq f(y) - f(x) + |\dot{x}|^2 = f(y) - e + \frac{3}{2}|\dot{x}|^2 \quad (7.71)$$

and that if $0 \leq s \leq t$, *one has*

$$\ddot{\varphi}(s) + \lambda \dot{\varphi}(s) \leq f(y) - e(t) + \frac{3}{2}|\dot{x}(s)|^2. \quad (7.72)$$

5) *Integrate once (7.72) between s and t to get*

$$\dot{\varphi}(t) \leq \dot{\varphi}(0)e^{-\lambda t} + \frac{f(y) - e(t)}{\lambda}(1 - e^{-\lambda t}) + \frac{3}{2}\int_0^t e^{-\lambda(t-s)}|\dot{x}(s)|^2\,ds.$$

(7.73)

6) *Integrate once again (7.73) using that e is non-increasing and $C :=$ $\int_0^{+\infty}|\dot{x}|^2\,dt < +\infty$ to get*

$$\varphi(t) \leq \varphi(0) + \frac{\dot{\varphi}(0)(1 - e^{-\lambda t})}{\lambda} + \frac{(f(y) - f(x(t)))(\lambda t - 1 + e^{-\lambda t})}{\lambda^2} + \frac{3C}{2\lambda}.$$

7) *Deduce that $\limsup_{t\to+\infty} f(x(t)) \leq f(y)$ and show that $f(x(t))$ converges to $\inf f$ and $\dot{x}(t)$ converges to 0 as $t \to +\infty$.*

8) *Show that if f admits a minimizer then $f(x(t)) - \inf f = O\left(\frac{1}{t}\right)$.*

Exercise 7.12 (Conjugate gradient). *Given $A \in \mathcal{S}_d^{++}(\mathbb{R})$ and $b \in \mathbb{R}^d$ our aim is to solve the linear system $Ax = b$, i.e., to compute $x := A^{-1}(b)$. We say that two vectors u and v are conjugate with respect to A if*

$$(Au) \cdot v = 0.$$

We also define

$$f(y) := \frac{1}{2}(Ay) \cdot y - b \cdot y.$$

1) *Show that x is the only minimizer of f.*

2) *Show that there exists a basis $(v_1, \ldots, v_d)$ of $\mathbb{R}^d$ of mutually conjugate vectors with respect to A and that x admits the explicit expression*

$$x = \sum_{i=1}^d \frac{b \cdot v_i}{Av_i \cdot v_i} v_i.$$

3) *The aim of the conjugate gradient algorithm is precisely to construct such a basis in an inductive way, as follows: start from $x_0 \in \mathbb{R}^d$, set $v_0 = g_0 = \nabla f(x_0) = Ax_0 - b$ and once x_{k-1} and v_{k-1} are known, if $v_{k-1} = 0$ stop the algorithm while if $v_{k-1} \neq 0$ set*

$$x_k := \mathrm{argmin}_{x_{k-1}+\mathbb{R}v_{k-1}} f$$

and

$$g_k := Ax_k - b, \quad v_k = g_k + \beta_k v_{k-1},$$

where β_k is such that v_k and v_{k-1} are conjugates with respect to A. Show that $g_k \cdot v_{k-1} = 0$ and compute β_k.

4) *Show that if $g_0, \ldots, g_l$ are all non-zero then $v_0, \ldots, v_l$ are mutually conjugate with respect to A. (Hint: compare $\mathrm{Span}\{v_0, \ldots, v_{k-1}\}$, $\mathrm{Span}\{g_0, \ldots, g_{k-1}\}$, $\mathrm{Span}\{v_0, Av_0, \ldots, Av_{k-2}\}$ and argue by induction.)*

5) *Show that the algorithm converges to x in at most d steps.*

6) *Show that x_k is the minimizer of f on the affine space $x_0 + \mathrm{span}\{v_0, \ldots, v_{k-1}\}$.*

Exercise 7.13. *Let $f_0, f_1, \ldots, f_m$ be convex functions on $\mathbb{R}^d$ (equipped with its usual Euclidean structure) such that*

- *f_0 is differentiable and, for some $\alpha > 0$, one has*

$$(\nabla f_0(y) - \nabla f_0(x)) \cdot (y - x) \geq \alpha |y - x|^2, \quad \forall (x, y) \in \mathbb{R}^d \times \mathbb{R}^d;$$

- *f_i is M-Lipschitz for $i = 1, \ldots, m$.*

We assume that the set $C := \{x \in \mathbb{R}^d : f_i(x) \leq 0, \ i = 1, \ldots, m\}$ is non-empty and are interested in the constrained problem

$$\inf_{x \in C} f_0(x). \tag{7.74}$$

Defining $F := (f_1, \ldots, f_m) : \mathbb{R}^d \to \mathbb{R}^m$ we recall that the Lagrangian of (7.74) is given by

$$\mathscr{L}(x, \lambda) := f_0(x) + \lambda \cdot F(x), \quad \forall (x, \lambda) \in \mathbb{R}^d \times \mathbb{R}^m.$$

1) *Show that (7.74) has a unique solution x^*. From now on we also assume that there exists a (non-necessarily unique) vector of KKT multipliers λ^* which means that*

$$\lambda^* \in \mathbb{R}_+^m, \quad \lambda^* \cdot F(x^*) = 0, \mathscr{L}(x^*, \lambda^*) \leq \mathscr{L}(x, \lambda^*), \forall x \in \mathbb{R}^d.$$

2) *Show that for every $\rho > 0$ one has*

$$\lambda^* = \mathrm{proj}_{\mathbb{R}_+^m}(\lambda^* + \rho F(x^*))$$

and that

$$\nabla f_0(x^*)(x - x^*) + \lambda^* \cdot (F(x) - F(x^*)) \geq 0, \quad \forall x \in \mathbb{R}^d.$$

3) *Given $\rho > 0$, Uzawa's algorithm with step ρ consists starting from $\lambda_0 \in \mathbb{R}_+^m$ to recursively set*

$$x_k = \mathrm{argmin}\mathscr{L}(., \lambda_k), \quad \lambda_{k+1} := \mathrm{proj}_{\mathbb{R}_+^m}(\lambda_k + \rho F(x_k)), \tag{7.75}$$

show that this sequence is well-defined and that

$$\nabla f_0(x_k)(x - x_k) + \lambda_k \cdot (F(x) - F(x_k)) \geq 0, \quad \forall x \in \mathbb{R}^d.$$

4) *Show that*

$$\alpha |x_k - x^*|^2 + (\lambda_k - \lambda^*) \cdot (F(x_k) - F(x^*)) \leq 0$$

and that

$$|\lambda_{k+1} - \lambda^*| \leq |\lambda_k - \lambda^* + \rho(F(x_k) - F(x^*))|.$$

5) *Show that*

$$|\lambda_{k+1} - \lambda^*|^2 + (2\rho\alpha - M^2\rho^2)|x_k - x^*|^2 \leq |\lambda_k - \lambda^*|^2.$$

6) *Deduce that x_k converges to x^* as soon as $\rho \in \left(0, \frac{2\alpha}{M^2}\right)$.*
7) *Show that Uzawa's algorithm is nothing but the projected gradient algorithm on the Lagrangian dual (see Chapter 6) of (7.74).*

Exercise 7.14. *Let C be a convex compact subset of $\mathbb{R}^d$, $f : \mathbb{R}^d \to \mathbb{R}$ be convex, differentiable with an M-Lipschitz gradient. We are interested in the constrained problem*

$$\inf_{x \in C} f(x). \tag{7.76}$$

Starting from $x_0 \in C$, the Frank–Wolfe algorithm builds inductively a sequence of points of C as follows: given x_k, let

$$y_{k+1} \in \text{argmin}_{y \in C} \nabla f(x_k) \cdot y \tag{7.77}$$

and

$$x_{k+1} = x_k + \lambda_k(y_{k+1} - x_k) \text{ with } \lambda_k := \frac{2}{2+k}. \tag{7.78}$$

1) *Show that (7.76) admits at least a solution x^* and give a necessary and sufficient condition characterizing the solutions of (7.76).*
2) *Show that the Frank–Wolfe algorithm is well-defined.*
3) *Give an explicit solution of (7.77) (of course depending on $\nabla f(x_k)$) in the following cases: C is the euclidean unit ball, $C := \{x \in \mathbb{R}_+^d, \sum_{i=1}^d x_i = 1\}$, $C := \prod_{i=1}^d [a_i, b_i]$.*
4) *Set $\theta_k := f(x_k) - f(x^*)$, and show that*

$$\theta_k \leq \nabla f(x_k) \cdot (x_k - y_{k+1}).$$

5) *Show that*

$$\theta_{k+1} - \theta_k \le \lambda_k \nabla f(x_k)(y_{k+1} - x_k) + \frac{\lambda_k^2 M}{2} |y_{k+1} - x_k|^2.$$

6) *Use the previous inequalities (and the explicit choice of the step λ_k in (7.78)) to deduce that*

$$\theta_{k+1} - \frac{k}{k+2}\theta_k \le \frac{2M \operatorname{diam}(C)^2}{(2+k)^2}$$

and then (multiplying the previous inequality by $(k+1)(k+2)$) that

$$\theta_k \le \frac{2M \operatorname{diam}(C)^2}{k+1}.$$

7) *Show that, if, in addition, f is strictly convex, then x_k converges to the solution of (7.76). Show that the iterates of the Frank–Wolfe algorithm need not converge without assuming strict convexity of f.*

Exercise 7.15. *Let $T \colon \mathbb{R}^d \to \mathbb{R}^d$ be a firmly non-expansive map, i.e.,*

$$(T(x) - T(y)) \cdot (x - y) \ge \|T(x) - T(y)\|^2, \quad \forall (x, y) \in \mathbb{R}^d \times \mathbb{R}^d.$$

Let us denote by $\operatorname{Fix}(T)$ the set of fixed-points of T:

$$\operatorname{Fix}(T) := \{x \in \mathbb{R}^d : x = T(x)\}$$

and assume that $\operatorname{Fix}(T) \neq \emptyset$. Finally, given $x_0 \in \mathbb{R}^d$ consider the sequence of iterates T defined inductively $x_{n+1} = T(x_n)$.

1) *Show that for every $x \in \operatorname{Fix}(T)$, the sequence $\|x_n - x\|$ is non-increasing.*
2) *Show that (x_n) converges to a point of $\operatorname{Fix}(T)$.*

Exercise 7.16. *Let $T \colon \mathbb{R} \to \mathbb{R}$. Show the equivalence*

$$T \text{ is firmly non-expansive} \iff T \text{ is monotone and 1-Lipschitz.}$$

Let $R \colon \mathbb{R}^2 \to \mathbb{R}^2$ be defined by $R(x, y) = (-y, x)$. Show that R is monotone and 1-Lipschitz but is not firmly non-expansive.

Exercise 7.17. *Let E be a Hilbert space and $A \colon E \to E$ be continuous and monotone (i.e., $(A(x) - A(y)) \cdot (x - y) \ge 0, \forall (x, y) \in E^2$).*

1) *Give an example in $E = \mathbb{R}^2$ of monotone and continuous map which is not a gradient.*

2) *Let v in A. Show that $A^{-1}(\{v\})$ is convex whenever non-empty.*
3) *Let E be a Hilbert space and $T : E \to E$ be 1-Lipschitz. Show that if Fix(T) is non-empty it is convex.*

Exercise 7.18. *For every $x \in \mathbb{R}^d$ set*

$$\|x\|_2 := \sqrt{\sum_{i=1}^{d} x_i^2}, \quad \|x\|_\infty := \max_{i=1,\dots,d} |x_i|,$$

compute $\text{prox}_{\|\cdot\|_2^2}$ *and* $\text{prox}_{\|\cdot\|_\infty}$.

Exercise 7.19 (Moreau's proximal identity). *Let E be a Hilbert space, $f \in \Gamma_0(E)$. Show that for every $x \in E$*

$$\text{prox}_f(x) + \text{prox}_{f^*}(x) = x.$$

Let $\varepsilon > 0$, f_ε be the Moreau–Yosida regularization of f (see Exercise 3.28) and let $(x, y) \in E^2$. Show that

$$f_\varepsilon(x) = \frac{1}{2\varepsilon} \|x - y\|^2 + f(y) \iff y = \text{prox}_{\varepsilon f}(x)$$

and that $\nabla f_\varepsilon = \frac{1}{\varepsilon}(\text{id} - \text{prox}_{\varepsilon f})$.

Exercise 7.20. *Let E be a Hilbert space, $f_i \in \Gamma_0(E)$ for $i = 1, \dots, m$, we are interested in*

$$\inf \left\{ \sum_{i=1}^{m} f_i(x), \ x \in E \right\}. \tag{7.79}$$

For $\overline{x} = (x_1, \dots, x_m) \in E^m$ (E^m being equipped with its standard product Hilbertian structure) define

$$f(\overline{x}) := \sum_{i=1}^{m} f_i(x_i)$$

and

$$D := \{ \overline{x} = (x_1, \dots, x_m) \in E^m : x_1 = x_2 \cdots = x_m \}, \quad g := \chi_D.$$

1) *Show that (7.79) is equivalent to*

$$\inf_{\overline{x} \in E^m} \{ f(\overline{x}) + g(\overline{x}) \}.$$

2) *Show that* $\mathrm{prox}_f(\overline{x}) = (\mathrm{prox}_{f_1}(x_1), \dots, \mathrm{prox}_{f_m}(x_m))$.
3) *Compute* prox_g.
4) *Describe a splitting scheme for (7.79) and give conditions which guarantee the convergence of this scheme.*

Exercise 7.21 (Proximal average, see [12]). *Let E be a Hilbert space,* $f_1, \dots, f_k \in \Gamma_0(E)^k$, $g := \frac{1}{2}\| \cdot \|^2$, $\lambda := (\lambda_1, \dots, \lambda_k) \in \mathbb{R}_+^k$ *such that* $\sum_{i=1}^k \lambda_i = 1$ *and set for all $x \in E$,*

$$g_\lambda(x) := \inf \left\{ \sum_{i=1}^k \lambda_i(f_i(x_i) + g(x_i)), \ (x_1, \dots, x_k) \in E^k : \sum_{i=1}^k \lambda_i x_i = x \right\}$$
$$-g(x).$$

1) *Show that g_λ never takes the value $-\infty$, is convex and proper.*
2) *Show that* $\mathrm{prox}_{g_\lambda} = \sum_{i=1}^k \lambda_i \mathrm{prox}_{f_i}$.

Chapter 8

When Optimization and Data Meet

The aim of this chapter is to introduce some classic (but still important in the active field of machine learning) methods for data processing. The emphasis will of course be on optimization (and matrix) techniques and, at the risk of disappointing some statistically oriented readers, we will almost never explicitly refer to any of the underlying probabilistic models (which are of chief importance in data analysis but beyond the scope of this book). The chapter is organized as follows. Section 8.1 is devoted to Principal-component analysis, a basic technique to represent data in high dimensions using only (but in an optimal way) few dimensions. In Section 8.2, we consider optimization methods for linear systems which may be either overde-termined (more independent equations than unknowns) or underdetermined (more unknowns than independent equations), such methods are basic tools for inverse problems and linear regressions. Finally, Section 8.3 presents two popular classification techniques, the first one is based on logistic regression and the second one, on a more geometric approach, called support-vector machines.

8.1. Principal component analysis

Imagine that for a population of n individuals, one has access to a large number m of data (for instance, income, expenditures in a large list of categories, e.g., food, clothing, cosmetics, insurance, public transport). One can think of such data sets as a cloud of n points in an m-dimensional space. Of course, one cannot visualize directly such point clouds if $m \geq 3$, and it is important to know how to represent these data using a (much) smaller number of components than m. Principal component analysis (PCA) is a classic method to represent datasets consisting of a cloud of points in $\mathbb{R}^m$ using only $l < m$ components of the observations without losing too much information. The principle is to look for an orthogonal projection of rank l which explains best the variability of the point cloud. This is an optimization problem which is related to the so-called *singular value decomposition* (SVD) of the matrix containing the data.

8.1.1. *Singular value decomposition*

Definition 8.1. Let $A \in \mathcal{M}_{n \times m}(\mathbb{R})$. The singular values of A are the square roots of the eigenvalues of the ($m \times m$ symmetric positive semidefinite) matrix $A^T A$.

Since $A^T A$ is symmetric, it has an orthonormal basis of eigenvectors, $u_1, \ldots, u_m$ so that denoting by U the $m \times m$ orthogonal matrix whose columns are $u_1, \ldots, u_m$, we have

$$A^T A = U^{-1} \text{diag}(\sigma_1^2, \ldots, \sigma_m^2) U = U^T \text{diag}(\sigma_1^2, \ldots, \sigma_m^2) U, \qquad (8.1)$$

where $\text{diag}(\sigma_1^2, \ldots, \sigma_m^2)$ is the diagonal matrix with entries $\sigma_1^2, \ldots, \sigma_m^2$ on its diagonal. The decomposition (8.1) can also be written in terms of the orthogonal projection $u_j u_j^T$ onto u_j as

$$A^T A = \sum_{j=1}^m \sigma_j^2 u_j u_j^T. \qquad (8.2)$$

Note that since A and $A^T A$ have the same nullspace, by the rank-nullity theorem they have the same rank so that the rank of A coincides with the number of non-zero singular values of A (in particular there are no more than n non-zero singular values of A). We shall always take the convention that the singular values are sorted in a non-increasing way: $\sigma_1 \geq \cdots \geq \sigma_m$ so that if we denote by $r \leq \min(n, m)$ the rank of A we have

$$\sigma_j > 0 \quad \text{for} \quad j \leq r, \ \sigma_j = 0, \ \text{for} \quad j > r. \qquad (8.3)$$

The matrix A then admits a singular value decomposition as expressed in the following:

Theorem 8.1 (Singular value decomposition). *Let $A \in \mathcal{M}_{n \times m}(\mathbb{R})$. Then there exist an $n \times n$ orthogonal matrix V, an $m \times m$ orthogonal matrix U such that*

$$A = V^T \Sigma U, \qquad (8.4)$$

where Σ is the $n \times m$ matrix with zero entries outside its diagonal and having the singular values of A as entries on its diagonal.

Proof. As above, we denote by $r \leq \min(m, n)$ the rank of A, take the convention (8.3) and choose an orthonormal basis $u_1, \ldots, u_m$ of eigenvectors of $A^T A$: $A^T A u_i = \sigma_i^2 u_i$. For $i \leq r$, set $v_i := \frac{1}{\sigma_i} A u_i$ so that $A u_i = \sigma_i v_i$ and for i, j in $\{1, \ldots, r\}^2$ observe that[a]

$$v_i \cdot v_j = v_i^T v_j = \frac{1}{\sigma_i \sigma_j} A u_i \cdot A u_j = \frac{1}{\sigma_i \sigma_j} A^T A u_i \cdot u_j$$

$$= \frac{\sigma_i}{\sigma_j} \sigma_i^2 u_i \cdot u_j = \frac{\sigma_i}{\sigma_j} \delta_{ij} = \delta_{ij}$$

so that $\{v_1, \ldots, v_r\}$ is an orthonormal family of vectors of $\mathbb{R}^n$. Let us complete, if necessary, this family so as to obtain an orthonormal basis $\{v_1, \ldots, v_n\}$ of $\mathbb{R}^n$. By construction for $i \leq r$, we have $A u_i = \sigma_i v_i$ and for $i > r$, $A^T A u_i = 0$ so that (taking the scalar product with u_i) $A u_i = 0 = \sigma_i v_i$. Denoting by V the $n \times n$ matrix having $\{v_1, \ldots, v_n\}$ as columns, V is orthogonal and the fact that $A u_i = \sigma_i v_i$ for $i = 1, \ldots, m$ directly gives the singular value decomposition (8.4). $\qquad \square$

Exercise 8.1 (Singular values as critical values). *Let $A \in \mathcal{M}_{n \times m}(\mathbb{R})$, consider the optimization problem*

$$\inf \left\{ A u \cdot v = u^T A^T v, \; u \in \mathbb{R}^m, \; \sum_{j=1}^{m} u_j^2 = 1, \; v \in \mathbb{R}^n, \; \sum_{i=1}^{n} v_i^2 = 1 \right\}. \quad (8.5)$$

1) *Show that (8.5) admits solutions and that if u, v solve (8.5) then u is an eigenvector of $A^T A$. (Hint: use Lagrange multipliers.)*
2) *Show that the value of (8.5) is the smallest singular value of A.*
3) *Find an optimization problem whose value gives the largest singular value of A.*

Exercise 8.2 (SVD and left/right inversion).

1) *Let $u \in \mathcal{L}(\mathbb{R}^m, \mathbb{R}^n)$ assume that u has rank m. Show that there exists $v \in \mathcal{L}(\mathbb{R}^n, \mathbb{R}^m)$ such that $v \circ u = \mathrm{id}_{\mathbb{R}^m}$. (Hint: take a basis of $\mathbb{R}^m$ and look at its image by u.)*

---

[a] δ_{ij} is the usual Kronecker symbol: $\delta_{ij} = \begin{cases} 1 & \text{if } i = j, \\ 0 & \text{otherwise.} \end{cases}$

2) Let $u \in \mathcal{L}(\mathbb{R}^m, \mathbb{R}^n)$ assume that u has rank n, show that there exists $v \in \mathcal{L}(\mathbb{R}^n, \mathbb{R}^m)$ such that $u \circ v = \mathrm{id}_{\mathbb{R}^n}$. (Hint: take a basis of $\mathbb{R}^n$ and look at its preimage by u.)

3) Let $A \in \mathcal{M}_{n \times m}(\mathbb{R})$ be a matrix a rank m, use the SVD of A, to find a left-inverse of A, i.e., a matrix $B \in \mathcal{M}_{m \times n}(\mathbb{R})$ such that $BA = I_m$. (Hint: find explicitly a left-inverse of the diagonal rectangular matrix Σ.)

4) Let $A \in \mathcal{M}_{n \times m}(\mathbb{R})$ be a matrix a rank n, use the SVD of A, to find a right-inverse of A, i.e., a matrix $B \in \mathcal{M}_{m \times n}(\mathbb{R})$ such that $AB = I_n$.

8.1.2. Principal component analysis

Let us now come back to our data processing purpose. Given a population of n individuals, we observe m real-valued characteristics or variables (age, weight, income, food, clothing, health products consumptions, etc.). These data can be stored in a matrix $X \in \mathcal{M}_{n \times m}(\mathbb{R})$ where each row i corresponds to an individual observation and each column corresponds to the observation of one variable in the population. The (empirical) mean of variable $j \in \{1, \dots, m\}$ is

$$\overline{x}_j = \frac{1}{n} \sum_{i=1}^{n} X_{ij}$$

and its (empirical) standard deviation is

$$s_j := \sqrt{\frac{1}{n} \sum_{i=1}^{n} (X_{ij} - \overline{x}_j)^2},$$

which captures the variability of the jth variable in our population of n individuals. The data can be expressed in different units for the different variables (for instance, some variables can be expressed in grams and others in tons). It is therefore often desirable to rescale them, or even to transform them into dimensionless and centered quantities. One possible (but not the only one of course) way to do so is to replace X_{ij} by

$$Y_{ij} = \frac{X_{ij} - \overline{x}_j}{s_j}$$

provided that $s_j > 0$ (which means that variable j is not constant in the population).

Once one has centered,[b] and possibly rescaled, X and obtained a new matrix $Y \in \mathcal{M}_{n \times m}(\mathbb{R})$ which is centered, i.e., has zero column sums:

$$\sum_{i=1}^{n} Y_{ij} = 0, \ j = 1, \ldots, m.$$

The goal of PCA is to approximate Y in a sparse way that is with l variables (which are linear combinations of the original variables) where $l < m$ is fixed. In other words, we have now, n vectors of $\mathbb{R}^m$. These vectors sum to zero, and we would like to project orthogonally these vectors onto an l-dimensional subspace of $\mathbb{R}^m$ so as to maximize the variability of the projected variables. Let $P \in \mathcal{M}_m(\mathbb{R})$ be the matrix of an orthogonal projector, i.e.,

$$P^T = P = PP^T$$

so that P is the matrix of the orthogonal projection onto its range and $I_m - P$ is the orthogonal projection onto its nullspace. Denoting by y_i the transpose of the ith row of Y, since Py_i and $y_i - Py_i$ are orthogonal, we have

$$\|Y\|_2^2 := \mathrm{tr}(YY^T) = \sum_{i=1}^{n} |y_i|^2 = \sum_{i=1}^{n} |Py_i|^2 + \sum_{i=1}^{n} |y_i - Py_i|^2, \qquad (8.6)$$

i.e., the squared norm of the matrix Y can be decomposed as the sum of the squared norm of the approximation PY^T

$$\sum_{i=1}^{n} |Py_i|^2 = \mathrm{tr}(PY^TYP^T) = \mathrm{tr}(P^TPY^TY) = \mathrm{tr}(PY^TY)$$

and the sum of the squared residuals (or errors)

$$\sum_{i=1}^{n} |y_i - Py_i|^2 = \mathrm{tr}((I_m - P)Y^TY(I_m - P)T) = \mathrm{tr}((I_m - P)Y^TY).$$

So, given a fixed rank $l < m$, PCA amounts to the maximization problem

$$\sup\{\mathrm{tr}(PY^TY) : P^TP = P = P^T, \ \mathrm{rank}(P) = l\} \qquad (8.7)$$

---

[b]We let the reader check that if we consider the variant of (8.8) where instead of looking for an optimal linear projection, we look for an optimal *affine* projection, centering the data is actually optimal.

which, with the decomposition formula (8.6), is equivalent to the best approximation problem

$$\inf \left\{ \sum_{i=1}^{n} |y_i - Py_i|^2 : P^T P = P = P^T, \ \text{rank}(P) = l \right\}. \tag{8.8}$$

Now it is convenient to represent orthogonal projection matrices of rank l as follows. If P is such a matrix and $\{a_1, \ldots, a_l\}$ is an orthonormal basis of the range of P, then

$$P = \sum_{k=1}^{l} a_k a_k^T. \tag{8.9}$$

Conversely, if $\{a_1, \ldots, a_l\}$ is an orthonormal family in $\mathbb{R}^m$, P given by (8.9) is the matrix of the orthogonal projector onto the vector space spanned by $\{a_1, \ldots, a_l\}$. One can therefore rewrite (8.7) as

$$\sup_{(a_1, \ldots, a_l) \in (\mathbb{R}^m)^l} \left\{ \sum_{i=1}^{n} \sum_{k=1}^{l} |a_k^T y_i|^2 : a_k^T a_l = \delta_{kl}, \ (k,l) \in \{1, \ldots, l\}^2 \right\}. \tag{8.10}$$

It is in particular clear from (8.10) that (8.7) admits solutions (in (8.10), we are maximizing a continuous function onto a non-empty compact subset of $(\mathbb{R}^m)^l$). One can explicitly solve (8.7) in terms of the singular values and the SVD of Y as follows.

Theorem 8.2. *Let $\sigma_1 \geq \cdots \geq \sigma_m$ be the singular values of Y written in decreasing order, and let $u_1, \ldots, u_m$ be orthonormal eigenvectors of $Y^T Y$ associated with the eigenvalues $\sigma_1^2, \ldots, \sigma_m^2$. The orthogonal projector on the space spanned by l first eigenvectors $\{u_1, \ldots, u_k\}$*

$$P_l^* := \sum_{k=1}^{l} u_k u_k^T$$

is optimal for (8.7). Moreover, the value of (8.7) is $\sum_{i=1}^{l} \sigma_i^2$.

Proof. Let us first write the diagonalization of $Y^T Y$ as

$$Y^T Y = \sum_{i=1}^{m} \sigma_i^2 u_i u_i^T$$

and the orthogonal projection matrix of rank l, P as

$$P = \sum_{k=1}^{l} a_k a_k^T,$$

where $(a_1, \ldots, a_m)$ is an arbitrary orthogonal basis of $\mathbb{R}^m$. Then the quantity we seek to maximize in (8.7) reads as

$$\operatorname{tr}(PY^TY) = \sum_{i=1}^{m} \sigma_i^2 \operatorname{tr}(Pu_i u_i^T) = \sum_{i=1}^{m} \sigma_i^2 \sum_{k=1}^{l} \operatorname{tr}(a_k a_k^T u_i u_i^T)$$

$$= \sum_{i=1}^{m} \sigma_i^2 \sum_{k=1}^{l} a_k^T u_i \operatorname{tr}(a_k u_i^T) = \sum_{i=1}^{m} \sigma_i^2 \sum_{k=1}^{l} (a_k^T u_i)^2.$$

Let us then set

$$\gamma_{ik} := (a_k \cdot u_i)^2 = (a_k^T u_i)^2, \quad (i,k) \in \{1, m\}^2$$

and observe that since both $(u_1, \ldots, u_m)$ and $(a_1, \ldots, a_m)$ are orthonormal bases, one has

$$\sum_{i=1}^{m} \gamma_{ik} = |a_k|_2^2 = 1, \quad \sum_{k=1}^{m} \gamma_{ik} = |u_i|_2^2 = 1$$

so that the matrix γ with entries γ_{ik} is *bistochastic*, i.e., has non-negative entries with its row and column sums[c] all equal to 1. Setting

$$\alpha_k := \begin{cases} 1 & \text{if } k = 1, \ldots, l, \\ 0 & \text{if } k = l+1, \ldots, m, \end{cases}$$

we can rewrite $\operatorname{tr}(PY^TY)$ as

$$\operatorname{tr}(PY^TY) = \sum_{i=1}^{m} \sum_{k=1}^{m} \sigma_i^2 \alpha_k \gamma_{ik}. \tag{8.11}$$

Next set

$$\varphi_i := (\sigma_i^2 - \sigma_l^2)_+, \quad i = 1, \ldots, m$$

---

[c]Recall that we have already studied these matrices in the context of the assignment problem in Chapter 6.

and

$$\psi_k := \begin{cases} \sigma_l^2 & \text{if } k = 1, \ldots, l, \\ 0 & \text{if } k = l+1, \ldots, m. \end{cases}$$

It is straightforward to check that for every i and k in $\{1, \ldots, m\}$, one has

$$\varphi_i + \psi_k \geq \sigma_i^2 \alpha_k.$$

Multiplying this inequality by γ_{ik}, summing and using the fact that γ is bistochastic, we get

$$\text{tr}(PY^TY) = \sum_{i=1}^{m} \sum_{k=1}^{m} \sigma_i^2 \alpha_k \gamma_{ik}$$

$$\leq \sum_{i=1}^{m} \varphi_i + \sum_{k=1}^{m} \psi_k = \sum_{i=1}^{l} \sigma_i^2.$$

But if we take the projector $P_l^* := \sum_{k=1}^{l} u_k u_k^T$, then

$$\text{tr}(P_l^* Y^T Y) = \sum_{i=1}^{l} \sigma_i^2,$$

which shows the optimality of P_l^* and the fact that the value of (8.7) is the sum of the l largest eigenvalues of Y^TY. $\qquad\square$

Exercise 8.3. *Give an alternative proof of Theorem 8.2, starting from (8.11) and using the Birkhoff–von Neumann theorem (see Theorem 6.15).*

Remark 8.1. The vectors $u_1, \ldots, u_l$ represent (orthogonal) linear combinations of the initial variables and are called the principal components. One could also have defined the principal components sequentially: u_1 is obtained by maximizing $u^T Y^T Y u$ among unit vectors, then u_2 is obtained by maximizing $u^T Y^T Y u$ among unit vectors, orthogonal to u_1, and at step k, u_k is found as a maximizer of $u^T Y^T Y u$ among unit vectors, orthogonal to $u_1, \ldots, u_{k-1}$.

Remark 8.2. The quality of the representation of the data using only an optimal rank l projector as described above can be measured by the ratio

$$\frac{\text{tr}(P_l^* Y^T Y)}{\text{tr}(Y^T Y)} = \frac{\sum_{i=1}^{l} \sigma_i^2}{\sum_{i=1}^{m} \sigma_i^2}.$$

This quantity expresses which percentage of the variance of the data is explained by the first l principal components. The choice of l in practice should somehow result from a balance between sparsity and the quality of the previous ratio.

8.2. Minimization for linear systems

Solving linear systems with m unknowns $x \in \mathbb{R}^m$ and n equations written in matrix form as

$$Ax = b \qquad (8.12)$$

for some $A \in \mathcal{M}_{n \times m}(\mathbb{R})$ and a right-hand side $b \in \mathbb{R}^n$ is of course a fundamental issue which is also of great importance in applications. Even in the ideal situation where $n = m$ and A is non-singular so that (8.12) admits $x = A^{-1}(b)$ a unique solution, it may be the case in practice that the right-hand side b is not known perfectly (because of measurement or rounding errors) and one has to wonder whether such small errors will result in small errors as well in the solution of the system. This is related to the conditioning number of the matrix A, which we will discuss in the next section. Now, if the system (8.12) has no solution, i.e., if b is not in the range of A (that is the vector space spanned by the columns of A), a reasonable way to find an approximate solution is by minimizing the error between Ax and b, and the most common way is to project b onto the range of A, this is the least squares method which is of particular importance in linear regressions (see Section 8.2.2) and very much related to the notions of Tikhonov regularization and Moore–Penrose inverses (see Section 8.2.3). Finally, in the opposite *underdetermined* situation where (8.12) has several (hence infinitely many) solutions, one has to select one according to some criterium, when m is much larger than n and A has rank n, one would like to select a *sparse* solution, i.e., a solution with few non-zero components. We shall see in Section 8.2.4 that a way to achieve this sparsity goal is by l^1-minimization which is at the heart of popular methods such as the Lasso and the Basis Pursuit.

8.2.1. *Matrix operator norms and conditioning numbers*

Let us first recall the definition of operator matrix norms (which are the matrix analog of operator norms for linear continuous maps, as defined in Proposition 1.12).

Definition 8.2. Let m and n be in $\mathbb{N}^*$, M and N be norms on $\mathbb{R}^m$ and $\mathbb{R}^n$ respectively. The operator norm on $\mathcal{M}_{n \times m}(\mathbb{R})$ (subordinated to the norms M and N) is defined by

$$|||A||| := \sup\{N(Ax),\ x \in \mathbb{R}^m,\ M(x) \leq 1\},\ \forall A \in \mathcal{M}_{n \times m}(\mathbb{R}).$$

It is straightforward that an operator norm is indeed a norm, but operator norms have special properties that arbitrary norms on $\mathcal{M}_{n \times m}(\mathbb{R})$ generally do not have. By definition, one has

$$N(Ax) \leq |||A|||M(x),\quad \forall (x, A) \in \mathbb{R}^m \times \mathcal{M}_{n \times m}(\mathbb{R}).$$

Let now $p \in \mathbb{N}^*$, P be a norm on $\mathbb{R}^p$, and $B \in \mathcal{M}_{m \times p}$, if we denote by $|||B|||$ the operator norm of B subordinated to P and M and by $|||AB|||$ the operator norm of AB subordinated to P and N, then one has

$$|||AB||| \leq |||A|||\ |||B|||.$$

One therefore says that operator norms are *submultiplicative*. In particular if $m = n$ and $M = N$, $|||I_n||| = 1$ and therefore whenever A is a non-singular $n \times n$ matrix, taking $B = A^{-1}$ and using submultiplicativity, one has

$$|||I_n||| = 1 \leq |||A|||\ |||A^{-1}|||. \tag{8.13}$$

Let us now fix a norm N in $\mathbb{R}^n$ and consider $A \in \mathcal{M}_n(\mathbb{R})$ non-singular. We are interested in comparing the solution of the linear system (8.12) $Ax = b$ with the solution of the slightly perturbed problem

$$A(x + \delta x) = b + \delta b,$$

where $\delta b \in \mathbb{R}^n$ is a small perturbation of the right-hand side of the initial system (8.12). More precisely, we would like to have a bound on the relative error on the solution in terms of the relative error in the right-hand side; that is, we look for a comparison between

$$\frac{N(\delta x)}{N(x)}\quad \text{and}\quad \frac{N(\delta b)}{N(b)}. \tag{8.14}$$

Put differently, we look for a constant C such that

$$N(\delta x)N(b) \leq CN(x)N(\delta b) \tag{8.15}$$

for every b and δb (note that (8.15) makes sense when $b = x = 0$, which is not the case of the relative errors defined in (8.14)). Since $\delta x = A^{-1}(\delta b)$ and $b = Ax$, we have

$$N(\delta x) \leq |||A^{-1}|||N(\delta b) \quad \text{and} \quad N(b) \leq |||A|||N(x).$$

Therefore (8.15) holds for the constant $C = c(A)$ equal to the *conditioning number* of A (subordinated to the norm N) defined by

$$c(A) := |||A^{-1}||| \, |||A|||. \tag{8.16}$$

It follows from (8.13) that

$$c(A) \geq 1.$$

Matrices with a conditioning number close to 1 are called well-conditioned whereas matrices with a very large conditioning number are called ill-conditioned (and one should be careful with such matrices when it comes to approximate and/or iterate their inversion!).

Example 8.1. Consider the elementary example of a (diagonal!) linear system in two dimensions (and use the Euclidean norm $|\cdot|_2$ in $\mathbb{R}^2$):

$$A := \begin{pmatrix} 10 & 0 \\ 0 & 0.1 \end{pmatrix}, \; b := \begin{pmatrix} 1 \\ 0 \end{pmatrix}, \; \delta b := \begin{pmatrix} 0 \\ 0.01 \end{pmatrix}.$$

Then

$$x = \begin{pmatrix} 0.1 \\ 0 \end{pmatrix}, \; \delta x := \begin{pmatrix} 0 \\ 0.1 \end{pmatrix},$$

hence

$$\frac{|\delta x|_2}{|x|_2} = 1 = 100 \times \frac{|\delta b|_2}{|b|_2}$$

that is a relative error of 1% in the right-hand side of the system may result in a relative error of 100% in the solution of the system.

Exercise 8.4 (Conditioning and singular values). *Equip $\mathbb{R}^n$ with its usual euclidean norm $|\cdot|_2$ and $\mathcal{M}_n(\mathbb{R})$ with*

$$\|A\|_2 := \sup\{|Ax|_2 : x \in \mathbb{R}^n, \; |x|_2 \leq 1\}.$$

Let $A \in \mathcal{M}_n(\mathbb{R})$ be non-singular, and consider the condition number

$$c_2(A) := \|A\|_2 \|A^{-1}\|_2,$$

show that

$$c_2(A) = \frac{\sigma_{\max}}{\sigma_{\min}},$$

where $\sigma_{\max}$ and $\sigma_{\min}$ denote, respectively, the largest and smallest singular value of A. What are the non-singular matrices for which $c_2(A) = 1$?

8.2.2. *Least squares and linear regressions*

We again consider the $n \times m$ linear system (8.12) but for the moment do not make any assumption about its exact solvability. Recall that the null space and range of A are, respectively, defined by

$$N(A) := \{x \in \mathbb{R}^m : Ax = 0\}, \ R(A) := \{Ax, \ x \in \mathbb{R}^m\}$$

so that $N(A)$ and $R(A)$ are linear subspaces of $\mathbb{R}^m$ and $\mathbb{R}^n$, respectively. By definition, $\operatorname{rank}(A) = \dim(R(A))$ and it follows from the rank-nullity theorem that

$$m = \operatorname{rank}(A) + \dim(N(A)).$$

Denoting by $|\cdot|_2$ the usual Euclidean norm, the least squares approach to (8.12) consists in the minimization

$$\inf_{x \in \mathbb{R}^m} f(x) := \frac{1}{2}|Ax - b|_2^2 = \frac{1}{2}\sum_{i=1}^{n}((Ax)_i - b_i)^2. \qquad (8.17)$$

Of course, if $b \in R(A)$, the minimum of f is 0 and achieved precisely when x solves the system $Ax = b$. If, on the contrary, the system is not solvable, i.e., $b \notin R(A)$, then the least squares problem (8.17) should be seen as a surrogate for (8.12). Note that (8.17) is nothing but an orthogonal projection problem (see Proposition 1.16) in the sense that

$$x \in \mathbb{R}^m \text{ solves } (8.17) \iff Ax \text{ solves the projection problem } \inf_{y \in R(A)} |y - b|_2^2$$
$$(8.18)$$

and since $R(A)$ is closed[d] and convex, this shows at once that:

- (8.17) admits solutions,
- if x and $\tilde{x}$ solve (8.17) then $x - \tilde{x} \in N(A)$ (because the projection of b onto $R(A)$ is unique).

A more explicit characterization of solutions of (8.17) is as follows.

Lemma 8.1. *Let $x \in \mathbb{R}^m$. Then x solves (8.17) if and only if it solves the following linear system (called the normal equations for (8.17)):*

$$A^T A x = A^T b. \tag{8.19}$$

Proof. Note that f is convex and differentiable and

$$\nabla f(x) = A^T A x - A^T b.$$

Since minimizers of f coincide with its critical points, the desired result follows. □

One can directly derive the normal equations from the orthogonal projection problem in (8.18).

Exercise 8.5. *Equip $\mathbb{R}^m$ and $\mathbb{R}^n$ with their usual scalar product and denote by $A^T \in \mathcal{M}_{m \times n}(\mathbb{R})$ the transpose of A.*

- *Show that $R(A)^\perp = N(A^T)$ and $N(A)^\perp = R(A^T)$, and deduce from Proposition 1.16 that $x \in \mathbb{R}^m$ solves (8.17) if and only if $Ax - b \in R(A)^\perp$.*
- *Deduce from the previous question that x solves (8.17) if and only if it solves the normal equations (8.19).*

In the particular case where $\text{rank}(A) = m$, i.e., $N(A) = \{0\}$, the $m \times m$ matrix $A^T A$ is non-singular[e] and we get a more explicit result.

Corollary 8.1. *If $N(A) = \{0\}$, then (8.17) has a unique solution which is given by*

$$x = (A^T A)^{-1} A^T b. \tag{8.20}$$

---

[d]Note that in infinite-dimensional Hilbert spaces, continuous linear maps need not have a closed range, so least squares problems do not necessarily possess solutions, see Exercise 8.18.

[e]This is because if $x \in N(A^T A)$ then $0 = x^T A^T A x = |Ax|_2^2$ so that $x \in N(A) = \{0\}$.

Linear regression

Corollary 8.1 is particularly useful in the framework of *linear regression*. In this context, one has at disposal n individual observations for a set of $m + 1$ (real-valued) variables and one wishes to estimate a certain linear relation between a variable of interest and the m other ones (called the explanatory variables). For $i = 1, \ldots, n$, denoting $(x_i^1, \ldots, x_i^m)$ the values of the explanatory variables and by y_i the value of the variable of interest for the ith observation, linear regression aims at fitting the observed values of $y = (y_1, \ldots, y_n)^T$ with a linear combination of the explanatory variables that is a vector of the form $X\beta$ where X is the $n \times m$ matrix whose ith row is $(x_i^1, \ldots, x_i^m)$ and $\beta \in \mathbb{R}^m$ is a vector of coefficients. Determining the coefficients β of the regression so as to minimize the sum of the squared errors between the observation y_i and the prediction of the linear model $\sum_{k=1}^m x_i^k \beta_k$ leads to the least squares problems[f]

$$\inf_{\beta \in \mathbb{R}^m} |y - X\beta|_2^2 := \sum_{i=1}^n (y_i - (X\beta)_i)^2. \tag{8.21}$$

It is often the case in practice that $n \geq m$ (more observations than variables) and that the columns of X are linearly independent (otherwise that would mean that some explanatory variables are redundant and thus could be removed), in which case, thanks to Corollary 8.1, (8.21) has a unique solution $\hat{\beta}$ (least square estimator of the regression coefficients) given by the following formula:

$$\hat{\beta} = (X^T X)^{-1} X^T y.$$

Least squares with SVD

We now return to the least squares problem (8.17) in the general case where $\text{rank}(A) = r \leq m$. The SVD of A from Theorem 8.1 reads $A = V^T \Sigma U$ where V is an $n \times n$ orthogonal matrix, U is an $m \times m$ orthogonal matrix and Σ is an $n \times m$ diagonal matrix having the singular values of A on its diagonal (with the convention $\sigma_1 \geq \cdots \geq \sigma_r > 0$ and $\sigma_j = 0$ for $j = r + 1, \ldots, m$). Let $x \in \mathbb{R}^m$, since V is orthogonal we have

$$|Ax - b|_2^2 = |V^T \Sigma U x - b|_2^2 = |V^T \Sigma U x - V^T V b|_2^2 = |\Sigma U x - V b|_2^2$$

---

[f] I have chosen on purpose notations which differ from those in (8.17).

so setting

$$\widetilde{x} = Ux, \ \widetilde{b} = Vb,$$

we have

$$|Ax - b|_2^2 = \sum_{i=1}^{r} (\sigma_i \widetilde{x}_i - \widetilde{b}_i)^2 + \sum_{i=r+1}^{n} \widetilde{b}_i^2$$

and the latter quantity is obviously minimized when

$$\widetilde{x}_i = \frac{\widetilde{b}_i}{\sigma_i} \quad \text{for } i = 1, \dots, r$$

(no condition on $\widetilde{x}_i$ for $i = r+1, \dots, m$). In particular, choosing $\widetilde{x}_i = 0$ for $i = r+1, \dots, m$, and denoting by u_i the ith column of U, we have that

$$x := \sum_{i=1}^{r} \frac{\widetilde{b}_i}{\sigma_i} u_i = \sum_{i=1}^{r} \frac{(Vb)_i}{\sigma_i} u_i \tag{8.22}$$

solves the least squares problem (8.17) (and, of course, any solution of (8.17) is obtained by adding to this particular x an element of N(A)).

8.2.3. *Tikhonov regularization, the Moore–Penrose inverse*

When the least squares problem (8.17) has several solutions, one may wonder if there is a canonical way to select a distinguished one. Given $\delta > 0$ a regularization parameter, consider the Tikhonov regularization of (8.17):

$$\inf_{x \in \mathbb{R}^m} \frac{1}{2} |Ax - b|_2^2 + \frac{\delta}{2} |x|_2^2, \tag{8.23}$$

which admits as unique solution:

$$x_\delta = A_\delta^+ b, \ A_\delta^+ := (\delta I_m + A^T A)^{-1} A^T. \tag{8.24}$$

The solution of (8.17) selected by Tikhonov regularization when letting the regularization parameter vanish is the one of minimal norm.

Lemma 8.2. *Let x_δ be the solution of (8.24). Then x_δ converges as $\delta \to 0^+$ to the solution of*

$$\inf_{x \in \mathbb{R}^m} \frac{1}{2} |x|_2^2 : |Ax - b|_2 = \text{dist}(b, \text{R}(A)), \tag{8.25}$$

where $\text{dist}(b, \text{R}(A)) := \min_{z \in \mathbb{R}^m} |Az - b|_2$.

Proof. Let us denote by $\overline{x}$ the solution of (8.25), using the fact that $|Ax_\delta - b|_2 \geq \mathrm{dist}(b, \mathrm{R}(A)) = |A\overline{x} - b|_2$, the optimality of x_δ for (8.24) gives $|x_\delta|_2 \leq |\overline{x}|_2$, in particular x_δ is bounded and $\delta|x_\delta|^2 \to 0$, so using

$$\delta|x_\delta|_2^2 + |Ax_\delta - b|_2^2 \leq \delta|\overline{x}|_2^2 + |A\overline{x} - b|_2^2$$

and $|x_\delta|_2 \leq |\overline{x}|_2$, we see that if y is a cluster point of x_δ as $\delta \to 0^+$, one should have $|Ay - b|_2 = \mathrm{dist}(b, \mathrm{R}(A))$ and $|y|_2 \leq |\overline{x}|_2$ which implies that $y = \overline{x}$ and convergence of x_δ to $\overline{x}$ as $\delta \to 0^+$.   □

The previous lemma in particular implies the convergence of the matrix A_δ^+, we then set:

$$A^+ := \lim_{\delta \to 0^+} (\delta I_m + A^T A)^{-1} A^T. \tag{8.26}$$

Note that, again by the previous lemma, A^+b is the orthogonal projection of 0 onto the set of solutions of (8.17) which is given by the normal equations (8.19). It is therefore immediate to check that A^+b is the only $x \in \mathbb{R}^m$ which:

- solves (8.19), i.e., $A^T A x = A^T b$, and
- belongs to $\mathrm{N}(A^T A)^\perp = \mathrm{R}(A^T A)$.

This yields the following characterization of A^+:

$$A^T A A^+ = A^T, \ \mathrm{R}(A^+) \subset \mathrm{R}(A^T A). \tag{8.27}$$

The matrix $A^+ \in \mathcal{M}_{m \times n}(\mathbb{R})$ is called the Moore–Penrose inverse of A and is fully characterized as follows.

Theorem 8.3 (Moore–Penrose inverse). *The matrix*

$$A^+ := \lim_{\delta \to 0^+} (\delta I_m + A^T A)^{-1} A^T$$

satisfies the following conditions:

$$AA^+A = A, \ A^+AA^+ = A^+, \ AA^+ \ and \ A^+A \ are \ symmetric. \tag{8.28}$$

Moreover, if $B \in \mathcal{M}_{m \times n}(\mathbb{R})$ satisfies

$$ABA = A, \ BAB = B, \ AB \ and \ BA \ are \ symmetric, \tag{8.29}$$

then $B = A^+$.

Proof. Since AA_δ^+ and $A_\delta^+ A = I_m - \delta(\delta I_m + A^T A)^{-1}$ are symmetric, the symmetry of AA_+ and A_+A is obtained by letting $\delta \to 0^+$. Taking the transpose of the identity $A^T AA^+ = A^T$ from (8.27) and using the symmetry of AA^+, we then obtain $AA^+ A = A$. Finally, to show that $A^+ AA^+ = A^+$, we first observe that $\mathrm{R}(A^+ AA^+) \subset \mathrm{R}(A^+) \subset \mathrm{R}(A^T A)$, we then use the identities $AA^+ A = A$ and $A^T AA^+ = A^T$ to get

$$A^T A(A^+ AA^+) = A^T AA^+ = A^T$$

so that $A^+ AA^+$ satisfies the two requirements in (8.27) and therefore coincides with A^+. Let now $B \in \mathcal{M}_{m \times n}(\mathbb{R})$ satisfy (8.29). Then, transposing $ABA = A$ and using the symmetry of AB, we get $A^T AB = A^T$. In a similar way, we obtain $B^T = B^T A^T B^T = B^T BA$ so that $\mathrm{N}(A) = \mathrm{N}(A^T A) \subset \mathrm{N}(B^T)$, passing to the orthogonals, this yields

$$\mathrm{R}(B) = \mathrm{N}(B^T)^\perp \subset \mathrm{N}(A^T A)^\perp = \mathrm{R}(A^T A).$$

Thanks to (8.27), we can conclude that $B = A^+$. $\qquad\square$

8.2.4. l^1-penalization and sparse solutions

Basis Pursuit for the underdetermined case

We now consider the linear system (8.12) $Ax = b$ with $A \in \mathcal{M}_{n \times m}(\mathbb{R})$, $b \in \mathbb{R}^n$ but in the undetermined case where

$$\mathrm{rank}(A) = n < m. \qquad (8.30)$$

So that (8.12) admits infinitely many solutions. As already mentioned, in this underdetermined situation, we would like to select sparse solutions, one could do so by minimizing the number of non-zero components of x:

$$\inf_{x \in \mathbb{R}^m : Ax = b} |x|_0 := \#\{i \in \{1, \ldots, m\} : x_i \neq 0\} \qquad (8.31)$$

(where $\#A$ denotes the cardinality of a finite set A). The problem is that $|\cdot|_0$ is highly non-convex (and not even continuous). As we shall see, a quite good surrogate (which can be viewed as some sort of convex relaxation) of $|\cdot|_0$ is given by the l^1-norm:

$$|x|_1 := \sum_{i=1}^m |x_i|.$$

The *Basis Pursuit* method (pioneered by Chen, Donoho and Saunders, see [34]) consists in looking for solutions of (8.12) of minimal l^1-norm, i.e,

$$\inf_{x \in \mathbb{R}^m \,:\, Ax=b} |x|_1. \tag{8.32}$$

Note that (8.32) is a convex minimization problem with (feasible, thanks to (8.30)) affine constraints, by coercivity of $|\cdot|_1$ it clearly possesses solutions (nonunique in general). In fact, (8.32) can be reformulated as a linear-programming problem.

Exercise 8.6. *Consider the LP problem*

$$\inf \left\{ \sum_{i=1}^{m} (y_i + z_i) : y_i \geq 0, \ z_i \geq 0, \ A(y - z) = b \right\}. \tag{8.33}$$

1) *Show that (8.33) admits solutions and that if (y, z) solves (8.33) then $\sum_{i=1}^{n} y_i z_i = 0$. (Hint: replace (y_i, z_i) by $((y_i - z_i)_+, (y_i - z_i)_-)$.)*
2) *Show that (8.32) and (8.33) are equivalent in the sense that x solves (8.32) if and only if (x_+, x_-) (the vector of componentwise positive/negative parts) solves (8.33).*
3) *Write the dual LP program of (8.33) and deduce necessary and sufficient optimality conditions for (8.32).*

The fact that l^1-minimization induces sparsity (see the recent work of Boyer *et al.* [20] for very interesting extensions) is expressed in the next.

Theorem 8.4. *Assume (8.30), then (8.32) admits a solution $\bar{x}$ such that*

$$|\bar{x}|_0 \leq n.$$

In particular, the system (8.12) admits a solution with no more than n non-zero components.

Proof. Let r be the value of (8.32), if $r = 0$, $b = 0$ and there is nothing to prove. We therefore assume that $r > 0$ and that x solves (8.32), in particular $|x|_1 = r > 0$ and by optimality

$$B_1(r) := \{ y \in \mathbb{R}^m : |y|_1 < r \} \text{ is disjoint from } x + N(A).$$

By the Hahn–Banach separation theorem, there exists $q \in \mathbb{R}^m \setminus \{0\}$ such that

$$\sup_{y \in B_1(r)} q^T y \leq \inf_{h \in N(A)} q^T(x + h). \tag{8.34}$$

In particular, this implies that the linear form $h \mapsto q^T h$ is lower bounded hence identically 0 on $N(A)$:

$$h \in N(A) \Rightarrow q^T h = 0, \text{ i.e., } q \in N(A)^\perp. \tag{8.35}$$

Now it is obvious that

$$\sup_{y \in B_1(r)} q^T y = \max_{y \in \overline{B}_1(r)} q^T y = r|q|_\infty \quad \text{where} \quad |q|_\infty := \max_{i=1,\ldots,m} |q_i| > 0.$$

Taking $h = 0$, $y_\varepsilon = (1 - \varepsilon)x$ in (8.34) and letting $\varepsilon \to 0$, we thus also get

$$q^T x = \sum_{i=1}^m q_i x_i = r|q|_\infty = \sum_{i=1}^m |q|_\infty |x_i|. \tag{8.36}$$

Since for every i, one has $q_i x_i \leq |q|_\infty |x_i|$ the previous equality yields

$$q_i x_i = |q|_\infty |x_i|, \quad \text{for} \quad i = 1, \ldots, m. \tag{8.37}$$

Thus, setting

$$I_+ := \{i \in \{1, \ldots, m\} : x_i > 0\}, \ I_- := \{i \in \{1, \ldots, m\} : x_i < 0\},$$

$$I := I_+ \cup I_-, \ I_0 := \{1, \ldots, m\} \setminus I,$$

there holds

$$x_i q_i > 0, \forall i \in I, \ q_i = |q|_\infty > 0, \forall i \in I_+, \ q_i = -|q|_\infty, \forall i \in I_-. \tag{8.38}$$

Define then

$$C := \{y \in \mathbb{R}^m : Ay = b, \ y_i = 0, \forall i \in I_0, \ q_i y_i \geq 0, \forall i \in I, \ q^T y = r|q|_\infty\}.$$

Note that $x \in C$, hence C is a non-empty convex compact subset of $\mathbb{R}^m$. Moreover, if $y \in C$, one deduces from the definition of C and (8.38)

$$|y|_1 = \sum_{i \in I} \frac{q_i}{|q_i|} y_i = \frac{q^T y}{|q|_\infty} = r,$$

hence every $y \in C$ solves (8.32). We know from Proposition 3.3 that $\mathrm{Ext}(C) \neq \emptyset$. Let then $\bar{x} \in \mathrm{Ext}(C)$, and let

$$J := \{i \in \{1,\ldots,m\} : \bar{x}_i \neq 0\}(\subset I), \ J_0 := \{1,\ldots,m\} \setminus J$$

and let $h \in \mathrm{N}(A)$ such that $h_j = 0$ whenever $j \in J_0$. Then it follows from (8.35) that $q^T h = 0$, hence for $\varepsilon > 0$ small enough, $\bar{x} \pm \varepsilon h$ belongs to C which contradicts the extremality of $\bar{x}$ unless $h = 0$. This shows that the linear map

$$h \in \mathbb{R}^m \mapsto (Ah, (h_j)_{j \in J_0}) \in \mathbb{R}^{n+\#J_0} = \mathbb{R}^{n+m-\#J}$$

is one-to-one hence

$$|\bar{x}|_0 = \#J \leq n,$$

which ends the proof since $\bar{x}$ being in C it solves (8.32). □

Exercise 8.7 (Dual form of Basis Pursuit). *Prove that*

$$\min(8.32) = \max\{b^T \lambda, \ \lambda \in \mathbb{R}^n, \ |A^T \lambda|_\infty \leq 1\}.$$

(Hint: use the proof above as well as the identity $\mathrm{N}(A)^\perp = \mathrm{R}(A^T)$, *see Exercise 8.5) and give the primal–dual extremality conditions between these two problems.*

Exercise 8.8. *Show that for every* $x \in \mathbb{R}^m$, *one has*

$$|x|_1 = \sup_{y \in \mathbb{R}^m, \, |y|_\infty \leq 1} y^T x \ \text{and} \ |x|_\infty = \sup_{y \in \mathbb{R}^m, \, |y|_1 \leq 1} y^T x$$

and compute the subdifferentials $\partial |\cdot|_1$ *and* $\partial |\cdot|_\infty$.

The LASSO

One drawback of the least squares method is that it leads to (approximate) solutions x for which $|x|_0$ can be as large as m. One way to overcome this is to use the LASSO (Least Absolute Shrinkage and Selection Operator) which consists in adding an l^1 penalization with parameter $\lambda > 0$ to the quadratic error:

$$\inf_{x \in \mathbb{R}^m} f_\lambda(x) := \frac{1}{2}|Ax - b|_2^2 + \lambda|x|_1 \tag{8.39}$$

(note that here we do not make any assumption on the rank of A). Observe that (8.39) is a (non-smooth) convex minimization problem and by coercivity, it possesses solutions[g]. One convenient way to characterize solutions of (8.39) is by duality. We leave the following (which is an easy application of the Fenchel–Rockafellar theorem) as an exercise to the reader:

Exercise 8.9. *Prove that* $\min$(8.39) *coincides with*

$$\max_{q \in \mathbb{R}^n \,:\, |A^T q|_\infty \le \lambda} b^T q - \frac{1}{2}|q|_2^2. \tag{8.40}$$

Show that (8.40) admits a unique solution $q \in \mathbb{R}^n$ *and that* $x \in \mathbb{R}^m$ *solves (8.39) if and only if*

$$q = b - Ax \text{ and } \lambda|x|_1 = x^T A^T q. \tag{8.41}$$

Using (8.41) and the fact that $|A^T q|_\infty \le \lambda$, defining

$$p := A^T q = A^T b - A^T Ax$$

(note that p_i represents how much x fails to solve the ith normal equation in the ordinary least squares problem), we have

$$|p_i| \le \lambda \text{ and } \lambda|x_i| = p_i x_i, \quad \forall i \in \{1, \dots, m\}$$

in particular

$$x_i > 0 \Rightarrow p_i = \lambda, \ x_i < 0 \Rightarrow p_i = -\lambda$$

so that

$$|p_i| < \lambda \Rightarrow x_i = 0.$$

This is the well-known thresholding effect induced by the LASSO: if the ith normal equation is satisfied up to an error strictly less than λ, then, the corresponding component x_i is set to 0. The choice of the parameter λ by the user, should therefore reflect a compromise between least squares accuracy and sparsity. Choosing λ too large leads to a trivial solution.

Exercise 8.10. *Prove that if* $\lambda \ge |A^T b|_\infty$ *then* 0 *solves (8.39).*

---

[g]Not necessarily unique if $\text{rank}(A) < m$, however if both x and $\tilde{x}$ solve (8.39) then $A(x - \tilde{x}) = 0$ (otherwise, one would have $f_\lambda(\frac{x+\tilde{x}}{2}) < \frac{1}{2}(f_\lambda(x) + f_\lambda(\tilde{x}))$).

Remark 8.3. In the context of linear regression, one can always add nonlinear (powers, logarithms, etc.) functions of the initial explanatory variables as new explanatory variables. Of course, adding such variables improves the goodness of fit of the model. But this is at the risk of over-fitting the model with variables which are in fact irrelevant. The LASSO is a popular method to select, within a large set of variables, those which are really relevant. As explained above, l^1 penalization acts as a sort of threshold on the coefficients of the regression so as to (hopefully) only keep the most meaningful ones.

Remark 8.4. Neither (8.32) nor (8.39) have closed-form solutions. One needs to use iterative methods, such as the one we saw in Chapter 7, to solve them in practice. In particular, the forward–backward splitting scheme from Section 7.4.1 enables one to solve (8.39). The Frank–Wolfe algorithm (see Exercise 7.14) is also well-suited to treat l^1 constraints (see Exercise 8.11) below.

Exercise 8.11. *Consider the variant of (8.39) where, instead of an l^1 penalization, we have an l^1 constraint:*

$$\inf\left\{\frac{1}{2}|Ax - b|_2^2 : x \in \mathbb{R}^m, \ |x|_1 \leq r\right\} \tag{8.42}$$

for some $r > 0$.

1) *Show that (8.42) admits solutions and give necessary and sufficient optimality conditions for these solutions.*
2) *Consider the Frank–Wolfe algorithm (see Exercise 7.14) for (8.42) initialized with $x_0 = 0$ (say). Give closed-form solutions of its iterates x_k. (Hint: look for a solution of (7.77) which is of the form $\pm r e_i$ where $(e_1, \ldots, e_m)$ is the canonical basis of $\mathbb{R}^m$.)*
3) *Show that $|x_k|_0 \leq k$ for every $k \in \mathbb{N}$ and discuss the convergence of $(f(x_k) - \inf(8.42))$ and of x_k.*

8.3.   Classification

Assume that we are given a set of n observations of the form $(y_i, x_i)_{i=1,\ldots,n}$ where y_i is a binary *label* which only takes two values, say $y_i \in \{-1, 1\}$ and $x_i \in \mathbb{R}^m$ represents observations of some real-valued explanatory variables (for instance, y can represent the failure/success of some project and x a list of characteristics of the project holder and the project itself). A typical classification problem is to use these data to estimate the value of the label for a new observation of the explanatory variables. This requires to deduce from the data a relation between y and x and of course, since y is binary, it makes little sense to perform a linear regression.

8.3.1. *Logistic regression*

A popular classification method is logistic regression, which assumes that the $(y_i, x_i)_i$ can be seen as i.i.d. realizations of a random pair (Y, X) with values in $\{-1, 1\} \times \mathbb{R}^m$ where the conditional probability of Y given $X = x$ takes the form

$$\mathbb{P}(Y = 1 | X = x) = \frac{e^{\beta^T x}}{1 + e^{\beta^T x}},$$

$$\mathbb{P}(Y = -1 | X = x) = \frac{1}{1 + e^{\beta^T x}} = \frac{e^{-\beta^T x}}{1 + e^{-\beta^T x}}$$

for some parameter $\beta \in \mathbb{R}^m$ to be determined. Note that the logistic model can be written in a more synthetic way as

$$\mathbb{P}(Y = y | X = x) = \frac{e^{y\beta^T x}}{1 + e^{y\beta^T x}} = \frac{1}{1 + e^{-y\beta^T x}}, \quad y \in \{-1, 1\}. \tag{8.43}$$

The *likelihood* of the model is given by

$$\prod_{i=1}^{n} \mathbb{P}(Y = y_i | X = x_i).$$

In view of (8.43), its log, called the *log-likelihood*, then takes the form

$$-\sum_{i=1}^{n} \log\left(1 + e^{-y_i \beta^T x_i}\right).$$

The *maximum-likelihood* estimator of the parameter β is then obtained (as its name indicates!) by maximizing the likelihood, or equivalently by solving

$$\inf_{\beta \in \mathbb{R}^m} F(\beta) := \sum_{i=1}^{n} \log\left(1 + e^{-y_i \beta^T x_i}\right), \tag{8.44}$$

note that F is smooth and it is convex, indeed it can be written as

$$F(\beta) := \sum_{i=1}^{n} g(y_i \beta^T x_i) \text{ where } g(t) := \log(1 + e^{-t})$$

and g is easily seen to be convex:

$$g'(t) = -\frac{1}{1+e^t}, \; g''(t) = \frac{e^t}{(1+e^t)^2} > 0.$$

So that β solves (8.44) if and only if $\nabla F(\beta) = 0$, i.e.,

$$\sum_{i=1}^{n} \frac{y_i x_i}{1 + e^{y_i \beta^T x_i}} = 0.$$

One should be slightly cautious when dealing with the existence of a minimizer of F, such existence cannot be taken for granted and actually depends on the data:

Proposition 8.1. *If the data are completely separated, i.e., if there exists* $b \in \mathbb{R}^m$ *such that*

$$y_i b^T x_i > 0, \quad \forall i = 1, \ldots, n, \tag{8.45}$$

then (8.44) does not have any solution. If the data overlap, i.e., for every $b \in \mathbb{R}^m \setminus \{0\}$, *there exists* j *in* $\{1, \ldots, n\}$ *such that*

$$y_j b^T x_j < 0, \tag{8.46}$$

then (8.44) admits at least one solution.

Proof. If (8.45) holds, then $F(\lambda b)$ tends to 0 as $\lambda \to \infty$. Since $F \geq 0$ this implies that $\inf_{\mathbb{R}^m} F = 0$ but this infimum is obviously not attained. Assume now that the data do not overlap and set

$$m(b) := \min\{y_j b^T x_j, \; j = 1, \ldots, n\},$$

m is continuous (it is actually concave and Lipschitz), it therefore achieves its maximum on the sphere and the value of this maximum is negative thanks to (8.46):

$$\max_{|b|=1} m(b) = -C < 0.$$

By homogeneity of m this yields:

$$\min\{y_j b^T x_j, \; j = 1, \ldots, n\} \leq -C|b|, \; \forall b \in \mathbb{R}^m.$$

We thus get

$$F(\beta) = \sum_{i=1}^{n} \log\left(1 + e^{-y_i \beta^T x_i}\right) \geq -\min\{y_i \beta^T x_i, \; i = 1, \ldots, n\} \geq C|\beta|$$

so that F is coercive hence attains its minimum. $\qquad \square$

Remark 8.5. As the proof above suggests, if the data are close to be completely separated or poorly overlap (it is typically the case in small samples), some coefficients of the logistic regression tend to $\pm\infty$. This separation problem may be important in practice and there are various regularization strategies which may prevent it. The penalization proposed by Firth [49] is particularly popular and interesting from a mathematical point of view.

8.3.2. *Support-vector machines*

As before, we are given data $(y_i, x_i) \in \{-1, 1\} \times \mathbb{R}^m$ for $i = 1, \ldots, n$ which split into two classes according to the value of the binary variable, we thus set

$$I_{\pm 1} := \{i \in \{1, \ldots, n\} : y_i = \pm 1\}, \quad C_{\pm 1} := \mathrm{co}\{x_i, \ i \in I_{\pm 1}\}.$$

The key idea behind support-vector (SVM) machines, pioneered by Vapnik [96], is to look for an hyperplane separating the sets $\{x_i, \ i \in I_{-1}\}$ and $\{x_i, \ i \in I_1\}$ which is the same as separating the (non-empty) convex sets C_{-1} and C_1 which is of course possible if and only if the data are *linearly separable*, i.e.,

$$C_{-1} \cap C_1 = \emptyset \tag{8.47}$$

which, by strict separation, is the same as

$$\exists (\beta, \beta_0) \in \mathbb{R}^m \setminus \{0\} \times \mathbb{R} : y_i(\beta^T x_i + \beta_0) > 0, \ \forall i = 1, \ldots, n. \tag{8.48}$$

For β, β_0 separating the two classes, i.e., satisfying (8.48), the hyperplane

$$H_{\beta, \beta_0} := \{x \in \mathbb{R}^m : \beta^T x + \beta_0 = 0\}$$

separates the two convex sets C_{-1} and C_1 and is called the decision frontier. Note that the distance from a point $x \in \mathbb{R}^m$ to the decision frontier is given by[h]

$$\mathrm{dist}(x, H_{\beta, \beta_0}) := \min_{y \in H_{\beta, \beta_0}} |x - y|_2 = \frac{|\beta^T x + \beta_0|}{|\beta|_2}.$$

---

[h]Indeed, the projection y of x onto H_{β, β_0} is characterized by $\beta^T y + \beta_0 = 0$ and $y - x = \lambda \beta$ for some $\lambda \in \mathbb{R}$; this implies $0 = \beta^T x + \beta_0 + \lambda|\beta|_2^2$ and then $|y - x|_2 = \frac{|\beta^T x + \beta_0|}{|\beta|_2}$.

Of course, if (8.48) holds, there are many separating hyperplanes, the *margin* is the least distance from the data to the decision frontier, i.e.,

$$\text{margin}(\beta, \beta_0) := \min_{i=1,\ldots,n} \text{dist}(x_i, H_{\beta,\beta_0}) = \min_{i=1,\ldots,n} \frac{y_i(\beta^T x_i + \beta_0)}{|\beta|_2}$$

and the principle behind support-vector machines methods is to look for a margin maximizing separating hyperplane. This may look complicated at first sight, but the key point is that $\text{margin}(\beta, \beta_0)$ is homogeneous of degree 0 (i.e., $\text{margin}(t\beta, t\beta_0) = \text{margin}(\beta, \beta_0)$ for every $t > 0$). We can therefore normalize (β, β_0) and the most convenient way to do so is by imposing

$$\min_{i=1,\ldots,n} \{y_i(\beta^T x_i + \beta_0)\} = 1. \tag{8.49}$$

Under this normalization, maximizing the margin is equivalent to minimizing $|\beta|$. This leads to

$$\inf_{(\beta,\beta_0)\in\mathbb{R}^{m+1}} \left\{ \frac{1}{2}|\beta|_2^2 : y_i(\beta^T x_i + \beta_0) \geq 1, \ \forall i = 1, \ldots, n \right\}. \tag{8.50}$$

Note that in (8.50) we have slightly relaxed (8.49) by an inequality. But this is without loss of generality, since the minimizer (β, β_0) of (8.50) (its existence and uniqueness is obvious under (8.48)) has to satisfy (8.49). If it was not the case, one could divide β, β_0 by a constant strictly larger than 1 and get a new admissible vector with strictly smaller norm than β. The solution of (8.50) is not explicit but it can be characterized by the existence of KKT multipliers $\lambda_i \geq 0$ (see Chapter 4) satisfying the first-order conditions

$$\beta = \sum_{i=1}^{n} \lambda_i y_i x_i, \ \sum_{i=1}^{n} \lambda_i y_i = 0, \tag{8.51}$$

together with the complementary slackness conditions

$$\lambda_i(y_i(\beta^T x_i + \beta_0) - 1) = 0, \quad \forall i = 1, \ldots, n. \tag{8.52}$$

In general, data are not linearly separable, i.e., (8.48) does not hold and it is not possible to separate the two classes by a linear classifier $x \mapsto \beta^T x + \beta_0$. However, allowing some misclassified data, it is still possible to use SVM so as to minimize misclassification errors. Introducing some slack variables $s_i \geq 0$ this leads to

$$\inf_{(\beta,\beta_0)\in\mathbb{R}^{m+1},\, s\in\mathbb{R}^n_+} \left\{ \frac{1}{2}|\beta|_2^2 + \mu \sum_{i=1}^n s_i : y_i(\beta^T x_i + \beta_0) \geq 1 - s_i, \; \forall i = 1,\ldots,n \right\},$$

where $\mu > 0$ is a penalization parameter which controls the tolerance on misclassified data.

8.4. Exercises

Exercise 8.12 (Variational characterization of the expectation). *Let X be a square integrable $\mathbb{R}^d$-valued random vector on some probability space $(\Omega, \mathcal{F}, \mathbb{P})$. Show that*

$$\inf_{x\in\mathbb{R}^d} \mathbb{E}(|X - x|_2^2)$$

admits $\mathbb{E}(X)$ as unique solution.

Exercise 8.13 (Variational characterization of the median). *Assume that X is an integrable random variable on some probability space $(\Omega, \mathcal{F}, \mathbb{P})$, consider the minimization problem*

$$\inf_{m\in\mathbb{R}} \mathbb{E}(|X - m|). \tag{8.53}$$

1) *Show that (8.53) admits solutions and that the set of solutions of (8.53) is an interval.*
2) *For $Y \in L^1(\Omega, \mathcal{F}, \mathbb{P})$ and $m \in \mathbb{R}$ set $\Phi(Y, m) := \mathbb{E}(|X - (Y + m)|)$ and $v(Y) := \inf_{m\in\mathbb{R}} \Phi(Y, m)$ so that $\inf(8.53) = v(0)$. Show that v is convex and continuous.*
3) *Recalling that $L^\infty(\Omega, \mathcal{F}, \mathbb{P})$ is the topological dual of $L^1(\Omega, \mathcal{F}, \mathbb{P})$, compute $\Phi^*(Z, 0)$ for every $Z \in L^\infty(\Omega, \mathcal{F}, \mathbb{P})$.*
4) *Deduce that (8.53) coincides with the value of its dual*

$$\sup\{\mathbb{E}(ZX) : Z \in L^\infty(\Omega, \mathcal{F}, \mathbb{P}), \; |Z| \leq 1 \; a.s., \; \mathbb{E}(Z) = 0\}, \tag{8.54}$$

and that (8.54) has at least one solution. (Hint: use the convex duality theorem (Theorem 6.2).)
5) *Show that if Z solves (8.54) then m solves (8.53) if and only if $Z = 1$ a.s. on $\{X > m\}$ and $Z = -1$ a.s. on $\{X < m\}$.*

6) *Show that m solves (8.53) if and only if m is a median of X, i.e.,*

$$\mathbb{P}(X < m) \leq \frac{1}{2} \ \text{and} \ \mathbb{P}(X > m) \leq \frac{1}{2}.$$

Exercise 8.14. *The Frobenius norm of a square matrix $M = (M_{ij})_{i,j} \in \mathcal{M}_n(\mathbb{R})$ is defined by*

$$\|M\|_F := \operatorname{tr}(M^T M)^{1/2} = \left(\sum_{i=1}^{n} \sum_{j=1}^{n} M_{ij}^2 \right)^{1/2}.$$

1) *Show that $\| \cdot \|_F$ is submultiplicative, i.e., $\|MN\|_F \leq \|M\|_F \|N\|_F$ for every $(M, N) \in \mathcal{M}_n(\mathbb{R})^2$.*
2) *Show that $\|I_n\|_F = \sqrt{n}$. Is $\| \cdot \|_F$ an operator norm?*

Exercise 8.15 ((Counter-)Examples of operator norms). *Given $A \in \mathcal{M}_{n \times m}(\mathbb{R})$, let us define*

$$N_\infty(A) := \max \{|A_{ij}|, \ (i,j) \in \{1, \ldots, n\} \times \{1, \ldots, m\}\},$$

$$N_1(A) := \sum_{i=1}^{n} \sum_{j=1}^{m} |A_{ij}|$$

as well as

$$|||A|||_1 := \max \left\{ \sum_{i=1}^{n} | \sum_{j=1}^{m} A_{ij} x_j| : x \in \mathbb{R}^m, \ \sum_{j=1}^{m} |x_j| \leq 1 \right\}$$

and

$$|||A|||_\infty := \max \left\{ \max_{i=1,\ldots,n} | \sum_{j=1}^{m} A_{ij} x_j| : x \in \mathbb{R}^m, \ \max_{j=1,\ldots,m} |x_j| \leq 1 \right\}.$$

1) *Show that N_∞ is a norm on $\mathcal{M}_{n \times m}(\mathbb{R})$ and that it is not an operator norm.*
2) *Show that N_1 is a norm on $\mathcal{M}_{n \times m}(\mathbb{R})$ and that it is not an operator norm.*
3) *Show that the operator norms $|||A|||_1$ and $|||A|||_\infty$ can be expressed as*

$$|||A|||_1 = \max_{j=1,\ldots,m} \sum_{i=1}^{n} |A_{ij}| \ \text{and} \ |||A|||_\infty = \max_{i=1,\ldots,n} \sum_{j=1}^{m} |A_{ij}|.$$

Exercise 8.16 (Regression line). *Let $n \in \mathbb{N}^*$, $(x_i, y_i) \in \mathbb{R}^2$, for $i = 1, \ldots, n$. The regression line for these data is the line with equation $y = \alpha + \beta x$ which solves the least squares problem:*

$$\inf_{(\alpha, \beta) \in \mathbb{R}^2} \sum_{i=1}^{n} (y_i - \alpha - \beta x_i)^2. \tag{8.55}$$

Let us define the (empirical) means

$$\overline{x} := \frac{1}{n} \sum_{i=1}^{n} x_i, \ \overline{y} := \frac{1}{n} \sum_{i=1}^{n} y_i.$$

1) *Show that*

$$\frac{1}{n} \sum_{i=1}^{n} (x_i - \overline{x})^2 = \frac{1}{n} \sum_{i=1}^{n} x_i^2 - \overline{x}^2 \ \text{and} \ \frac{1}{n} \sum_{i=1}^{n} (x_i - \overline{x})(y_i - \overline{y}) = \frac{1}{n} \sum_{i=1}^{n} x_i y_i - \overline{xy}.$$

2) *Show (by a direct computation and a convexity argument) that (α, β) solves (8.55) if and only if*

$$\overline{y} = \alpha + \beta \overline{x}$$

(which means that the empirical mean of the data belongs to the regression line) and

$$\frac{1}{n} \sum_{i=1}^{n} x_i y_i = \alpha \overline{x} + \frac{\beta}{n} \sum_{i=1}^{n} x_i^2.$$

3) *Show that if $\sum_{i=1}^{n} (x_i - \overline{x})^2 \neq 0$ (i.e., x_i is not constant) then (8.55) has a unique solution given by*

$$\beta = \frac{\sum_{i=1}^{n} (x_i - \overline{x})(y_i - \overline{y})}{\sum_{i=1}^{n} (x_i - \overline{x})^2}, \ \alpha = \overline{y} - \frac{\sum_{i=1}^{n} (x_i - \overline{x})(y_i - \overline{y})}{\sum_{i=1}^{n} (x_i - \overline{x})^2} \overline{x}.$$

4) *Solve (8.55) when $\sum_{i=1}^{n} (x_i - \overline{x})^2 = 0$.*

Exercise 8.17. *Let $A \in \mathcal{M}_{n \times m}(\mathbb{R})$, express the Moore–Penrose inverse of A using the SVD of A. Show that if $\text{rank}(A) = n$ (respectively, $\text{rank}(A) = m$), A^+ is a right (respectively, left) inverse of A. Show that $(A^T)^+ = (A^+)^T$. Compute A^+ when*

$$A = \begin{pmatrix} 1 & 1 \\ 1 & 1 \end{pmatrix}, \ A = \begin{pmatrix} 1 & 1 \\ 1 & 1 \\ 1 & 1 \end{pmatrix}.$$

Exercise 8.18 (A least squares problem without solutions in l^2).
Equip $l^2 := \{(x_n)_n \in \mathbb{R}^{\mathbb{N}} : \sum_{n=0}^{+\infty} x_n^2 < +\infty\}$ with is usual Hilbertian structure. For $x \in l^2$, define $A(x) \in l^2$ by $(A(x))_n = \frac{1}{n+1}x_n$ for every $n \in \mathbb{N}$.

1) Show that $A \in \mathcal{L}_c(l^2)$. Is the range of A closed? (Hint: consider $A(x^k)$ where $x_n^k = 1$ if $n \leq k$ and $x_n^k = 0$ if $n > k$.)
2) Let $b \in l^2$ be given by $b_n = \frac{1}{n+1}$ for every n. Show that the value of the least squares problem

$$\inf_{x \in l^2} \|A(x) - b\|_{l^2}^2$$

is 0 but is not attained.

Exercise 8.19 (Regularization of the problem from Exercise 8.18).
Let $A \in \mathcal{L}_c(l^2)$ be as in Exercise 8.18 and let $b \in l^2$. For $\lambda > 0$ consider:

$$\inf_{x \in l^2} \frac{1}{2}\|A(x) - b\|_{l^2}^2 + \frac{\lambda}{2}\|x\|_{l^2}^2 \tag{8.56}$$

and

$$\inf_{x \in l^2} \frac{1}{2}\|A(x) - b\|_{l^2}^2 + \lambda\|x\|_{l^1}, \tag{8.57}$$

where $\|x\|_{l^1} := \sum_{n=0}^{+\infty} |x_n|$.

1) Show that (8.56) has a unique solution and compute it.
2) Show that (8.57) has a unique solution and compute it.

Exercise 8.20. Let x and $\tilde{x}$ be two solutions of the LASSO problem (8.39). Prove that

$$Ax = A\tilde{x} \text{ and } \lambda|x|_1 = \lambda|\tilde{x}|_1$$

defining v as the value of the LASSO problem:

$$v(\lambda) := \inf(8.39).$$

Prove that v is concave non-decreasing on $\mathbb{R}_+$, differentiable on $\mathbb{R}_+^*$ and constant on $[\|A^T b\|_\infty, +\infty)$.

Exercise 8.21 (Elastic net). Let $A \in \mathcal{M}_{n \times m}(\mathbb{R})$, $b \in \mathbb{R}^n$, $\lambda > 0$, $\mu > 0$ consider

$$v(\lambda, \mu) := \inf_{x \in \mathbb{R}^m} f_{\lambda,\mu}(x), \quad f_{\lambda,\mu}(x) := \frac{1}{2}|Ax - b|_2^2 + \lambda|x|_1 + \frac{\mu}{2}|x|_2^2. \tag{8.58}$$

1) *Show that (8.58) has a unique solution $x(\lambda, \mu)$.*
2) *Show that $f_{\lambda, \mu}$ is convex and compute its subgradient. Deduce a necessary and sufficient optimality condition for (8.58).*
3) *Give a dual formulation of (8.58).*
4) *Show that v is concave and differentiable on $(0, +\infty)^2$ and study the continuity and monotonicity of $(\lambda, \mu) \mapsto (|x(\lambda, \mu)|_1, |x(\lambda, \mu)|_2^2)$.*

Chapter 9

An Invitation to the Calculus
of Variations

The calculus of variations is one of the oldest[a] and most fascinating branches of mathematical analysis. In part because of its intimate connections with physics (classical mechanics, optics, elasticity, etc.) and geometry (harmonic maps, minimal hypersurfaces, etc.), it has attracted the attention of giants of mathematics (Bernoulli, Euler, Lagrange, Dirichlet, Weierstrass, Hilbert, Noether to name a few) over the last four centuries. The modest goal of this chapter is to give a short introduction to this wide topic for which there are excellent textbooks such as Buttazzo, Giaquinta and Hildebrandt [28], Cesari [33], Dacorogna [39, 40], Ekeland and Temam [46], Giusti [60], Giaquinta and Hildebrandt [56, 57], Morrey [77], Young [99].

9.1. Preliminaries

The standard problem in the calculus of variations in one independent variable (say t interpreted as a time variable belonging to some interval $[a, b]$) consists in finding curves γ in $\mathbb{R}^m$ which minimize an integral of the form

$$E(\gamma) := \int_a^b L(t, \gamma(t), \dot{\gamma}(t)) \mathrm{d}t,$$

---

[a]At the risk of being a bit anachronic, it may be traced back to the foundation of Carthage (around 810 BC). Indeed, Queen Dido, when she founded Carthage (after her brother, Pygmalion, murdered her husband) was ironically offered by the locals an area that an ox hide would cover, a problem intimately related to the isoperimetric inequality (see Exercise 9.21).

where $L : [a, b] \times \mathbb{R}^m \times \mathbb{R}^m$ is called the Lagrangian, and $\dot{\gamma}(t)$ is the velocity of the curve $\gamma = (\gamma_1, \ldots, \gamma_m)$ at time t, i.e., $\dot{\gamma}(t) = (\dot{\gamma}_1(t), \ldots, \dot{\gamma}_m(t))$, possibly fixing the endpoints $\gamma(a) = \alpha$, $\gamma(b) = \beta$ or imposing other constraints, a cost depending also on the free-endpoints $\gamma(0)$, $\gamma(1), \ldots$

We are here in the realm of optimization in infinite dimensions which relies very much on functional analytic arguments. The choice of a good functional space where to look for optimal curves is an important issue. In the first place, we need the functional E which depends on the velocity of the curve to be well-defined. A first choice would be to work in a space of smooth functions (C^1 say) but this is too naive. On the one hand, spaces of smooth functions lack good weak compactness properties (they are not reflexive). On the other hand, because E being a Lebesgue integral, one does not really need the velocity to be defined everywhere but only a.e. which suggests to consider spaces of functions whose derivative (in some weak sense to be made precise) lie in some Lebesgue space. The good framework for such *weakly differentiable* functions is that of the Sobolev spaces about which we will recall some basic facts in the following sections, for proofs and fine properties of Sobolev spaces, we advise the reader to consult the classic textbooks of Brezis [23] or Adams [2].

As we shall see, variational problems in one independent variable are a bit special.[b] Indeed, variational problems in dimension one can be efficiently attacked by the dynamic programming approach and the Hamilton–Jacobi theory (see Section 9.5.2). In view of their importance in applications (in physics, geometry but also image processing), we will also consider the case of several independent variables say $x = (x_1, \ldots, x_d) \in \Omega$ where Ω is an open bounded subset of $\mathbb{R}^d$, satisfying suitable regularity assumptions, and one looks for functions $u = (u_1, \ldots, u_m) : \Omega \to \mathbb{R}^m$ minimizing the integral functional

$$J(u) := \int_\Omega L(x, u(x), Du(x))\mathrm{d}x,$$

where Du is the Jacobian matrix of u, possibly subject to a Dirichlet boundary condition which prescribes u on the boundary of Ω, $\partial\Omega$.

The following sections assume the reader is already familiar with measure and integration theory and L^p spaces, if this is not the case, an excellent

---

[b]This is why I have chosen to use the notation $t \mapsto \gamma(t)$ and to speak about curves (and velocities denoted $\dot{\gamma}$) in the case of one independent variable and to rather use the notation $x \mapsto u(x)$ when there are several variables $x = (x_1, \ldots, x_d)$.

reference is the book by Hewitt and Stromberg [63], see also Rudin [88]. In what follows, given $A \subset \mathbb{R}^m$ measurable^c we will denote by $\mathbf{1}_A$ the characteristic function (1 on A and 0 outside) of A and by $|A|$ the (m-dimensional) Lebesgue measure of A.

9.1.1. *On weak derivatives*

Let Ω be an open subset of $\mathbb{R}^d$. Given $u \in L^1_{\text{loc}}(\Omega)$ one can define the linear form $\{u\}$ over the space of test-functions $C_c^\infty(\Omega)$ by

$$\langle \{u\}, \varphi \rangle := \int_\Omega u\varphi, \ \forall \varphi \in C_c^\infty(\Omega). \tag{9.1}$$

It follows from Proposition 2.9 that if u and v are in $L^1_{\text{loc}}(\Omega)$ and $\{u\} = \{v\}$ then $u = v$ a.e. so that $\{u\}$ fully characterizes u. Now, note that if $u \in C^1(\Omega)$ and $i \in \{1, \ldots, d\}$ then, by integration by parts, one has

$$\int_\Omega \partial_i u\varphi = -\int_\Omega u\partial_i\varphi, \text{ i.e., } \langle \{\partial_i u\}, \varphi \rangle = -\langle \{u\}, \partial_i\varphi \rangle, \quad \forall \varphi \in C_c^\infty(\Omega),$$

which suggests to take the previous relation as a definition for the partial derivatives of u when u is only $L^1_{\text{loc}}(\Omega)$, this is the point of view of the theory of distributions and leads to the definition.

Definition 9.1. Given $u \in L^1_{\text{loc}}(\Omega)$ and $i \in \{1, \ldots, d\}$, the ith partial derivative $\partial_i\{u\}$ in the weak sense (or in the sense of distributions) is the linear form on $C_c^\infty(\Omega)$ defined by

$$\langle \partial_i\{u\}, \varphi \rangle = -\langle \{u\}, \partial_i\varphi \rangle, \quad \forall \varphi \in C_c^\infty(\Omega). \tag{9.2}$$

As already mentioned above, if $u \in C^1$ then $\partial_i\{u\} = \{\partial_i u\}$ that is weak derivatives of smooth functions coincide with classical derivatives. In the definition above, we could have as well considered test-functions in $C_c^1(\Omega)$ instead of $C_c^\infty(\Omega)$. Of course, we can define higher partial derivatives in the weak sense by

$$\langle \partial^\alpha\{u\}, \varphi \rangle = (-1)^{|\alpha|}\langle \{u\}, \partial^\alpha\varphi \rangle,$$

for every multi-index α and every $\varphi \in C_c^\infty(\Omega)$ (or simply $C_c^{|\alpha|}(\Omega)$).

---

^{c}Throughout this chapter, measurable has to be understood as measurable for the d-dimensional Lebesgue measure, likewise *almost everywhere* will mean everywhere except perhaps on a set of zero Lebesgue measure.

Let us now assume that Ω is an open bounded subset of $\mathbb{R}^d$ with a C^1 boundary (see Section 2.4.2), and denote by $C^k(\overline{\Omega})$ the space of restrictions of functions of $C_c^k(\mathbb{R}^d)$ to Ω. Let $f \in C(\overline{\Omega})$ and $g \in C(\partial\Omega)$ a classic solution of

$$\operatorname{div}(\psi) = f \text{ in } \Omega, \ \psi \cdot \nu = g \text{ on } \partial\Omega \tag{9.3}$$

is a $\psi \in C^1(\Omega, \mathbb{R}^d) \cap C(\overline{\Omega}, \mathbb{R}^d)$ which satisfies pointwise the conditions in (9.3). It then follows from the integration by parts formula (2.54) that if ψ is a classical solution of (9.3) then

$$\int_\Omega \psi \cdot \nabla\varphi = -\int_\Omega f\varphi + \int_{\partial\Omega} g\varphi d\sigma, \ \forall\varphi \in C^1(\overline{\Omega}). \tag{9.4}$$

Conversely (again thanks to Proposition 2.9), if $\psi \in C^1(\Omega, \mathbb{R}^d) \cap C(\overline{\Omega}, \mathbb{R}^d)$ satisfies (9.4) then it solves (9.3). Therefore, one can take (9.4) as a definition for ψ to be a solution of (9.3). Given $f \in L^1(\Omega)$ and $g \in L^1(\partial\Omega, \sigma)$ a weak solution of (9.3) is by definition a $\psi \in L^1(\Omega, \mathbb{R}^d)$ such that (9.4) holds (note that an obvious condition for (9.4) to be solvable is that $\int_\Omega f = \int_{\partial\Omega} g d\sigma$). One also says that (9.4) is the weak formulation of (9.3).

Given Ω a non-empty open subset of $\mathbb{R}^d$ and $p \in [1, +\infty]$, the Sobolev space $W^{1,p}(\Omega)$ consists of all L^p functions whose first derivatives also belong to L^p:

$$W^{1,p}(\Omega) := \{u \in L^p(\Omega) : \partial_i\{u\} \in L^p(\Omega), \ \forall i = 1, \ldots, d\}. \tag{9.5}$$

In other words, $u \in L^p(\Omega)$ belongs to $W^{1,p}(\Omega)$ if, for every $i = 1, \ldots, d$, there exists $g_i \in L^p(\Omega)$ such that

$$\int_\Omega u\partial_i\varphi = -\int_\Omega g_i\varphi, \ \forall\varphi \in C_c^1(\Omega),$$

the functions g_i are then unique (up to a.e. equivalence) and they are simply denoted $g_i = \partial_i u$. The gradient of u is by definition $\nabla u := (\partial_1 u, \ldots, \partial_d u)^T \in L^p(\Omega)^d$. The space $W^{1,p}(\Omega)$ clearly is a vector space (and partial derivatives are linear operators), and it is usually equipped with the norm:

$$\|u\|_{W^{1,p}(\Omega)} := \|u\|_{L^p} + \|\nabla u\|_{L^p(\Omega)}. \tag{9.6}$$

For $p = 2$, one usually denotes $W^{1,2}(\Omega)$ as $H^1(\Omega)$, its norm $\|\cdot\|_{W^{1,2}}$ is equivalent to the norm deriving from the scalar product:

$$\|u\|_{H^1(\Omega)}^2 = \langle u, u \rangle \text{ with } \langle u, v \rangle := \int_\Omega (uv + \nabla u \cdot \nabla v). \tag{9.7}$$

A first classical result is the following theorem.

Theorem 9.1. $W^{1,p}(\Omega)$ *is a Banach space, it is reflexive whenever* $p \in (1, +\infty)$. *The space* $H^1(\Omega)$ *equipped with the scalar product from (9.7) is a Hilbert space.*

9.1.2. Sobolev functions in dimension 1

Let I be an open interval of the real line and $p \in [1, +\infty]$. For $\gamma \in W^{1,p}(I)$ we will denote by $\dot{\gamma} \in L^p(I)$ the derivative of γ. An important result is that functions in $W^{1,p}(I)$ admit a continuous representative and, roughly speaking, the next result expresses the fact that, in dimension 1, $W^{1,p}$ functions simply are primitive of L^p functions:

Theorem 9.2. *Let* $\gamma \in W^{1,p}(I)$. *Then* γ *admits a continuous representative that we will again denote* $\gamma \in C(\overline{I})$ *and is such that*

$$\gamma(t) - \gamma(s) = \int_s^t \dot{\gamma}(\tau)d\tau, \forall (s,t) \in I^2. \tag{9.8}$$

Note that if $p \in (1, +\infty)$ then for $\gamma \in W^{1,p}(I)$ and s and t in I, it follows from (9.8) and Hölder's inequality that

$$|\gamma(t) - \gamma(s)| \leq \int_s^t |\dot{\gamma}| \leq |t - s|^{1 - \frac{1}{p}} \|\dot{\gamma}\|_{L^p}, \tag{9.9}$$

hence functions in $W^{1,p}(I)$ are Hölder continuous with exponent $1 - \frac{1}{p}$. Likewise, elements of $W^{1,\infty}$ are Lipschitz.

A consequence of Theorem 9.2 is the following corollary.

Corollary 9.1 (Du Bois-Reymond's lemma). *If* $\gamma \in W^{1,p}(I)$ *is such that* $\dot{\gamma} \in C(I)$, *then* γ *is of class* C^1 *(and weak and strong derivatives of course agree).*

Theorem 9.3 (Density of test functions). *Let* $p \in [1, +\infty)$ *and* $\gamma \in W^{1,p}(I)$. *Then there exists a sequence* $\gamma_n \in C_c^\infty(\mathbb{R})$ *such that the restrictions* $\gamma_{n|I}$ *converge in* $W^{1,p}(I)$ *to* γ.

Theorem 9.4 (Sobolev embedding in dimension 1). *Let* I *be an open interval of* $\mathbb{R}$. *Then there exists a constant* $M = M(I)$ *(independent of* p*) such that for every* $p \in [1, +\infty]$, *and every* $\gamma \in W^{1,p}(I)$ *one has*

$$\|\gamma\|_{L^\infty} \leq M \|\gamma\|_{W^{1,p}}. \tag{9.10}$$

In other words, the embedding $W^{1,p}(I) \subset L^\infty(I)$ is continuous (which we denote as $W^{1,p}(I) \hookrightarrow L^\infty(I)$).

Theorem 9.5 (Rellich–Kondrachov compactness theorem). *Let I be a bounded open interval of $\mathbb{R}$.*

- *If $p > 1$ and γ_n is bounded in $W^{1,p}(I)$, then γ_n has a subsequence which converges in $C(\overline{I})$.*
- *If γ_n is bounded in $W^{1,1}(I)$, then for every $q \in [1, +\infty)$, γ_n has a subsequence which converges in $L^q(I)$.*

Combining the Rellich–Kondrachov theorem with the reflexivity of $L^p(I)$ for $p \in (1, +\infty)$, we deduce that if I is bounded and γ_n is a bounded sequence in $W^{1,p}(I)$ with $p \in (1, +\infty)$, then, one can find a subsequence γ_{n_j} and some $\gamma \in W^{1,p}(I)$ such that

- γ_{n_j} converges uniformly to γ on $\overline{I}$ as $j \to +\infty$,
- $\dot{\gamma}_{n_j} \rightharpoonup \dot{\gamma}$ in $L^p(I)$ as $j \to +\infty$.

Proposition 9.1. *Let I be an open interval of $\mathbb{R}$ and $p \in [1, +\infty]$. If γ and σ are in $W^{1,p}(I)$ then $\gamma\sigma \in W^{1,p}(I)$ and $(\gamma\sigma) = \gamma\dot{\sigma} + \dot{\gamma}\sigma$. Hence, for every s and t in I one has the integration by parts formula*

$$\int_s^t \dot{\gamma}\sigma = -\int_s^t \gamma\dot{\sigma} + \gamma(t)\sigma(t) - \gamma(s)\sigma(s).$$

For $p \in [1, +\infty)$ we denote by $W_0^{1,p}(I)$ the closure of $C_c^1(I)$ in $(W^{1,p}, \|\cdot\|_{W^{1,p}})$. Equipped with the topology of $W^{1,p}(I)$, $W_0^{1,p}(I)$ is by definition a closed subspace of $W^{1,p}(I)$, hence a Banach space. We also denote $H_0^1(I) := W_0^{1,2}(I)$. A simple characterization of $W_0^{1,p}(I)$ is the following theorem.

Theorem 9.6. *Let $\gamma \in W^{1,p}(I)$. Then $\gamma \in W_0^{1,p}(I)$ if and only if $\gamma = 0$ on ∂I. In particular, $W_0^{1,p}(\mathbb{R}) = W^{1,p}(\mathbb{R})$.*

Proposition 9.2 (Poincaré's inequality). *Assume that I is bounded. Then there exists a constant C such that*

$$\|\gamma\|_{W^{1,p}} \le C\|\dot{\gamma}\|_{L^p}, \quad \forall u \in W_0^{1,p}(I).$$

In particular, on $W_0^{1,p}(I)$, $\gamma \mapsto \|\dot{\gamma}\|_{L^p}$ is a norm which is equivalent to the $W^{1,p}$-norm. In a similar way, on $H_0^1(I)$, $(\gamma, \sigma) \mapsto \int_I \dot{\gamma}\dot{\sigma}$ is a scalar product which defines a norm equivalent to the H^1-norm.

We denote by $W^{1,p}(I,\mathbb{R}^m)$ or $W^{1,p}(I)^m$ the space of vector-valued functions $\gamma = (\gamma_1,\ldots,\gamma_m) : I \to \mathbb{R}^m$, such that $\gamma_j \in W^{1,p}(I)$ for $j = 1,\ldots,m$. Of course, arguing componentwise all the theorems above apply to $W^{1,p}(I,\mathbb{R}^m)$. We also set $H^1(I,\mathbb{R}^m) := W^{1,2}(I,\mathbb{R}^m)$ and equip it with the scalar product $(\gamma,\sigma) \mapsto \int_I (\gamma \cdot \sigma + \dot{\gamma} \cdot \dot{\sigma})$.

In practice, one barely uses the weak topology of $W^{1,p}$ but rather uses simple characterizations of weak convergence as follows.

Exercise 9.1. *Let $p \in (1,+\infty)$, $m \in \mathbb{N}^*$, $a < b$ be two reals, γ_n be a sequence in $W^{1,p}((a,b),\mathbb{R}^m)$ and $\gamma \in W^{1,p}((a,b),\mathbb{R}^m)$. Prove the equivalence between:*

- *γ_n converges weakly to γ in $W^{1,p}((a,b),\mathbb{R}^m)$,*
- *$\gamma_n(a)$ converges to $\gamma(a)$ and $\dot{\gamma}_n$ converges weakly in L^p to $\dot{\gamma}$,*
- *γ_n converges uniformly to γ and $\dot{\gamma}_n$ converges weakly in L^p to $\dot{\gamma}$.*

9.1.3. *Sobolev functions in higher dimensions*

In higher dimensions, the major difference with the scalar case is that Sobolev functions need not be continuous (or even bounded), as examples of negative powers of the norm easily show. To simplify the exposition, we will mostly work in this section in an open subset with a C^1 boundary.

Theorem 9.7 (Density of smooth functions). *If Ω is an open subset of $\mathbb{R}^d$ with a C^1 boundary, $p \in [1,+\infty[$ and $u \in W^{1,p}(\Omega)$, then there exists $(u_n)_n \in C_c^\infty(\mathbb{R}^d)^{\mathbb{N}}$ such that $u_n|_\Omega$ converges to u in $W^{1,p}(\Omega)$.*

Theorem 9.8 (Sobolev and Morrey inequalities). *Let $d \geq 2$, let Ω be an open subset of $\mathbb{R}^d$ with a C^1 boundary such that $\partial\Omega$ is bounded and let $p \in [1,+\infty]$.*

1) *If $p < d$, $W^{1,p}(\Omega) \hookrightarrow L^q(\Omega)$ for every $q \in [p,p^*]$ where p^* is given by*

$$\frac{1}{p^*} = \frac{1}{p} - \frac{1}{d}. \tag{9.11}$$

2) *If $p = d$, $W^{1,p}(\Omega) \hookrightarrow L^q(\Omega)$ for every $q \in [d,\infty)$.*
3) *If $p > d$, $W^{1,p}(\Omega) \hookrightarrow L^\infty(\Omega)$. Moreover, for $p > d$, functions in $W^{1,p}(\Omega)$ admit a continuous representative and there is a constant C such that for every $u \in W^{1,p}(\Omega)$ (still denoting by u this continuous representative) one has a Hölder continuity estimate given by Morrey's inequality:*

$$|u(x) - u(y)| \leq C\|u\|_{W^{1,p}}|x - y|^{1-\frac{d}{p}}, \quad \forall (x,y) \in \Omega \times \Omega. \tag{9.12}$$

Theorem 9.9 (Rellich–Kondrachov). *Let $d \geq 2$, let Ω be an open bounded subset of $\mathbb{R}^d$ with a C^1 boundary and let $p \in [1, +\infty]$. Assume that (u_n) is bounded in $W^{1,p}(\Omega)$.*

1) *If $p < d$, for every $q \in [1, p^*)$ with p^* given by (9.11), (u_n) has a subsequence which converges strongly in $L^q(\Omega)$.*
2) *If $p = d$, for every $q \in [1, \infty)$, (u_n) has a subsequence which converges strongly in $L^q(\Omega)$.*
3) *If $p > d$, (u_n) has a subsequence which converges in $C(\overline{\Omega})$.*

Even though $W^{1,p}(\Omega)$ functions need not be continuous they have a well-defined trace on the boundary of Ω.

Theorem 9.10 (Trace theorem). *Let Ω be an open bounded subset of $\mathbb{R}^d$ with a C^1 boundary and let $p \in [1, +\infty)$. Then there exists a linear continuous (trace) operator $\mathrm{tr} \in \mathcal{L}_c(W^{1,p}(\Omega), L^p(\partial\Omega, \sigma))$ such that*

$$\mathrm{tr}\, u = u|_{\partial\Omega}, \quad \forall u \in C_c^1(\mathbb{R}^d).$$

By density of smooth functions, the trace operator tr is unique and we shall simply denote $\mathrm{tr}\, u$ by $u|_{\partial\Omega}$ for $u \in W^{1,p}(\Omega)$. Arguing by density, one can use the trace theorem for various integration by parts formulas for Sobolev functions. For instance, if $\psi = (\psi_1, \ldots, \psi_d)$ with each $\psi_i \in W^{1,1}(\Omega)$, one has the divergence formula:

$$\int_\Omega \mathrm{div}(\psi) = \int_{\partial\Omega} \psi \cdot \nu d\sigma.$$

If u and v are in $H^1(\Omega)$, one has

$$\int_\Omega \partial_i u\, v = -\int_\Omega u\, \partial_i v + \int_{\partial\Omega} (uv)\nu_i d\sigma.$$

If $1 \leq p < \infty$, we denote by $W_0^{1,p}(\Omega)$ the closure of $C_c^\infty(\Omega)$ in $W^{1,p}(\Omega)$ for $\|\cdot\|_{W^{1,p}(\Omega)}$. Equipped with the norm of $W^{1,p}(\Omega)$, $W_0^{1,p}(\Omega)$ is a closed (hence also weakly closed) subspace of $W^{1,p}(\Omega)$, the connection with the trace map when Ω has a smooth boundary is as follows.

Proposition 9.3. *Let Ω be an open bounded subset of $\mathbb{R}^d$ with a C^1 boundary and let $p \in [1, +\infty)$ and let tr be the trace map from $W^{1,p}(\Omega)$ to $L^p(\partial\Omega, \sigma)$. Then $W_0^{1,p}(\Omega)$ is the null space of tr:*

$$W_0^{1,p}(\Omega) = \{u \in W^{1,p}(\Omega) : u|_{\partial\Omega} = 0\}.$$

A key inequality is as follows.

Theorem 9.11 (Poincaré's inequality). *Let $p \in [1, \infty)$ and Ω be a bounded open subset of $\mathbb{R}^d$. Then there exists a constant C such that*

$$\|u\|_{L^p(\Omega)} \leq C\|\nabla u\|_{L^p(\Omega)}, \quad \forall u \in W_0^{1,p}(\Omega).$$

In other words, $u \mapsto \|\nabla u\|_{L^p(\Omega)}$ defines on $W_0^{1,p}(\Omega)$ a norm which is equivalent to the $W^{1,p}(\Omega)$-norm. When Ω is bounded and has a C^1 boundary, a useful variant of the Poincaré inequality above is the Poincaré inequality with a trace term:

$$\|u\|_{L^p(\Omega)} \leq C(\|u_{|\partial\Omega}\|_{L^p(\partial\Omega,\sigma)} + \|\nabla u\|_{L^p(\Omega)}), \quad \forall u \in W^{1,p}(\Omega),$$

which does not impose that $u_{|\partial\Omega} = 0$.

9.2. On integral functionals

9.2.1. *Continuity, semicontinuity*

Let $p \geq 1$, Ω be an open subset of $\mathbb{R}^d$ and $F : \Omega \times \mathbb{R}^m \to \mathbb{R} \cup \{+\infty\}$ be measurable, such that

$$\text{for a.e. } x \in \Omega, \ u \mapsto F(x,u) \text{ is lsc on } \mathbb{R}^m, \tag{9.13}$$

and there exists some $\alpha \in L^1(\Omega, \mathbb{R})$ and some $M \in \mathbb{R}$ (possibly negative) such that

$$\text{for a.e. } x \in \Omega \text{ and every } u \in \mathbb{R}^m, F(x,u) \geq \alpha(x) + M|u|^p. \tag{9.14}$$

Then, given $u \in L^p(\Omega, \mathbb{R}^m)$ define

$$J_F(u) := \int_\Omega F(x, u(x))\mathrm{d}x. \tag{9.15}$$

These assumptions ensure that J_F is well-defined, possibly taking the value $+\infty$.

Lemma 9.1. *Let $p \geq 1$, F satisfy (9.13)–(9.14). Then, the functional J_F defined in (9.15) is lsc for the L^p topology. If, in addition $F(x,.)$ is convex for a.e. $x \in \Omega$, then J_F is lsc for the weak L^p topology.*

Proof. Let u_n converge to u in L^p, then passing to a subsequence if necessary, we may assume that u_n converges a.e. to u (see [23, Theorem IV.9]). Now observe that $F(x, u_n) - \alpha - M|u_n|^p$ is non-negative, Fatou's lemma and (9.14) then gives

$$\liminf_n \int_\Omega (F(x, u_n(x)) - \alpha(x) - M|u_n(x)|^p)\mathrm{d}x$$

$$= \liminf_n J_F(u_n) - \int_\Omega \alpha - M \int_\Omega |u|^p$$

$$\geq \int_\Omega \liminf_n (F(x, u_n(x)) - \alpha(x) - M|u_n(x)|^p)\mathrm{d}x$$

$$\geq J_F(u) - \int_\Omega \alpha - M \int_\Omega |u|^p.$$

This shows that J_F is lsc for the L^p topology. If in addition F is convex in its second argument, J_F is convex hence also lsc for the weak L^p topology, thanks to Theorem 3.8. □

Corollary 9.2. *Let $p \geq 1$, $F : \Omega \times \mathbb{R}^m \to \mathbb{R}$ be measurable such that $F(x,.)$ is continuous for a.e. x and such that there exist $\alpha \in L^1(\Omega, \mathbb{R})$ and a constant M such that*

$$|F(x, u)| \leq M|u|^p + \alpha(x), \text{ for a.e. } x \in \Omega \text{ and every } u \in \mathbb{R}^m. \quad (9.16)$$

Then, the functional J_F defined in (9.15) is continuous for the L^p topology.

Proof. Both F and $-F$ satisfy the assumptions of Lemma 9.1 so that both J_F and $-J_F$ are lsc. □

9.2.2. *The importance of being convex*

When $d = 1$, a remarkable fact is that the convexity of the integrand is not only a sufficient condition for weak lsc but also a necessary one. To see this, let us first recall the classic Riemann–Lebesgue lemma.

Lemma 9.2. *Let (a, b) be a finite open interval, let $\gamma_0 \in L^\infty((a, b), \mathbb{R}^m)$ and extend γ_0 to $\mathbb{R}$ by periodicity (i.e., set $\gamma_0(t + k(b - a)) = \gamma_0(t)$ for every $k \in \mathbb{Z}$ and $t \in (a, b)$) and for $n \in \mathbb{N}^*$ set*

$$\gamma_n(t) := \gamma_0(a + n(t - a)), \quad \forall t \in (a, b).$$

Then γ_n converges weakly $*$ in $L^\infty((a,b),\mathbb{R}^m)$ to[d] $\bar\gamma$ the average of γ_0:

$$\bar\gamma := \frac{1}{b-a}\int_a^b \gamma_0.$$

Proof. Making the change of variable $t \in (a,b) \mapsto \frac{t-a}{b-a}$ we may assume that $a = 0$ and $b = 1$, arguing componentwise, we may also assume that $m = 1$. Then, we observe that $\|\gamma_n\|_{L^\infty} = \|\gamma_0\|_{L^\infty}$ so that γ_n is bounded in L^∞, hence passing to a subsequence, if necessary, we may assume that γ_n weakly-$*$ converges to some $\gamma \in L^\infty$. Let now $\varphi \in C_c((0,1),\mathbb{R})$ and compute:

$$\int_0^1 \gamma_n\varphi = \int_0^1 \gamma_0(nt)\varphi(t)\mathrm{d}t = \sum_{k=0}^{n-1}\int_{\frac{k}{n}}^{\frac{k+1}{n}}\gamma_0(nt)\varphi(t)dt$$

$$= \sum_{k=0}^{n-1}\int_0^{\frac{1}{n}}\gamma_0(ns+k)\varphi\left(s+\frac{k}{n}\right)ds = \sum_{k=0}^{n-1}\int_0^{\frac{1}{n}}\gamma_0(ns)\varphi\left(s+\frac{k}{n}\right)ds$$

$$= \sum_{k=0}^{n-1}\frac{1}{n}\int_0^1\gamma_0(t)\varphi\left(\frac{t+k}{n}\right)dt = \int_0^1\gamma_0(t)\left(\frac{1}{n}\sum_{k=0}^{n-1}\varphi\left(\frac{t+k}{n}\right)\right)dt.$$

Then, we observe that, uniformly in $t \in [0,1]$, we have

$$\frac{1}{n}\sum_{k=0}^{n-1}\varphi\left(\frac{t+k}{n}\right) \to \int_0^1\varphi \text{ as } n \to +\infty.$$

Hence

$$\lim_n \int_0^1 \gamma_n\varphi = \int_0^1\gamma_0\int_0^1\varphi = \int_0^1\bar\gamma\varphi.$$

In particular $\int_0^1(\gamma - \bar\gamma)\varphi = 0$ for every $\varphi \in C_c((0,1),\mathbb{R})$ but since $C_c((0,1),\mathbb{R})$ is dense in L^1 this gives $\gamma = \bar\gamma$ and the whole sequence γ_n weakly $*$ converges to $\bar\gamma$. $\qquad\square$

---

[d]Recall that $L^\infty((a,b),\mathbb{R}^m)$ can be identified with the dual of $L^1((a,b),\mathbb{R}^m)$ so a sequence $\gamma_n \in L^\infty(((a,b),\mathbb{R}^m)$ weakly-$*$ converges to some $\gamma \in L^\infty$ if and only if $\int_a^b \gamma_n \cdot \varphi$ converges to $\int_a^b \gamma \cdot \varphi$ for every $\varphi \in L^1((a,b),\mathbb{R}^m)$. It then follows from the Banach–Aaloglu theorem and the fact that $L^1(\Omega)$ is separable (see [23, Chapter III]) that every bounded sequence in L^∞ possesses a subsequence which converges weakly-$*$.

Theorem 9.12. *Let $F \in C(\mathbb{R}^m, \mathbb{R})$ and set*

$$J_F(\gamma) := \int_a^b F(\gamma(t))dt, \forall \gamma \in L^\infty((a,b), \mathbb{R}^m).$$

If J_F is weakly-$*$ sequentially lsc in L^∞, then F is convex.

Proof. Let $(x,y) \in \mathbb{R}^m$, $\lambda \in [0,1]$ and set

$$\lambda_0 := \mathbf{1}_{(a, a+\lambda(b-a))}$$

and extend λ_0 to $\mathbb{R}$ by periodicity to the whole of $\mathbb{R}$. For $n \in \mathbb{N}^*$ and $t \in (a,b)$ set:

$$\lambda_n(t) := \lambda_0(a + n(t-a))$$

and then define the sequence of curves

$$\gamma_n(t) := (1 - \lambda_n(t))x + \lambda_n(t)y.$$

It follows from Lemma 9.2 that γ_n converges weakly $*$ in L^∞ to $(1-\lambda)x+\lambda y$. Now since by construction λ_n takes values in $\{0,1\}$ we have

$$F(\gamma_n(t)) = (1 - \lambda_n(t))F(x) + \lambda_n(t)F(y)$$

so, thanks to Lemma 9.2 again, $F(\gamma_n)$ converges weakly-$*$ in L^∞ to $(1-\lambda)F(x)+\lambda F(y)$, In particular, $\int_a^b F(\gamma_n(t))$ converges to $(b-a)((1-\lambda)F(x)+\lambda F(y))$. If J_F is weakly $*$ sequentially lsc in L^∞, we thus have

$$(b-a)((1-\lambda)F(x) + \lambda F(y)) = \lim_n \int_a^b F(\gamma_n)$$

$$\geq J_F((1-\lambda)x + \lambda y)$$

$$= (b-a)F((1-\lambda)x + \lambda y). \qquad \square$$

Remark 9.1. Observe that being lsc for the weak-$*$ topology of L^∞ is a weaker condition than being weakly lsc in L^p, hence convexity of F is also necessary for J_F to be weakly lsc in L^p.

An immediate consequence of Theorem 9.12 is the following corollary.

Corollary 9.3. *Let $p \geq 1$, $F \in C(\mathbb{R}^m, \mathbb{R})$ and set*

$$I_F(\gamma) := \int_a^b F(\dot{\gamma}(t))dt, \quad \forall \gamma \in W^{1,p}((a,b), \mathbb{R}^m).$$

If I_F is weakly sequentially lsc in $W^{1,p}((a,b), \mathbb{R}^m)$, then F is convex.

Proof. If θ_n converges weakly $*$ to θ in L^∞ and we define $\gamma_n(t) := \int_a^t \theta_n$ and $\gamma := \int_a^t \theta$ for every $t \in (a, b)$, then γ_n converges weakly to γ in $W^{1,p}((a, b), \mathbb{R}^m)$ (see Exercise 9.1). Hence, if I_F is weakly sequentially lsc in $W^{1,p}((a, b), \mathbb{R}^m)$, then $\theta \in L^\infty \mapsto \int_a^b F(\theta(t))dt$ is weakly $*$ sequentially lsc in L^∞ which, by Theorem 9.12, implies that F is convex. $\qquad\square$

Remark 9.2. The proof of Corollary 9.3 follows from Theorem 9.12 by simple integration. In higher dimensions, i.e., for functionals of the form

$$(u_1, \ldots, u_m) \mapsto \int_\Omega F(Du),$$

where Ω is an open subset of $\mathbb{R}^d$, the picture is more complex. Convexity is still a necessary condition for weak lsc when $m = 1$ (see [39, 46]) but it is no longer the case in the *vectorial* setting where both m and d are larger than 2 (see Exercise 9.15 for an example). It is Charles Morrey who first identified the right condition, which he called quasiconvexity,[e] for weak lower semicontinuity in the vectorial case, see [1, 39, 77, 81].

In the proof of Theorem 9.12, for simplicity we took F depending only on γ but adding a (continuous, say) dependence on time does not create extra difficulties, we let the reader check the following exercise.

Exercise 9.2. *Let* $F \in C([a, b] \times \mathbb{R}^m, \mathbb{R})$ *and set*

$$J_F(\gamma) := \int_a^b F(t, \gamma(t))dt, \forall \gamma \in L^\infty((a, b), \mathbb{R}^d).$$

Show that if J_F *is weakly-$*$ sequentially lsc in* L^∞ *then* $F(t, \cdot)$ *is convex for every* $t \in [a, b]$. *(Hint: show for every subinterval* $[t_0, t_0 + \varepsilon] \subset [a, b]$, $\gamma \mapsto \frac{1}{\varepsilon} \int_{t_0}^{t_0+\varepsilon} F(t, \gamma(t))dt$ *is weakly-$*$ lsc, argue as in the proof of Theorem 9.12 and let* $\varepsilon \to 0$.)

9.2.3. *Differentiability*

Given Ω on open bounded subset of $\mathbb{R}^d$, m and N in $\mathbb{N}^*$ consider a Lagrangian: $L : \Omega \times \mathbb{R}^m \times \mathbb{R}^N \to \mathbb{R}$ measurable and such that

$$\text{for a.e. } x \in \Omega, (u, v) \in \mathbb{R}^m \times \mathbb{R}^N \mapsto L(x, u, v) \text{ is of class } C^1, \qquad (9.17)$$

---

[e] Not to be confused with the notion of quasiconvexity we saw in Exercise 2.16.

we denote by $\nabla_u L(x, u, v)$ and $\nabla_v L(x, u, v)$ the corresponding partial gradients[f] and assume that for some $p \geq 1$, $q \geq 1$ and $C \geq 0$, one has for almost every $x \in \Omega$ and every $(u, v) \in \mathbb{R}^m \times \mathbb{R}^N$:

$$|\nabla_u L(x, u, v)| \leq C(1 + |u|^{p-1}), \quad |\nabla_v L(x, u, v)| \leq C(1 + |v|^{q-1}) \quad (9.18)$$

and for some $\alpha \in L^1(\Omega, \mathbb{R})$

$$|L(x, u, 0)| \leq \alpha(x) + C|u|^p. \quad (9.19)$$

Combining (9.18) and (9.19), we see that for almost every $x \in \Omega$ and every $(u, v) \in \mathbb{R}^m \times \mathbb{R}^N$

$$|L(x, u, v)| \leq \alpha(x) + C|u|^p + C(|v| + |v|^q) \quad (9.20)$$

so that the functional

$$I(u, v) := \int_\Omega L(x, u(x), v(x))dx, \ (u, v) \in L^p(\Omega, \mathbb{R}^m) \times L^q(\Omega, \mathbb{R}^N) \quad (9.21)$$

is well-defined (and actually continuous for the strong topology of $L^p(\Omega, \mathbb{R}^m) \times L^q(\Omega, \mathbb{R}^N)$ by the same reasoning as in Corollary 9.2). Regarding the differentiability of I, we have the following proposition.

Proposition 9.4. *Let $p \geq 1$, $q \geq 1$, L satisfy (9.17)–(9.19) and let I be defined by (9.21). Then I is Gateaux-differentiable on $L^p(\Omega, \mathbb{R}^m) \times L^q(\Omega, \mathbb{R}^N)$ and for every $(\varphi, \psi) \in L^p(\Omega, \mathbb{R}^m) \times L^q(\Omega, \mathbb{R}^N)$ one has*

$$I'(u, v)(\varphi, \psi) = \int_\Omega \nabla_u L(x, u(x), v(x)) \cdot \varphi(x)dx$$

$$+ \int_\Omega \nabla_v L(x, u(x), v(x)) \cdot \psi(x)dx.$$

Proof. Let $(\varphi, \psi) \in L^p(\Omega, \mathbb{R}^m) \times L^q(\Omega, \mathbb{R}^N)$ and $\varepsilon \in \mathbb{R} \setminus \{0\}$, then let us write

$$\frac{1}{\varepsilon}[I(u + \varepsilon\varphi, v + \varepsilon\psi) - I(u, v)] = \int_\Omega \eta_\varepsilon,$$

---

[f]That is, $\nabla_u L(x, u, v) = (\partial_{u_1} L(x, u, v), \ldots, \partial_{u_m} L(x, u, v))^T \in \mathbb{R}^m$ and $\nabla_v L(x, u, v) = (\partial_{v_1} L(x, u, v), \ldots, \partial_{v_N} L(x, u, v))^T \in \mathbb{R}^N$.

where

$$\eta_\varepsilon(x) := \frac{1}{\varepsilon}[L(x, u(x) + \varepsilon\varphi(x), v(x) + \varepsilon\psi(x)) - L(x, u(x), v(x))].$$

It follows from (9.17) that for almost every $x \in \Omega$ one has

$$\lim_{\varepsilon \to 0} \eta_\varepsilon(x) = \nabla_u L(x, u(x), v(x)) \cdot \varphi(x) + \nabla_v L(x, u(x), v(x)) \cdot \psi(x). \quad (9.22)$$

It now follows from the mean value theorem that there is some $\theta \in (0, 1)$ (depending on ε and x) such that

$$\eta_\varepsilon(x) = \nabla_u L(x, u(x) + \varepsilon\theta\varphi(x), v(x) + \varepsilon\theta\psi(x)) \cdot \varphi(x)$$
$$+ \nabla_v L(x, u(x) + \varepsilon\theta\varphi(x), v(x) + \varepsilon\theta\psi(x)) \cdot \psi(x).$$

Using (9.18), we thus get

$$|\eta_\varepsilon| \leq \eta := C(1 + (|u| + |\varphi|)^{p-1})|\varphi| + C(1 + (|v| + |\psi|)^{q-1})|\psi|.$$

But since u and φ are in L^p, $(|u|+|\varphi|)^{p-1} \in L^{p'}$ where $p' = \frac{p}{p-1}$ is the conjugate exponent of p (with the convention $p' = \infty$ if $p = 1$) and since $\varphi \in L^p$, we deduce from Hölder's inequality that $(1 + (|u| + |\varphi|)^{p-1})|\varphi| \in L^1(\Omega)$. Likewise $(1 + (|v| + |\psi|)^{q-1})|\psi| \in L^1(\Omega)$. Hence η is an integrable majorant of $|\eta_\varepsilon|$. It therefore follows from Lebesgue's dominated convergence theorem and (9.22) that I is differentiable at (u, v) in the direction (φ, ψ), the corresponding directional derivative being given by the linear expression:

$$DI((u, v); (\varphi, \psi)) = \int_\Omega A \cdot \varphi + B \cdot \psi,$$

where $(A, B)(x) := (\nabla_u L(x, u(x), v(x)), \nabla_v L(x, u(x), v(x)))$. Thanks to (9.18), we thus have $(A, B) \in L^{p'}(\Omega, \mathbb{R}^m) \times L^{q'}(\Omega, \mathbb{R}^N)$. Thanks to Hölder's inequality, $(\varphi, \psi) \mapsto DI((u, v); (\varphi, \psi))$ is a continuous linear form on $L^p(\Omega, \mathbb{R}^m) \times L^q(\Omega, \mathbb{R}^N)$, which establishes the desired Gateaux-differentiability claim. $\qquad \square$

9.3. The direct method

The first question when facing a variational problem is whether a minimizer exists (in a suitable Sobolev space, say), we shall give some sufficient conditions in Section 9.3.2 which ensure existence. This will be based on the so-called *direct method of the calculus of variations* which consists in taking a minimizing sequence, to get some bound in a reflexive Sobolev space, to take a weakly convergent subsequence and then to invoke some weak lsc argument to show that the corresponding weak limit is indeed a minimizer.

9.3.1. *Obstructions to existence*

Before presenting the direct method, we wish to give some classic examples of simple variational problems which do not possess solutions.

Example 9.1 (Bolza, oscillating minimizing sequences). Consider

$$\inf_{\gamma \in W^{1,4}((0,1),\mathbb{R}),\, \gamma(0)=\gamma(1)=0} E(\gamma) := \int_0^1 [(1 - \dot{\gamma}^2(t))^2 + \gamma^2(t)]\mathrm{d}t. \qquad (9.23)$$

Since E is non-negative, so is the infimum of (9.23). Now consider $\gamma_0(t) := |t-1/2|-1/2$ for $t \in [0,1]$ extended by periodicity to the whole of $\mathbb{R}$ and for $n \in \mathbb{N}^*$ set $\gamma_n(t) := \frac{1}{n}\gamma_0(nt)$, note that γ_n is admissible that $\dot{\gamma}_n \in \{-1,1\}$ a.e. and $\|\gamma_n\|_\infty \leq \frac{1}{2n}$ a.e. so that $E(\gamma_n) \leq \frac{1}{4n^2}$, from which we deduce that the infimum of (9.23) is 0. But if $E(\gamma) = 0$, then, we should have at the same time $\gamma = 0$ a.e. hence $\gamma(t) = 0$ for every $t \in [0,1]$ (because γ being $W^{1,4}$ is continuous) and $\dot{\gamma} \in \{-1,1\}$, which clearly is impossible. What goes wrong in this example comes from the non-convexity of the potential $(1-\xi^2)^2$ which is minimal at 1 and -1 (this is a so-called double-well potential), the minimizing sequence constructed above oscillates very fast using the two slopes ± 1 which make the potential minimal and the fast oscillations of the velocity enables us to make the other term $\gamma^2(t)$ arbitrarily small.

A variant with free endpoints is left to the reader as an exercise.

Exercise 9.3. *Show that the problem*

$$\inf_{\gamma \in W^{1,8}((0,1),\mathbb{R})} E(\gamma) := \int_0^1 [(\dot{\gamma}^2(t) - 2\dot{\gamma}(t))^4 + (\gamma(t) - t)^8]\mathrm{d}t \qquad (9.24)$$

does not admit any solution.

Another obstruction for existence of minimizers is when the growth of the energy is linear. In this case the natural space to work with[g] is $W^{1,1}$

---

[g]In fact, one way to recover some sort of weak compactness is to work in the larger space BV of functions of bounded variation, but we won't develop it here even though this space plays a key role in many variational problems and in particular in minimal surfaces or isoperimetric problems, see [5, 47, 59].

which is not reflexive and one cannot hope for existence of a minimizer in general as the next example shows.

Example 9.2. Consider now

$$\inf_{\gamma \in W^{1,1}((0,1),\mathbb{R}),\, \gamma(0)=1,\gamma(1)=0} E(\gamma) := \int_0^1 [|\dot{\gamma}(t)| + |\gamma(t)|]dt. \qquad (9.25)$$

Observe that if γ is admissible then

$$E(\gamma) \geq \int_0^1 |\dot{\gamma}(t)|dt \geq |\gamma(1) - \gamma(0)| = 1$$

so that the infimum in (9.25) is larger than 1. Now observe that if $n \in \mathbb{N}^*$, and $\gamma_n(t) := (1 - nt)_+$ for $t \in [0,1]$, γ_n is admissible and

$$E(\gamma_n) = 1 + \frac{1}{2n}$$

so that the infimum in (9.25) is in fact 1. But if γ was admissible with $E(\gamma) = 1$ we should have $\int_0^1 |\gamma| = 0$ hence $\gamma(t) = 0$ for every $t \in [0,1]$ since γ is $W^{1,1}$ hence continuous on $[0,1]$, this obviously contradicts the boundary condition $\gamma(0) = 1$.

We leave as an exercise the next example (due to Weierstrass) of non-existence.

Exercise 9.4. *Consider*

$$\inf_{\gamma \in H^1((0,1)),\mathbb{R}),\, \gamma(0)=1,\, \gamma(1)=0} E(\gamma) := \int_0^1 t\dot{\gamma}^2(t)\, dt. \qquad (9.26)$$

1) *For $n \in \mathbb{N}^*$ set*

$$\gamma_n(t) := \begin{cases} 1 & \text{if } t \in [0, 1/n], \\ -\dfrac{\log(t)}{\log(n)} & \text{if } t \in [1/n, 1], \end{cases}$$

compute $E(\gamma_n)$.

2) *Deduce the value of the infimum in (9.26) and that (9.26) does not have any solution.*

9.3.2. *Existence in the separable case*

For the sake of simplicity, we consider in this section a separable problem. There are many variants and extensions that deal with more general Lagrangians and we will give another existence result that deals with non-separable Lagrangians in Section 9.4.2. Given Ω, an open bounded subset of $\mathbb{R}^d$ with a C^1 boundary, $p \in (1, +\infty)$, $F : \Omega \times \mathcal{M}_{m \times d} \to \mathbb{R} \cup \{+\infty\}$, $G : \Omega \times \mathbb{R}^m \to \mathbb{R} \cup \{+\infty\}$ measurable such that

$$\text{for a.e. } x \in \Omega, \ F(x, .) \text{ is convex, lsc on } \mathcal{M}_{m \times d}, \ G(x, .) \text{ is lsc on } \mathbb{R}^m. \tag{9.27}$$

We also assume that there exist α and β in $L^1(\Omega)$, $q \in [1, p)$ and $C \in \mathbb{R}$ such that

$$\text{for a.e. } x \in \Omega, \ \forall \xi \in \mathcal{M}_{m \times d}, \ F(x, \xi) \geq \alpha(x) + M|\xi|^p \tag{9.28}$$

and

$$\text{for a.e. } x \in \Omega, \ \forall u \in \mathbb{R}^m, \ G(x, u) \geq \beta(x) - C|u|^q. \tag{9.29}$$

We are finally given $u_0 \in W^{1,p}(\Omega)^m$ and consider

$$\inf_{u \in u_0 + W_0^{1,p}(\Omega)^m} J(u) := \int_\Omega [F(x, Du(x)) + G(x, u(x))] \mathrm{d}x, \tag{9.30}$$

where the condition $u \in u_0 + W_0^{1,p}(\Omega)^m$ has to be understood as the Dirichlet boundary condition $u = u_0$ on $\partial\Omega$ in the sense of traces and $Du(x)$ is the Jacobian matrix of u at x, i.e., the $m \times d$ matrix whose rows are $\nabla u_1(x)^T, \ldots, \nabla u_m(x)^T$. We finally assume that

$$J(u_0) < +\infty. \tag{9.31}$$

Then we have the following theorem:

Theorem 9.13. *Under assumptions (9.27)–(9.29) and (9.31), problem (9.30) admits at least one solution.*

Proof. Let us first remark that $u_0 + W_0^{1,p}(\Omega)^m$ is a closed affine subspace of $W^{1,p}(\Omega)^m$. It is therefore closed and convex hence also weakly closed. Let us prove now that $J + \chi_{u_0 + W_0^{1,p}(\Omega)^m}$ is coercive. In what follows, A will be a positive constant which may change from one line to another. Firstly,

Poincaré's inequality gives that for every $u \in u_0 + W_0^{1,p}(\Omega)^m$

$$\|u\|_{L^p} \le \|u_0\|_{L^p} + A(\|Du\|_{L^p} + \|Du_0\|_{L^p}). \tag{9.32}$$

Assumptions (9.28)–(9.29) give

$$J(u) \ge \int_\Omega (\alpha + \beta) + M\|Du\|_{L^p}^p - C\|u\|_{L^q}^q. \tag{9.33}$$

It follows from Hölder's inequality that

$$\|u\|_{L^q} \le |\Omega|^{\frac{1}{q} - \frac{1}{p}} \|u\|_{L^p}. \tag{9.34}$$

Combining (9.32)–(9.34) and $q < p$ then gives

$$J(u) \ge A(\|Du\|_{L^p}^p - \|Du\|_{L^p}^q - 1)$$
$$\ge A(\|Du\|_{L^p}^p - 1).$$

With Poincaré's inequality, this implies that $J + \chi_{u_0 + W_0^{1,p}(\Omega)^m}$ is coercive in $W^{1,p}(\Omega)^m$. Since $u_0 + W_0^{1,p}(\Omega)^m$ is weakly closed, $\chi_{u_0 + W_0^{1,p}(\Omega)^m}$ is sequentially weakly lsc. Let us now prove that J is sequentially weakly lsc: if u_n converges weakly in $W^{1,p}(\Omega)^m$ to some u (hence is bounded in $W^{1,p}(\Omega)^m$ thanks to the Banach–Steinhaus theorem), then by Rellich–Kondrachov's theorem, taking a subsequence if necessary, u_n converges strongly to u in L^p and Du_n converges weakly to Du in L^p. It then follows from assumptions (9.27)–(9.29) and Lemma 9.1 that

$$\liminf_n \int_\Omega F(x, Du_n(x))\mathrm{d}x \ge \int_\Omega F(x, Du(x))\mathrm{d}x,$$

$$\liminf_n \int_\Omega G(x, u_n(x))\mathrm{d}x \ge \int_\Omega G(x, u(x)),$$

which proves that J is sequentially weakly lsc in $W^{1,p}(\Omega)^m$. Finally recalling that $W^{1,p}(\Omega)^m$ is reflexive and (9.31), the existence of a minimizer follows from Theorem 1.29. $\qquad \square$

Remark 9.3. We let the reader check that the same existence result holds if $q = p$ in (9.29) and the constant C is small enough compared to the constant M in (9.28), namely $C < \frac{M}{\nu_p(\Omega)}$ where $\nu_p(\Omega)$ is the Poincaré constant:

$$\nu_p(\Omega) := \sup \left\{ \int_\Omega |v|^p, \; v \in W_0^{1,p}(\Omega)^m, \; \int_\Omega |Dv|^p \le 1 \right\}.$$

Remark 9.4. We have used Poincaré's inequality and taken advantage of the fact that $u - u_0 \in W_0^{1,p}(\Omega)^m$. If there is no boundary condition, $C < 0$ and $p = q$ in (9.29), then J is coercive in $W^{1,p}(\Omega)^m$ and then existence of a minimizer can be proved in a similar way as above.

Let us consider a variant when $d = 1$, $\Omega = (a, b)$ and instead of prescribing $\gamma(a)$ and $\gamma(b)$ there is a cost depending on these boundary values.

Exercise 9.5. *Let $F : (a, b) \times \mathbb{R}^m \to \mathbb{R} \cup \{+\infty\}$ be measurable, convex and lsc in its second argument and such that for some $p > 1$, some $M > 0$ and some $\alpha \in L^1(a, b)$*

$$F(t, v) \geq \alpha(t) + M|v|^p.$$

Let $G : (a, b) \times \mathbb{R}^m \to \mathbb{R} \cup \{+\infty\}$ be measurable, bounded from below and lsc in its second argument and let $\varphi, \psi : \mathbb{R}^m \to \mathbb{R} \cup \{+\infty\}$ be lsc and bounded from below with $\varphi(v) \to +\infty$ as $|v| \to +\infty$. Then, define

$$E(\gamma) := \int_a^b [F(t, \dot{\gamma}(t)) + G(t, \gamma(t))]dt + \varphi(\gamma(a)) + \psi(\gamma(b)),$$

$$\forall \gamma \in W^{1,p}((a, b), \mathbb{R}^m),$$

and assume that there exists $\gamma_0 \in W^{1,p}((a, b), \mathbb{R}^m)$ such that $E(\gamma_0) < +\infty$. Show that E has a minimizer on $W^{1,p}((a, b), \mathbb{R}^m)$.

9.3.3. Relaxation

Let $p \in (1, +\infty)$, $m \in \mathbb{N}^*$, $a < b$ be given $F \in C(\mathbb{R}^m, \mathbb{R})$ such that for some constant $\Lambda \geq 1$ one has

$$\frac{1}{\Lambda}|\xi|^p - \Lambda \leq F(\xi) \leq \Lambda(|\xi|^p + 1), \ \forall \xi \in \mathbb{R}^m \tag{9.35}$$

and $G : (a, b) \times \mathbb{R}^m \to \mathbb{R}$ measurable such that $\gamma \in \mathbb{R}^m \mapsto G(t, \gamma)$ is continuous for every $t \in (a, b)$, and such that there exist α and $\dot{\beta}$ in $L^1((a, b), \mathbb{R})$ constants $C \geq c > 0$ such that

$$\alpha(t) + c|\gamma|^p \leq G(t, \gamma) \leq \beta(t) + C|\gamma|^p, \ \forall (t, \gamma) \in (a, b) \times \mathbb{R}^m. \tag{9.36}$$

We are interested in

$$\inf_{\gamma \in W^{1,p}((a,b),\mathbb{R}^m)} J(\gamma) := \int_a^b [F(\dot{\gamma}(t)) + G(t, \gamma(t))]dt. \tag{9.37}$$

The assumptions above ensure, using Corollary 9.2, that J is continuous for the strong $W^{1,p}$ topology, but unless F is convex, the existence of a minimizer is not guaranteed since J need not be weakly lsc (if necessary, have a second look at Example 9.1 or Exercise 9.3). This is why one needs to introduce a relaxation of (9.37) in which, F^{**}, the convex envelope of F, plays a crucial role. By definition F^{**} is the largest convex minorant of F:

$$F^{**} := \sup\{f : f \text{ is convex and } f \leq F \text{ on } \mathbb{R}^m\}.$$

Thanks to (9.35), $\xi \mapsto \frac{1}{\Lambda}|\xi|^p - \Lambda$ is a convex minorant of F, hence

$$\frac{1}{\Lambda}|\xi|^p - \Lambda \leq F^{**}(\xi) \leq \Lambda(|\xi|^p + 1), \ \forall \xi \in \mathbb{R}^m. \tag{9.38}$$

In particular this implies that F^{**} is continuous by Corollary 3.6. A useful characterization of the convex envelope is as follows.

Lemma 9.3. *Let $F \in C(\mathbb{R}^m, \mathbb{R})$ satisfy (9.35) and F^{**} be its convex envelope. Then for every $\xi \in \mathbb{R}^m$ one has*

$$F^{**}(\xi) = \min\left\{\sum_{k=1}^{m+1} \lambda_k F(\xi_k), \ \lambda_k \geq 0, \xi_k \in \mathbb{R}^m, \ \sum_{k=1}^{m+1} \lambda_k(1, \xi_k) = (1, \xi)\right\}. \tag{9.39}$$

Proof. First of all, it is easy to check that $F^{**}(\xi)$ equals:

$$\inf\left\{\sum_{k=1}^{N} \lambda_k F(\xi_k), \ N \in \mathbb{N}^*, \ \lambda_k \geq 0, \xi_k \in \mathbb{R}^m, \ \sum_{k=1}^{N} \lambda_k(1, \xi_k) = (1, \xi)\right\}. \tag{9.40}$$

Indeed, the infimum in (9.40) is clearly less than $F(\xi)$ and it is easy to check that it is a convex function of ξ, hence it is smaller than $F^{**}(\xi)$. Moreover, if f is a convex minorant of F and $\lambda_k \geq 0, \xi_k \in \mathbb{R}^m$, $\sum_{k=1}^{N} \lambda_k(1, \xi_k) = (1, \xi)$, then

$$f(\xi) \leq \sum_{k=1}^{N} \lambda_k f(\xi_k) \leq \sum_{k=1}^{N} \lambda_k F(\xi_k)$$

so that $f(\xi)$ is less than the infimum of (9.40), hence $F^{**}(\xi)$ is smaller than the infimum of (9.40).

It now follows from (9.40) that $(\xi, F^{**}(\xi))$ is in the closure of the convex hull of the graph of F, $\mathrm{co}(\mathrm{graph}(F))$, but thanks to Carathéodory's theorem, every point in $\mathrm{co}(\mathrm{graph}(F))$ can be written as convex combinations of at most $m+2$ points of the form $(\xi_k, F(\xi_k))$. This shows that (9.40) can be improved by imposing the bound $m+2$ on N and enables us to deduce

$$F^{**}(\xi) = \inf \left\{ \sum_{k=1}^{m+2} \lambda_k F(\xi_k), \ \lambda_k \geq 0, \xi_k \in \mathbb{R}^m, \ \sum_{k=1}^{m+2} \lambda_k (1, \xi_k) = (1, \xi) \right\}.$$

$$(9.41)$$

Now we claim that the infimum in (9.41) is attained. Indeed take a minimizing sequence (λ_k^n, ξ_k^n)

$$\lambda_k^n \geq 0, \xi_k^n \in \mathbb{R}^m, \ \sum_{k=1}^{m+2} \lambda_k^n (1, \xi_k^n) = (1, \xi), \ \lim_n \sum_{k=1}^{m+2} \lambda_k^n F(\xi_k^n) = F^{**}(\xi).$$

Since λ_k^n is bounded, taking a subsequence if necessary, we may assume that λ_k^n converges to some λ_k. Moreover, thanks to (9.35), $\sum_{k=1}^{m+2} \lambda_k^n |\xi_k^n|^p$ is bounded. Let $K := \{k \in \{1, \ldots, m+2\} : \lambda_k > 0\}$, if $k \in K$, ξ_k^n is bounded hence can be assumed to converge to some ξ_k, if $k \notin K$ then (passing to subsequences if necessary) it is easy to see that $\lambda_k^n \xi_k^n$ tends to 0 (this is obvious if ξ_k^n is bounded and if $|\xi_k^n|$ tends to $+\infty$ it follows from the fact that $\lambda_k^n |\xi_k^n|^p$ is bounded and $p > 1$), we thus have

$$\lambda_k \geq 0, \forall k \in K, \ \sum_{k \in K} \lambda_k (1, \xi_k) = (1, \xi).$$

Of course $F^{**}(\xi) \leq \sum_{k \in K} \lambda_k F(\xi_k)$, and the converse inequality follows from the fact that F is bounded from below (thanks to (9.35)):

$$F^{**}(\xi) = \lim_n \sum_{k=1}^{m+2} \lambda_k^n F(\xi_k^n)$$

$$\geq \lim_n \sum_{k \in K} \lambda_k^n F(\xi_k^n) + \inf F \lim_n \sum_{k \notin K} \lambda_k^n = \sum_{k \in K} \lambda_k F(\xi_k),$$

which shows that $F^{**}(\xi) = \sum_{k \in K} \lambda_k F(\xi_k)$ so that the infimum in (9.41) is a minimum.

The last step consists in showing that a convex combination of $m+1$ points is enough in the minimization (9.41). We have just seen that there exist $(\lambda_k, \xi_k)_{k=1,\dots,m+2}$ such that $\lambda_k \geq 0$, $\sum_{k=1}^{m+2} \lambda_k(1, \xi_k) = (1, \xi_k)$ and $F^{**}(\xi) = \sum_{k=1}^{m+2} \lambda_k F(\xi_k)$. Assume that $\lambda_k > 0$ for $k = 1, \dots, m+2$. By convexity of F^{**} we have

$$\sum_{k=1}^{m+2} \lambda_k F(\xi_k) = F^{**}(\xi) \leq \sum_{k=1}^{m+2} \lambda_k F^{**}(\xi_k)$$

but since $F^{**}(\xi_k) \leq F(\xi_k)$ we in fact have $F(\xi_k) = F^{**}(\xi_k)$. Now let $q \in \partial F^{**}(\xi)^{\text{h}}$, i.e., $F^{**}(\xi) + F^*(q) = q \cdot \xi$ so that

$$\sum_{k=1}^{m+2} \lambda_k (F^{**}(\xi_k) + F^*(q)) = \sum_{k=1}^{m+2} \lambda_k F(\xi_k) + F^*(q) = \sum_{k=1}^{m+2} \lambda_k q \cdot \xi_k$$

together with Young's inequality $F^{**}(\xi_k) + F^*(q) \geq q \cdot \xi_k$ and $\lambda_k > 0$ this yields $F^{**}(\xi_k) + F^*(q) = q \cdot \xi_k$ for every k, i.e.,

$$q \in \partial F^{**}(\xi_k), \quad \forall k = 1, \dots, m+2. \tag{9.42}$$

From Carathédory's theorem, there exists $I \subset \{1, \dots, m+2\}$ with cardinality at most $m+1$ and $(\alpha_i)_{i \in I}$, $\alpha_i \geq 0$ such that $\sum_{i \in I} \alpha_i = 1$ and $\xi = \sum_{i \in I} \alpha_i \xi_i$. Let us define the affine function $l(x) := q \cdot x - F^*(q)$, for all $x \in \mathbb{R}^m$. Thanks to the fact that $q \in \partial F^{**}(\xi)$, (9.42) and $F(\xi_i) = F^{**}(\xi)$, we have

$$F^{**}(\xi) = l(\xi) = \sum_{i \in I} \alpha_i l(\xi_i) = \sum_{i \in I} \alpha_i F^{**}(\xi_i) = \sum_{i \in I} \alpha_i F(\xi_i),$$

which completes the proof of (9.39) since I has cardinality less than $m+1$. $\square$

Remark 9.5. Note that in the proof above, we have only used the lower bound in (9.35). The last part of the proof which finds an optimal convex combination with at most $m+1$ (instead of $m+2$) points is not important for relaxation purposes but it is a good occasion for the reader to check her/his knowledge of the convex analysis material encountered so far. More importantly, it reveals further geometric information on convex envelopes; namely the fact that the points ξ_k which achieve the minimum in (9.39) with a positive weight λ_k are contact points between F and F^{**} and all share a common subgradient (see the exercise below for an interesting consequence).

$^{\text{h}}F^{**}$ is convex and continuous hence subdifferentiable everywhere.

Exercise 9.6. *Assume $F \in C^1(\mathbb{R}^m, \mathbb{R})$ satisfies (9.35). Prove that $F^{**} \in C^1(\mathbb{R}^m, \mathbb{R})$. (Hint: take $q \in \partial F^{**}(\xi)$, show that if $\lambda_k > 0$, $\sum_k \lambda_k (1, \xi_k) = (1, \xi)$ and $F^{**}(\xi) = \sum_k \lambda_k F(\xi_k)$ then $q = \nabla F(\xi_k)$.) Find $F \in C^\infty(\mathbb{R}, \mathbb{R})$ that satisfies (9.35), such that $F^{**} \notin C^2(\mathbb{R}, \mathbb{R})$.*

We are now ready to state a relaxation theorem for (9.37).

Theorem 9.14. *Let F and G and J be as above. Then*

$$\inf_{\gamma \in W^{1,p}((a,b),\mathbb{R}^m)} J(\gamma) = \min_{\gamma \in W^{1,p}((a,b),\mathbb{R}^m)} \bar{J}(\gamma), \qquad (9.43)$$

where $\bar{J}$ is the relaxed functional defined for every $\gamma \in W^{1,p}((a,b), \mathbb{R}^m)$ by

$$\bar{J}(\gamma) := \int_a^b [F^{**}(\dot{\gamma}(t)) + G(t, \gamma(t))]dt.$$

Proof. Since $\bar{J} \leq J$ the inequality $\inf_{\gamma \in W^{1,p}} \bar{J} \leq \inf_{\gamma \in W^{1,p}} J$ is obvious. Now, thanks to (9.38)–(9.36) and the convexity of F^{**}, the direct method (see Remark 9.4) shows that $\bar{J}$ admits a minimizer so that the right-hand side of (9.43) is indeed a minimum. It remains to show that $\inf_{\gamma \in W^{1,p}} J \geq \min_{\gamma \in W^{1,p}} \bar{J}$.

It follows from our assumptions on F and G, (9.38) and Corollary 9.2 that both $\bar{J}$ and J are continuous for the strong $W^{1,p}$ topology. In particular, since $C^1([a,b])$ is dense in $W^{1,p}$, the infimum of $\bar{J}$ over $W^{1,p}$ is the same as over $C^1([a,b])$. It is easy to check that any $\gamma \in C^1([a,b])$ is the limit in $W^{1,p}$ (actually even[i] on $W^{1,\infty}$) of a sequence of continuous and piecewise affine curves. So the infimum of $\bar{J}$ over $W^{1,p}$ is the same as its infimum over continuous piecewise affine curves. Let γ be such a curve so that there exists a subdivision $a = t_0 < t_1 < \cdots < t_{N-1} < t_N = b$ of $[a,b]$ such that for every $i = 0, \ldots, N-1$ and every $t \in I_i := [t_i, t_{i+1})$ one has

$$\gamma(t) = \gamma(t_i) + \alpha_i(t - t_i) \text{ with } \alpha_i := \frac{\gamma(t_{i+1}) - \gamma(t_i)}{t_{i+1} - t_i}. \qquad (9.44)$$

It follows from Lemma 9.3 that for every i there exist $(\lambda_k^i, \alpha_k^i)_{k=1,\ldots,m+1} \in (\mathbb{R}_+ \times \mathbb{R}^m)^{m+1}$ such that

$$\sum_{k=1}^{m+1} \lambda_k^i = 1, \quad \sum_{k=1}^{m+1} \lambda_k^i \alpha_k^i = \alpha_i, \quad \sum_{k=1}^{m+1} \lambda_k^i F(\alpha_k^i) = F^{**}(\alpha_i). \qquad (9.45)$$

---

[i]Indeed if $\gamma \in C^1([a,b])$ and $a = t_0 < t_1 < \cdots < t_{N-1} < t_N = b$ is a subdivision of $[a,b]$ with $\max_{i=0,N-1} |t_{i+1} - t_i| \leq \varepsilon$ and $\gamma_n(t) = \gamma(t_i) + \frac{(t-t_i)(\gamma(t_{i+1})-\gamma(t_i))}{t_{i+1}-t_i}$ for $t \in [t_i, t_{i+1})$ then one has $\|\dot{\gamma}_n - \dot{\gamma}\|_{L^\infty} \leq \omega(\varepsilon)$ where ω is a modulus of continuity of $\dot{\gamma}$.

For $i = 0, \ldots, N - 1$ take a partition of I_i into $m + 1$ disjoint intervals $(I_i^k)_{k=1,\ldots,m+1}$ in such a way that

$$|I_i^k| = \lambda_k^i |I_i| = \lambda_k^i (t_{i+1} - t_i). \tag{9.46}$$

Now define on I_i, $\bar{\lambda}_k^i := \mathbf{1}_{I_i^k}$, extend $\bar{\lambda}_k^i$ to $\mathbb{R}$ by periodicity and set for every $n \in \mathbb{N}^*$ and every $t \in I_i$:

$$\lambda_k^{i,n}(t) := \bar{\lambda}_k^i(t + n(t - t_i)), \ \alpha^{i,n}(t) := \sum_{k=1}^{m+1} \lambda_k^{i,n}(t)\alpha_k^i$$

and define the (Lipschitz) curve γ_n by

$$\gamma_n(t) := \gamma(t_i) + \int_0^t \alpha^{i,n}(s)ds, \ \forall t \in I_i, \ i = 0, \ldots, N - 1 \tag{9.47}$$

so that $\dot{\gamma}_n := \sum_{i=0}^{N-1} \mathbf{1}_{I_i} \alpha^{i,n}$ and, by virtue of Lemma 9.2, we have

$$\dot{\gamma}_n \text{ converges weakly} * \text{ to } \dot{\gamma} = \sum_{i=0}^{N-1} \mathbf{1}_{I_i} \alpha_i \text{ in } L^\infty \tag{9.48}$$

and

$$\|\gamma_n - \gamma\|_{L^\infty} \to 0 \text{ as } n \to \infty. \tag{9.49}$$

From (9.49) and Lebesgue's dominated convergence theorem, we first get

$$\lim_n \int_a^b G(t, \gamma_n(t))dt = \int_a^b G(t, \gamma(t))dt. \tag{9.50}$$

Now observing that for every i, $(\lambda_k^{i,n})_{k=1,\ldots,m+1}$ are characteristic functions of disjoint sets, we have

$$\int_a^b F(\dot{\gamma}_n) = \sum_{i=0}^{N-1} \int_{I_i} F\left(\sum_{k=1}^{m+1} \lambda_k^{i,n}(t)\alpha_k^i\right) dt$$

$$= \sum_{i=0}^{N-1} \int_{I_i} \sum_{k=1}^{m+1} \lambda_k^{i,n}(t) F\left(\alpha_k^i\right) dt.$$

Thanks to Lemma 9.2 and (9.45), we thus get, for every i

$$\lim_n \int_{I_i} \sum_{k=1}^{m+1} \lambda_k^{i,n}(t) F\left(\alpha_k^i\right) dt = \int_{I_i} \sum_{k=1}^{m+1} \lambda_k^i F\left(\alpha_k^i\right) dt$$

$$= \int_{I_i} F^{**}(\dot{\gamma})$$

together with (9.50) this yields

$$\bar{J}(\gamma) = \lim_n J(\gamma_n) \geq \inf_{W^{1,p}} J$$

but this holds for any continuous piecewise affine curve γ, taking the infimum and using the density of such curves in $W^{1,p}$ we thus get

$$\inf_{W^{1,p}} \bar{J} \geq \inf_{W^{1,p}} J,$$

which concludes the proof. □

It is worth mentioning that the relaxation result above contains a necessary optimality condition for the unrelaxed problem.

Corollary 9.4. *Under the assumptions of this section if γ is a minimizer of J over $W^{1,p}((a,b), \mathbb{R}^m)$, then one has*

$$F^{**}(\dot{\gamma}(t)) = F(\dot{\gamma}(t)) \text{ for a.e. } t \in (a, b).$$

The relaxation result we presented above can be generalized to more general Lagrangians, higher dimensions (but when both d and m are larger than 2, it is another envelope than the convex one which appears in the relaxed functional, see [1]). For more on relaxation, we refer the interested reader to [27, 39, 46]. Relaxation results are also very much related to the theory of Young measures pioneered by Young [99], also see [15] or [81].

9.4. Euler–Lagrange equations and other necessary conditions

9.4.1. *Euler–Lagrange equations*

Given Ω an open bounded subset of $\mathbb{R}^d$ with a C^1 boundary, $m \in \mathbb{N}^*$ consider a Lagrangian: $L : \Omega \times \mathbb{R}^m \times \mathcal{M}_{m \times d} \to \mathbb{R}$ measurable and such that

$$\text{for a.e. } x \in \Omega, \ (u, \xi) \in \mathbb{R}^m \times \mathcal{M}_{m \times d} \mapsto L(x, u, \xi) \text{ is of class } C^1. \quad (9.51)$$

We then denote by $\nabla_u L(x, u, \xi) = (\partial_{u_1} L(x, u, v), \dots, \partial_{u_m} L(x, u, v))^T \in \mathbb{R}^m$ the partial gradient of L with respect to u. We denote the matrix ξ either as $\xi = [\xi_{ij}]_{i=1,\dots,m, \ j=1,\dots,d}$ or $\xi = (\xi_1, \dots, \xi_m)^T$ where $\xi_i \in \mathbb{R}^d$ is the ith row of ξ. For $i = 1, \dots, m$ we then denote by $\nabla_{\xi_i} L(x, u, \xi)$ the partial gradient with respect to ξ_i:

$$\nabla_{\xi_i} L(x, u, \xi) = (\partial_{\xi_{i1}} L(x, u, v), \dots, \partial_{\xi_{id}} L(x, u, v))^T \in \mathbb{R}^d.$$

We assume that there are some exponents $p \geq 1$ and $q \geq 1$ such that

$$q \in \begin{cases} [1, +\infty) & \text{if } p \geq d, \\ [1, p^*] & \text{with } p^* = \dfrac{pd}{d-p} \text{ if } p < d \end{cases} \qquad (9.52)$$

as well as some $C \geq 0$ such that, for almost every $x \in \Omega$, every $(u, \xi) \in \mathbb{R}^m \times \mathcal{M}_{m \times d}$ and every $i = 1, \ldots, m$:

$$|\nabla_{\xi_i} L(x, u, \xi)| \leq C(1 + |\xi|^{p-1}) \qquad (9.53)$$

and

$$|\nabla_u L(x, u, \xi)| \leq C(1 + |u|^{q-1}) \qquad (9.54)$$

and for some $\alpha \in L^1(\Omega, \mathbb{R})$

$$|L(x, u, 0)| \leq \alpha(x) + C|u|^q. \qquad (9.55)$$

We then define for every $u = (u_1, \ldots, u_m) \in W^{1,p}(\Omega, \mathbb{R}^m)$

$$J(u) := \int_\Omega L(x, u(x), Du(x))dx, \qquad (9.56)$$

where, as before $Du \in \mathcal{M}_{m \times d}$ is the Jacobian matrix of u, i.e., $Du_{ij} := \partial_j u_i$.

Condition (9.52) and the Sobolev embedding theorem ensure that $W^{1,p}$ embeds continuously into L^q. Recalling Corollary 9.2, these conditions imply that J is continuous for the strong $W^{1,p}$ topology. It directly follows from Proposition 9.4 and the fact that

$$u \in W^{1,p}(\Omega, \mathbb{R}^m) \mapsto (u, Du) \in L^q(\Omega, \mathbb{R}^m) \times L^p(\Omega, \mathcal{M}_{m \times d})$$

is linear and continuous that J is Gateaux–differentiable and that for every $(u, \varphi) \in W^{1,p}(\Omega, \mathbb{R}^m) \times W^{1,p}(\Omega, \mathbb{R}^m)$, one has

$$J'(u)(\varphi) = \int_\Omega \nabla_u L(x, u(x), Du(x)) \cdot \varphi(x)dx$$

$$+ \sum_{i=1}^m \int_\Omega \nabla_{\xi_i} L(x, u(x), Du(x)) \cdot \nabla\varphi_i(x)dx.$$

Theorem 9.15. *Let $p \geq 1$, $q \geq 1$, L satisfy (9.51)–(9.55) and J be defined by (9.56) and $u_0 \in W^{1,p}(\Omega)$. If u is a local minimizer[j] of J on $u_0 + W_0^{1,p}(\Omega)$ then u solves the Euler–Lagrange system of partial differential equations*

$$\text{div}(\nabla_{\xi_i} L(x, u(x), Du(x))) = \partial_{u_i} L(x, u(x), Du(x)), \ x \in \Omega, i = 1, \ldots, m$$

in the weak sense, which means that for every $i = 1, \ldots, m$ and every $v \in W_0^{1,p}(\Omega, \mathbb{R})$, one has

$$\int_\Omega [\partial_{u_i} L(x, u(x), Du(x))v(x) + \nabla_{\xi_i} L(x, u(x), Du(x)) \cdot \nabla v(x)] dx = 0.$$

$$(9.57)$$

Proof. Since u is a local minimizer of J on $u_0 + W_0^{1,p}(\Omega)$, it satisfies $J'(u)(\varphi) = 0$ for all $\varphi = (\varphi_1, \ldots, \varphi_m) \in W_0^{1,p}(\Omega)^m$. Let $v \in W_0^{1,p}(\Omega, \mathbb{R})$, $i \in \{1, \ldots, m\}$ taking $\varphi_k = 0$ if $k \neq i$ and $\varphi_i = v$ thus exactly gives (9.57). $\square$

As usual, if J is convex, being a critical point is sufficient for global minimality, we thus have the following proposition.

Proposition 9.5. *Let L satisfy (9.51)–(9.55) and J be defined by (9.56) and $u_0 \in W^{1,p}(\Omega)$. If J is convex and $u \in u_0 + W_0^{1,p}(\Omega)$ satisfies the Euler–Lagrange system (9.57) then*

$$J(u) \leq J(v), \forall v \in u_0 + W_0^{1,p}(\Omega).$$

Remark 9.6. The growth assumptions (9.53)–(9.54) may look heavy (they are indeed!) and purely technical. The key point in justifying the validity of the Euler–Lagrange equations is to be able to differentiate under the integral sign as we did in the proof of Proposition 9.4, which requires some L^1 bound on the norms of the gradient of the Lagrangian along the solution in order to apply Lebesgue's dominated convergence theorem. Of course, if we knew that a certain minimizer is Lipschitz then it would be bounded as well as its derivatives and justifying that it solves the Euler–Lagrange equations would be much easier. Unfortunately, it might be the

---

[j]That is, there is some $\delta > 0$ such that $J(u) \leq J(v)$ whenever $v \in u_0 + W_0^{1,p}(\Omega)$ satisfies $\|v - u\|_{W^{1,p}} \leq \delta$.

case that there is no Lipschitz minimizer[k] and, even worse, that the minimum of the functional over merely $W^{1,1}$ function is achieved but strictly less than among Lipschitz functions. The occurrence of such (a bit weird) situations is called the Lavrentiev phenomenon (see, e.g., [8, 26, 39]).

Let us now consider a variant where instead of minimizing J while prescribing the boundary values $u = u_0$ on $\partial\Omega$, the boundary values $u_{|\partial\Omega}$ are free but there is a certain boundary cost. Then let $\psi : \partial\Omega \times \mathbb{R}^m \to \mathbb{R}$ be measurable and such that

$$\text{for } \sigma\text{-a.e. } x \in \partial\Omega, \, u \in \mathbb{R}^m \mapsto \psi(x,u) \text{ is of class } C^1 \qquad (9.58)$$

and for some $C > 0$, for σ-a.e. $x \in \partial\Omega$ and every $u \in \mathbb{R}^m$, one has

$$|\nabla_u \psi(x,u)| \le C(1 + |u|^{p-1}), \psi(.,0) \in L^1(\partial\Omega, \sigma). \qquad (9.59)$$

Define then for every $u \in W^{1,p}(\Omega, \mathbb{R}^m)$,

$$j(u) := \int_{\partial\Omega} \psi(x, u(x))d\sigma(x) \qquad (9.60)$$

and observe that j is Gateaux-differentiable with

$$j'(u)(v) = \int_{\partial\Omega} \nabla_u \psi(x, u(x)) \cdot v(x)d\sigma(x), \, \forall(u,v) \in W^{1,p}(\Omega, \mathbb{R}^m),$$

where all the boundary terms above have to be understood in the sense of traces. Minimizers of $J + j$ over $W^{1,p}(\Omega)$ not only satisfy the Euler–Lagrange equations but also additional boundary conditions called transversality conditions.

Proposition 9.6 (Transversality conditions). *Let $p \ge 1$, $q \ge 1$, L satisfy (9.51)–(9.55), let J be defined by (9.56), let ψ satisfy (9.58)–(9.59) and j be defined by (9.60). If u is a local minimizer of $J + j$ on $W^{1,p}(\Omega, \mathbb{R}^m)$, u is a weak solution of the Euler–Lagrange equations (9.57) and satisfies for $i = 1, \dots, m$ the transversality condition*

$$\nabla_{\xi_i} L(x, u(x), Du(x)) \cdot \nu(x) + \partial_{u_i} \psi(x, u(x)) = 0 \text{ on } \partial\Omega, \qquad (9.61)$$

where $\nu(x)$ denotes the outward unit normal at x to $\partial\Omega$.

---

[k]For instance, it is easy to check that the infimum of $\int_0^1 (\gamma(t)\dot\gamma(t))^2 dt$ among $W^{1,1}((0,1), \mathbb{R})$ functions γ such that $\gamma(0) = 0$ and $\gamma(1) = 1$ is achieved for $\gamma(t) = \sqrt{t}$ which is not Lipschitz.

Proof. If u is a local minimizer of $J + j$ on $W^{1,p}(\Omega, \mathbb{R}^m)$ then for every $v = (v_1, \ldots, v_m) \in W^{1,p}(\Omega, \mathbb{R}^m)$, one has $J'(u)(v) + j'(u)(v) = 0$, i.e.,

$$0 = \sum_{i=1}^{m} \int_{\Omega} \nabla_{\xi_i} L(x, u(x), Du(x)) \cdot \nabla v_i(x) dx$$

$$+ \sum_{i=1}^{m} \int_{\Omega} \partial_{u_i} L(x, u(x), Du(x)) v_i(x) dx$$

$$+ \sum_{i=1}^{m} \int_{\partial\Omega} \partial_{u_i} \psi(x, u(x)) v_i(x) d\sigma(x),$$

which is exactly the weak formulation of (9.57)–(9.61) (see the discussion after (9.3)). $\square$

When $d = 1$ and $\Omega = (a, b)$ we may write the boundary cost j as

$$j(\gamma) = \psi_b(\gamma(b)) + \psi_a(\gamma(a))$$

and the transversality conditions take the form

$$\nabla_v L(b, \gamma(b), \dot{\gamma}(b)) + \nabla \psi_b(\gamma(b)) = 0, \ \nabla_v L(a, \gamma(a), \dot{\gamma}(a)) - \nabla \psi_a(\gamma(a)) = 0.$$

9.4.2. *On existence of minimizers again*

The reader may feel a bit disappointed at this point by the fact that we stated an existence result in the separable case in Theorem 9.13 and obtained Euler–Lagrange equations for non-separable Lagrangians in Theorem 9.15. Our aim now is to show that under the assumptions of the previous section and additional convexity and coercivity conditions, the existence of minimizers (which solve the Euler–Lagrange equations) can be taken for granted.

The setting is the same as in Section 9.4.1. To shorten notations we set $N := m \times d$ and identify $\mathcal{M}_{m \times d}$ to $\mathbb{R}^N$. We consider a Lagrangian L: $\Omega \times \mathbb{R}^m \times \mathbb{R}^N$ measurable in all its arguments such that

$$\text{for a.e. } x \in \Omega, \ (u, \xi) \in \mathbb{R}^m \mapsto L(x, u, \xi) \text{ is continuous,} \qquad (9.62)$$

$$\text{for a.e. } x \in \Omega, \ \forall u \in \mathbb{R}^m, \ \xi \in \mathbb{R}^N \mapsto L(x, u, \xi) \text{ is convex and } C^1, \quad (9.63)$$

and, for some pair of exponents $p \geq 1$ and $q \geq 1$, it satisfies the coercivity condition

$$\text{for a.e. } x \in \Omega, \ \forall (u, \xi) \in \mathbb{R}^m \times \mathbb{R}^N, \ L(x, u, \xi) \geq \beta(x) - \mu |u|^q + \nu |\xi|^p, \tag{9.64}$$

for some $\nu > 0$, $\mu \in \mathbb{R}$ and $\beta \in L^1$. We also assume that L satisfies the conditions (9.53) and (9.55) from Section 9.4.1. In the sequel, A will denote a positive constant that may vary from one line to another.

Note that the assumptions above imply a bound

$$|L(x, u, \xi)| \leq \theta(x) + A(|u|^q + |\xi|^p), \tag{9.65}$$

for a suitable constant $A > 0$ and $\theta \in L^1(\Omega)$. The functional

$$I(u, \xi) := \int_\Omega L(x, u(x), \xi(x)) dx, \forall (u, \xi) \in L^q(\Omega, \mathbb{R}^m) \times L^p(\Omega, \mathbb{R}^N) \tag{9.66}$$

therefore is well defined and continuous for the strong topology thanks to (9.62)–(9.65) and arguing as in Corollary 9.2. More interestingly, the assumptions above ensure that I is strong–weak sequentially lsc.

Theorem 9.16. *Let $p > 1$, $q \geq 1$, L satisfy (9.62)–(9.64), (9.53) and (9.55). Then if $u_n \to u$ in $L^q(\Omega, \mathbb{R}^m)$ and $\xi_n \rightharpoonup \xi$ in $L^p(\Omega, \mathbb{R}^N)$ one has*

$$\liminf_n I(u_n, \xi_n) \geq I(u, \xi).$$

Proof. Let us define for every $(x, u, v) \in \Omega \times \mathbb{R}^m \times \mathbb{R}^N$ the (partial) Legendre transform

$$L^*(x, u, v) := \sup_{\xi \in \mathbb{R}^N} \{\xi \cdot v - L(x, u, \xi)\}. \tag{9.67}$$

Thanks to (9.65) and (9.64), L^* satisfies for almost every $x \in \Omega$ and for every $(u, v) \in \mathbb{R}^m \times \mathbb{R}^N$,

$$|L^*(x, u, v)| \leq \gamma(x) + A(|u|^q + |v|^{p'}), \tag{9.68}$$

for some positive constant A, some $\gamma \in L^1(\Omega)$ and $p' = \frac{p}{p-1}$. In particular $L^*(., u(\cdot), v(\cdot)) \in L^1(\Omega)$ whenever $(u, v) \in L^q(\Omega, \mathbb{R}^m) \times L^{p'}(\Omega, \mathbb{R}^N)$. Since

$L(x, u, \cdot)$ is convex and continuous, the Fenchel–Moreau theorem gives

$$L(x, u, \xi) = \sup_{v \in \mathbb{R}^N} \{ v \cdot \xi - L^*(x, u, v) \}$$

and the supremum above is attained for $v = V(x, u, \xi) := \nabla_\xi L(x, u, \xi)$ which thanks to (9.53) satisfies the growth condition

$$|V(x, u, \xi))| \leq C(1 + |\xi|^{p-1}). \qquad (9.69)$$

For $(u, \xi) \in L^q(\Omega, \mathbb{R}^m) \times L^p(\Omega, \mathbb{R}^N)$, it follows from the Young–Fenchel inequality that

$$I(u, \xi) \geq \sup_{v \in L^{p'}(\Omega, \mathbb{R}^N)} \int_\Omega [v(x) \cdot \xi(x) - L^*(x, u(x), v(x))] dx.$$

Define then for every $x \in \Omega$, $\bar{v}(x) := V(x, u(x), \xi(x)) = \nabla_\xi L(x, u(x), \xi(x))$, it follows from (9.69) that $\bar{v} \in L^{p'}(\Omega)$ and by construction $L(x, u(x), \xi(x)) = \bar{v}(x) \cdot \xi(x) - L^*(x, u(x), \bar{v}(x))$ for every $x \in \Omega$. Hence I can be represented as a supremum of integral functionals which are affine with respect to the ξ variable:

$$I(u, \xi) = \sup_{v \in L^{p'}(\Omega, \mathbb{R}^N)} \int_\Omega [v(x) \cdot \xi(x) - L^*(x, u(x), v(x))] dx. \qquad (9.70)$$

Since a supremum of strong–weak sequentially lsc functionals is itself strong-weak sequentially lsc, we are left to show that for any *fixed* $v \in L^{p'}(\Omega, \mathbb{R}^N)$, whenever $u_n \to u$ in $L^q(\Omega, \mathbb{R}^m)$ and $\xi_n \rightharpoonup \xi$ in $L^p(\Omega, \mathbb{R}^N)$, one has

$$\liminf_n \int_\Omega [v(x) \cdot \xi_n(x) - L^*(x, u_n(x), v(x))] dx$$

$$\geq \int_\Omega [v(x) \cdot \xi(x) - L^*(x, u(x), v(x))] dx.$$

Since $\xi_n \rightharpoonup \xi$ in $L^p(\Omega, \mathbb{R}^N)$ and $v \in L^{p'}(\Omega, \mathbb{R}^N)$, $\int_\Omega v \cdot \xi_n$ converges to $\int_\Omega v \cdot \xi$. As for the convergence of the other term, we first remark that $u \mapsto L^*(x, u, v(x))$ is lsc (as a supremum of continuous functions). We claim that it is also usc, to see this, fix x for a moment and simply set $G(u) := L^*(x, u, v(x)) = \sup_{\xi \in \mathbb{R}^N} \{ v(x) \cdot \xi - L(x, u, \xi) \}$, the coercivity condition (9.64) ensures that the supremum is attained for some $\bar{\xi}$, and of course $v \cdot \bar{\xi} - L(x, u, \bar{\xi}) \geq -L(x, u, 0)$ which, with (9.55), yields the bound (with A depending on x):

$$|\bar{\xi}|^p \leq A(1 + |u|^q). \tag{9.71}$$

Now (again for a fixed x) let $u_n \to u$ in $\mathbb{R}^m$ and let $\bar{\xi}_n \in \mathbb{R}^N$ be such that $L^*(x, u_n, v(x)) = v(x) \cdot \bar{\xi}_n - L(x, u_n, \bar{\xi}_n)$. The convergence of u_n and the bound (9.71) implies that $\bar{\xi}_n$ is bounded, passing to subsequences if necessary, we may therefore assume that $\bar{\xi}_n$ converges to some $\bar{\xi}$ and that $L^*(x, u_n, v(x))$ converges which, since $L(x, \cdot, \cdot)$, is continuous gives

$$\lim_n L^*(x, u_n, v(x)) = v(x) \cdot \bar{\xi} - L(x, u, \bar{\xi}) \leq L^*(x, u, v(x)).$$

This shows that $G(\cdot) = L^*(x, \cdot, v(x))$ is continuous. With the bound (9.68) and Corollary 9.2, we deduce that if $u_n \to u$ in $L^q(\Omega, \mathbb{R}^m)$ then

$$\lim_n \int_\Omega L^*(x, u_n(x), v(x))dx = \int_\Omega L^*(x, u(x), v(x))dx,$$

which concludes the proof. $\qquad\square$

Remark 9.7. In fact, the lsc result of Theorem 9.16 holds under much more general conditions. The differentiability assumption on the Lagrangian with respect to ξ can be removed and continuity in (9.62) can be replaced by lower semicontinuity, we refer to Theorem 2.3.1 in Buttazzo's book [27] for such a general result. The strategy of proof is roughly the same and consists in writing I as a supremum of affine integral functionals in the variable ξ. But there are two technical difficulties (which we avoided above thanks to (9.53)): the interchange of suprema and integrals has to be justified as well as the measurability of the coefficients in the affine functions of ξ.

We deduce the following existence result.

Corollary 9.5. *Assume that Ω is an open bounded subset of $\mathbb{R}^d$ with a C^1 boundary, that $p > 1$ and $q \in [1, p)$, that L is a Lagrangian which satisfies (9.62)–(9.64), (9.53) and (9.55). Given $u_0 \in W^{1,p}(\Omega, \mathbb{R}^m)$ the functional*

$$J(u) := \int_\Omega L(x, u(x), \nabla u(x))dx$$

admits at least one minimizer on $u_0 + W_0^{1,p}(\Omega, \mathbb{R}^m)$. If, in addition, L satisfies (9.51) and (9.54), then, every minimizer of J on $u_0 + W_0^{1,p}(\Omega, \mathbb{R}^m)$ satisfies the Euler–Lagrange equations (9.57).

Proof. First note that $J(u_0) < +\infty$ thanks to (9.65). Let $u_n \in u_0 + W_0^{1,p}(\Omega, \mathbb{R}^m)$ be a minimizing sequence of J over $W_0^{1,p}(\Omega, \mathbb{R}^m)$. Using (9.64) and arguing as in the proof of Theorem 9.13, we get a bound on $\|u_n\|_{W^{1,p}}$. Thanks to Rellich–Kondrachov's theorem, passing to a subsequence, we may therefore assume that $u_n \to u$ in $L^q(\Omega, \mathbb{R}^m)$ and $Du_n \rightharpoonup Du$ in $L^p(\Omega, \mathbb{R}^{m \times d})$ for some $u \in u_0 + W_0^{1,p}(\Omega, \mathbb{R}^m)$. Thanks to theorem 9.16, we thus get

$$\liminf_n J(u_n) \geq J(u)$$

so that u minimizes J over $u_0 + W_0^{1,p}(\Omega, \mathbb{R}^m)$. Further assuming (9.51) and (9.54), the fact that minimizers satisfy the Euler–Lagrange equations (9.57) follows from Theorem 9.15. $\qquad\square$

9.4.3. *Examples*

Let us now consider some examples of Euler–Lagrange equations and transversality conditions. Note that for $d = 1$, the Euler–Lagrange equations are second-order ODEs in $\mathbb{R}^m$, if $m = 1$ and $d \geq 2$, they take the form of a second-order PDE and for $d \geq 2$, $m \geq 2$, they are systems of second-order PDEs. Note also that either boundary values are prescribed or boundary values are free, in which case, the Euler–Lagrange are supplemented with transversality boundary conditions.

Example 9.3 (Newton's law as an Euler–Lagrange ODE). Let $V \in C^1(\mathbb{R}^m, \mathbb{R})$ bounded from below, $a < b$. For $\gamma \in H^1((a,b), \mathbb{R}^m)$, define

$$E(\gamma) := \int_a^b \left[\frac{1}{2} |\dot{\gamma}(t)|^2 + V(\gamma(t)) \right] dt = \int_a^b \left[\frac{1}{2} \sum_{i=1}^m \dot{\gamma}_i(t)^2 + V(\gamma(t)) \right] dt \tag{9.72}$$

and given $x \in \mathbb{R}^m$, consider

$$\inf_{\gamma \in H^1((a,b), \mathbb{R}^m),\, \gamma(0)=x} E(\gamma). \tag{9.73}$$

Applying the direct method immediately gives the existence of (at least) a solution γ of (9.73). Of course, if γ is a solution of (9.73) it minimizes E among curves $\sigma \in H^1$ which satisfy the prescribed initial condition $\sigma(0) = x$

and which agree with γ at the terminal time, i.e., $\sigma(b) = \gamma(b)$, in particular γ solves the Euler–Lagrange ODE:

$$\frac{d}{dt}\dot{\gamma} = \nabla V(\gamma) \text{ on } (a, b) \tag{9.74}$$

in the weak H^1 sense. But since γ is H^1, it is continuous (actually $C^{0,\frac{1}{2}}$), hence the right-hand side of (9.74) is a continuous function, we deduce from du Bois–Reymond's lemma that $\dot{\gamma}$ is C^1, i.e., γ is C^2 so that γ is in fact a solution of (9.74)

$$\ddot{\gamma}(t) = \nabla V(\gamma(t)), \ t \in [a, b], \gamma(t) = 0.$$

The previous reasoning is typical from the calculus of variations: we start with a minimizer which a priori only has Sobolev regularity, we justify that it is a weak solution of a second-order equation which gives higher regularity. In the present example, we really gain a lot of regularity: $\gamma \in H^1$ means it only has one derivative in L^2, whereas $\gamma \in C^2$ gives two continuous derivatives. Of course, if V is C^2, we get that $\gamma \in C^3$ and one can iterate the argument as many times as the regularity of V allows. In particular, if V is C^∞ so is γ.

Note now that (9.74) has a very natural mechanical interpretation. Imagine that the curve γ represents the motion of a single point body (with mass normalized to 1) subject to the potential force field $\vec{F} = \nabla V$, then (9.74) is nothing but Newton's law of motion. The functional E represents the mechanical (kinetic plus potential) energy of the body and Newton's law precisely says that the motion of the body has to be a stationary point of the mechanical energy.

The Euler–Lagrange equation (9.74) has been obtained by the minimality of the action among all curves which agree with γ at both endpoints, but only $\gamma(a)$ is prescribed, we have one extra degree of freedom which is the terminal condition $\gamma(b)$. In other words, $E(\gamma) \leq E(\gamma + \varepsilon\varphi)$ for every $\varepsilon \in \mathbb{R}$ and every $\varphi \in C^1([a, b], \mathbb{R}^m)$ for which $\varphi(a) = 0$, for any such φ we thus have

$$0 = E'(\gamma)(\varphi) = \int_a^b (\dot{\gamma} \cdot \dot{\varphi} + \nabla V(\gamma) \cdot \varphi)$$

$$= \int_a^b (-\ddot{\gamma} + \nabla V(\gamma)) \cdot \varphi + [\dot{\gamma} \cdot \varphi]_a^b = \dot{\gamma}(b) \cdot \varphi(b),$$

where the second line has been obtained by integration by parts and using (9.74). Since we may prescribe any value in $\mathbb{R}^m$ for $\varphi(b)$, we discover another optimality condition for the free endpoint namely

$$\dot{\gamma}(b) = 0,$$

which is a special instance of transversality condition for free endpoints.

Example 9.4 (Dirichlet energy and Poisson equation). Let Ω be an open bounded subset of $\mathbb{R}^d$ with a C^1 boundary, $f \in L^2(\Omega)$ and $u_0 \in H_0^1(\Omega)$. Consider the energy

$$J_f(u) := \frac{1}{2} \int_\Omega |\nabla u|^2 + \int_\Omega fu$$

and the variational problem

$$\inf_{u \in u_0 + H_0^1(\Omega)} J_f(u). \tag{9.75}$$

It is easy to see that J_f strictly convex $u_0 + H_0^1(\Omega)$, by the direct method (9.75) it has a solution $u \in H^1(\Omega)$ which is unique and characterized by the Poisson equation

$$\Delta u = f, \text{ in } \Omega, \ u = u_0 \text{ on } \partial\Omega$$

in the weak sense (where we recall that $\Delta u = \sum_{i=1}^d \partial_{ii} u = \operatorname{div}(\nabla u)$ denotes the Laplacian of u). In particular, the minimizer of the Dirichlet energy J_0 over $u_0 + H_0^1(\Omega)$ is nothing but the harmonic extension of u_0. A natural question is whether being a solution of the Poisson equation gives improved regularity for u, this is indeed the case but more complicated to prove than when $d = 1$. This relies on the theory of elliptic regularity for which we refer the reader for instance to the textbook of Gilbarg and Trudinger [58].

Example 9.5 (p-Laplace equation). Let $p \in (1, +\infty)$, $f \in L^{p'}(\Omega)$ and $u_0 \in W_0^{1,p}(\Omega)$, consider the energy

$$J(u) := \frac{1}{p} \int_\Omega |\nabla u|^p + \int_\Omega fu$$

and the variational problem

$$\inf_{u \in u_0 + W_0^{1,p}(\Omega)} J(u). \tag{9.76}$$

It is not difficult to check that (9.76) has a unique solution u which is the only weak solution in $W^{1,p}$ of the p-Laplace equation:

$$\operatorname{div}(|\nabla u|^{p-2}\nabla u) = f \text{ in } \Omega, \ u = u_0 \text{ on } \partial\Omega.$$

Example 9.6 (Equation of minimal hypersurfaces). Given $u_0 \in W^{1,1}(\Omega)$ consider the minimal hypersurface problem

$$\inf_{u \in u_0 + W_0^{1,1}(\Omega)} S(u) := \int_\Omega \sqrt{1 + |\nabla u|^2}. \tag{9.77}$$

Since $W^{1,1}(\Omega)$ is not reflexive, nothing guarantees that (9.77) has solutions (in general it does not). However, if it does the solution is unique[1] which is characterized by the minimal surface equation:

$$\operatorname{div}\left(\frac{\nabla u}{\sqrt{1 + |\nabla u|^2}}\right) = 0, \; u = u_0 \text{ on } \partial\Omega, \tag{9.78}$$

which, geometrically, means that the level sets of u have a vanishing mean curvature. For more on minimal hypersurfaces and related problems we refer the reader to [59].

Example 9.7 (An example of system of Euler–Lagrange equations). Let us now consider a vectorial case $m = d = 2$ and consider

$$\inf_{(u,v) \in H_0^1(\Omega)^2} J(u,v) := \int_\Omega \left(\frac{1}{2}|\nabla u|^2 + \frac{1}{2}|\nabla v|^2 + u\cos(u + v^5)\right)$$

applying the direct method one easily finds a minimizer (not necessarily unique) (u,v) (which being H^1 is also L^q for every q since $d = 2$) and such minimizers all solve the system

$$-\Delta u - u\sin(u + v^5) + \cos(u + v^5) = 0, \; -\Delta v - 5uv^4 \sin(u + v^5) = 0, \text{ in } \Omega$$

together with the homogeneous Dirichlet boundary conditions $u = v = 0$ on $\partial\Omega$.

Exercise 9.7. *Let $f \in L^2(\Omega)$, $\alpha \in C^1(\mathbb{R})$ and $\frac{1}{C} \leq \alpha \leq C$ for some $C > 1$. Consider*

$$\inf_{u \in H_0^1(\Omega)} \int_\Omega \left(\frac{1}{2}\alpha(u)|\nabla u|^2 + fu\right). \tag{9.79}$$

Prove that (9.79) admits solutions and that the latter satisfy a certain Euler–Lagrange equation to determine.

---

[1]Since $p \mapsto \sqrt{1 + |p|^2}$ is strictly convex, if u and v are two minimizers one should have $\nabla u = \nabla v$ hence $u = v$ since $u_{|\partial\Omega} = v_{|\partial\Omega}$.

9.5. A focus on the case $d = 1$

We consider the case $d = 1$ and a Lagrangian $(t, x, v) \in \mathbb{R} \times \mathbb{R}^m \times \mathbb{R}^m \mapsto L(t, x, v)$. We assume that $L \in C^2(\mathbb{R} \times \mathbb{R}^m \times \mathbb{R}^m)$ and satisfies

$$D_{vv}L(t, x, v)\text{is positive definite}, \ \forall (t, x, v) \in \mathbb{R} \times \mathbb{R}^m \times \mathbb{R}^m \qquad (9.80)$$

and

$$L(t, x, v) \geq g(|v|), \ \forall (t, x, v) \in \mathbb{R} \times \mathbb{R}^m \times \mathbb{R}^m, \qquad (9.81)$$

where $g : \mathbb{R}_+ \to \mathbb{R}$ is superlinear, i.e.,

$$\lim_{\alpha \to +\infty} \frac{g(\alpha)}{\alpha} = +\infty.$$

9.5.1. Hamiltonian systems

Our goal is to reformulate the Euler–Lagrange system of ODEs

$$\frac{d}{dt}\nabla_v L(t, \gamma(t), \dot\gamma(t)) = \nabla_x L(t, \gamma(t), \dot\gamma(t)), \qquad (9.82)$$

which is of the second order and not explicit in the acceleration $\ddot\gamma$ as a system of first-order ODEs called a Hamiltonian system. To do so, define the Hamiltonian $H : \mathbb{R} \times \mathbb{R}^m \times \mathbb{R}^m \to \mathbb{R}$ by

$$H(t, x, p) := \sup_{v \in \mathbb{R}^m} \{-p \cdot v - L(t, x, v)\}, \ \forall (t, x, p) \in \mathbb{R} \times \mathbb{R}^m \times \mathbb{R}^m. \qquad (9.83)$$

Assumptions (9.80)–(9.81) ensure that the supremum in (9.83) is attained for a unique $v = V(t, x, p)$ which is characterized by the optimality condition

$$v = V(t, x, p) \iff p + \nabla_v L(t, x, v) = 0. \qquad (9.84)$$

One easily checks that $(t, x, p) \in \mathbb{R} \times \mathbb{R}^m \times \mathbb{R}^m \mapsto V(t, x, p)$ is continuous, it is even C^1 thanks to (9.80) and the implicit function theorem. This, in turn, implies that H is of class C^2 and its derivatives can be computed by the envelope theorem (see Theorem 5.2) as

$$\nabla_p H(t, x, p) = -V(t, x, p), \ \nabla_x H(t, x, p) = -\nabla_x L(t, x, V(t, x, p)), \qquad (9.85)$$

$$\partial_t H(t, x, p) = -\partial_t L(t, x, V(t, x, p)). \qquad (9.86)$$

Also note that, by definition, the Hamiltonian H is convex (and even strictly convex since its Legendre transform is differentiable) with respect to the p

variable. Now assume that γ solves the Euler–Lagrange equations (9.82) on some time interval (a, b) and introduce the so-called adjoint or momentum variable

$$p(t) := -\nabla_v L(t, \gamma(t), \dot{\gamma}(t)), \; t \in (a, b). \tag{9.87}$$

We then have the following equivalence theorem.

Theorem 9.17. *Given $\gamma : (a, b) \to \mathbb{R}^m$ a solution of the Euler–Lagrange equations (9.82) on (a, b), define the momentum variable p by (9.87) then the pair (γ, p) solves the Hamiltonian system*

$$\dot{\gamma}(t) = -\nabla_p H(t, \gamma(t), p(t)), \; \dot{p}(t) = \nabla_x H(t, \gamma(t), p(t)), \; t \in (a, b). \tag{9.88}$$

Conversely, if (γ, p) solves (9.88) then γ solves the Euler–Lagrange equations (9.82) and p and γ are related by (9.87).

Proof. Assume first that γ solves Euler–Lagrange equations (9.82), then from (9.87), (9.84) and (9.85), we deduce that

$$\dot{\gamma}(t) = V(t, \gamma(t), p(t)) = -\nabla_p H(t, \gamma(t), p(t))$$

and using the second identity in (9.85) and (9.82), we obtain

$$-\dot{p}(t) = \nabla_x L(t, \gamma(t), \dot{\gamma}(t)) = -\nabla_x H(t, \gamma(t), p(t)).$$

Assume now that (γ, p) solves (9.88) then $\dot{\gamma}(t) = V(t, \gamma(t), p(t))$. With (9.84) this gives $-p(t) = \nabla_v L(t, \gamma(t), \dot{\gamma}(t))$ so that (9.87) holds. We thus get, thanks to (9.85)

$$\begin{aligned}
\dot{p}(t) &= -\frac{d}{dt} \nabla_v L(t, \gamma(t), \dot{\gamma}(t)) \\
&= \nabla_x H(t, \gamma(t), p(t)) = -\nabla_x L(t, \gamma(t), V(t, \gamma(t), p(t))) \\
&= -\nabla_x L(t, \gamma(t), \dot{\gamma}(t)),
\end{aligned}$$

which shows that γ solves the Euler–Lagrange equations (9.82). $\qquad\square$

Under the assumptions of this section, let us briefly indicate how to obtain regularity of weak solutions of the Euler–Lagrange system (9.82). Indeed, assume that $\gamma \in W^{1,1}((a, b), \mathbb{R}^m)$ is such that

- $t \mapsto p(t) := -\nabla_v L(t, \gamma(t), \dot{\gamma}(t))$ belongs to $L^1((a, b), \mathbb{R}^m)$;
- $t \mapsto \nabla_x L(t, \gamma(t), \dot{\gamma}(t))$ belongs to $L^1((a, b), \mathbb{R}^m)$;

- γ is a weak solution of the Euler–Lagrange equation in the sense that

$$\int_a^b (\nabla_v L(t, \gamma(t), \dot\gamma(t)) \cdot \dot\varphi(t) + \nabla_x L(t, \gamma(t), \dot\gamma(t)) \cdot \varphi(t))dt = 0,$$

for every $\varphi \in C_c^1((a, b), \mathbb{R}^m))$.

Then p is $W^{1,1}$ hence continuous, using the Hamiltonian system (9.88) we deduce that γ and p are of class C^1, but since H is of class C^2 both p and γ are in fact of class C^2 and γ solves (9.82) in the classical sense. Note that assumption (9.80) is essential to obtain C^2 regularity (and it is quite intuitive since, formally, developing the Euler–Lagrange equation the only term containing $\ddot\gamma$ is $D_{vv}L(t, \gamma(t), \dot\gamma(t))\ddot\gamma(t))$.

If one considers a C^1 terminal cost ψ and critical points of the functional

$$\inf_{\gamma \in W^{1,1}((a,b),\mathbb{R}^m),\ \gamma(a)=x} \int_a^b L(t, \gamma(t), \dot\gamma(t))dt + \psi(\gamma(a)),$$

one can conveniently express the terminal transversality condition using the momentum variable p as

$$p(b) = \nabla\psi(\gamma(b)).$$

In other words, one has to supplement the Hamiltonian system (9.88) with the fixed initial condition $\gamma(a) = x$ and the terminal condition $p(b) = \nabla\psi(\gamma(b))$. Such a system of ODEs is forward–backward (note the difference with a Cauchy problem where one fixes initial conditions for both γ and p).

9.5.2. *Dynamic programming, Hamilton–Jacobi equations*

Given a Lagrangian L satisfying the same conditions as above, as well as terminal cost $\psi \in C(\mathbb{R}^m)$ bounded from below, a time interval (a, b), $x \in \mathbb{R}^m$ and $t \in [a, b]$, we are interested in optimizing the total cost starting from the initial condition $\gamma(t) = x$ at time t:

$$v(t, x) := \inf_{\gamma \in W^{1,1}((t,b),\mathbb{R}^m)),\gamma(t)=x} \left\{ \int_t^b L(s, \gamma(s), \dot\gamma(s))ds + \psi(\gamma(b)) \right\}.$$

$$(9.89)$$

The function $v \colon [a, b] \times \mathbb{R}^m \to \mathbb{R}$ is called the value (or Bellman). Of course, the value function coincides with the terminal cost at the terminal time b:

$$v(b, x) = \psi(x), \quad \forall x \in \mathbb{R}^m. \tag{9.90}$$

If a curve γ is optimal for the problem $v(t, x)$ and if we consider an intermediate time say $t + h \in (t, b]$ then the restriction of this curve to $[t + h, b]$ should of course remain optimal starting from $\gamma(t + h)$ at time $t + h$. This simple optimality principle implies consistency relations between the value functions as different dates known as the dynamic programming principle:

Proposition 9.7. *Let v be the value function defined in (9.89), let $t \in [a, b]$ and let $h > 0$ with $t + h \leq b$. Then for every $x \in \mathbb{R}^m$ one has*

$$v(t, x) = \inf_{\gamma \in W^{1,1}((t, t+h), \mathbb{R}^m)), \, \gamma(t)=x} \left\{ \int_t^{t+h} L(s, \gamma(s), \dot{\gamma}(s)) ds \right.$$

$$\left. + v(t + h, \gamma(t + h)) \right\}. \tag{9.91}$$

Proof. Let $\varepsilon > 0$ and $\gamma \in W^{1,1}(t, b), \mathbb{R}^m)$ satisfy $\gamma(t) = x$ and

$$v(t, x) \geq \int_t^b L(s, \gamma(s), \dot{\gamma}(s)) ds + \psi(\gamma(b)) - \varepsilon.$$

Then

$$v(t, x) \geq \int_t^{t+h} L(s, \gamma(s), \dot{\gamma}(s)) ds + \int_{t+h}^b L(s, \gamma(s), \dot{\gamma}(s)) ds + \psi(\gamma(b)) - \varepsilon$$

$$\geq \int_t^{t+h} L(s, \gamma(s), \dot{\gamma}(s)) ds + v(t + h, \gamma(t + h)) - \varepsilon,$$

which shows that $v(t, x)$ is larger than the right-hand side of (9.91) by letting ε tend to 0. To show the converse inequality, we choose a $\gamma \in W^{1,1}((t, t + h), \mathbb{R}^m)$ which realizes the infimum in the right-hand side of (9.91) up to $\frac{\varepsilon}{2}$ and $\sigma \in W^{1,1}((t + h, b), \mathbb{R}^m)$ such that $\sigma(t + h) = \gamma(t + h)$ and

$$v(t + h, \gamma(t + h)) \geq \int_{t+h}^b L(s, \sigma(s), \dot{\sigma}(s)) ds + \psi(\sigma(b)) - \frac{\varepsilon}{2}.$$

Since the curve obtained by gluing γ and σ is $W^{1,1}$ and starts at x at time t we deduce that $v(t, x)$ is less than the infimum of (9.91) plus ε, which enables us to conclude. $\qquad \square$

What is more striking is the fact that when one lets $h \to 0^+$ in (9.91), one discovers (under some conditions that we will discuss later on) that the value function satisfies a certain PDE called the Hamilton–Jacobi equation.

Theorem 9.18. *Let $t \in (a, b)$ and $x \in \mathbb{R}^m$ be such that v is differentiable at (t, x) and there exists an optimal curve γ for $v(t, x)$ which is C^1 in a neighborhood of time t. Then*

$$\partial_t v(t, x) = H(t, x, \nabla_x v(t, x)), \tag{9.92}$$

where H is the Hamiltonian associated with the Lagrangian L by (9.83).

Proof. Let $\xi \in \mathbb{R}^d$, it follows from the dynamic programming principle that for $0 < h < b - t$ one has

$$v(t, x) \leq \int_t^{t+h} L(s, x + (s - t)\xi, \xi)ds + v(t + h, x + h\xi)$$

but since we assumed that v is differentiable at (t, x) we obtain

$$v(t, x) \leq L(t, x, \xi)h + v(t, x) + \partial_t v(t, x)h + h\nabla_x v(t, x) \cdot \xi + o(h)$$

diving by h and letting $h \to 0^+$ thus yields

$$\partial_t v(t, x) \geq -L(t, x, \xi) - \nabla_x v(t, x) \cdot \xi$$

taking the supremum with respect to ξ, we arrive at

$$\partial_t v(t, x) \geq H(t, x, \nabla_x v(t, x)).$$

To show the converse inequality, we use the optimal curve γ for $v(t, x)$ to obtain

$$v(t, x) \geq \int_t^{t+h} L(s, \gamma(s), \dot{\gamma}(s))ds + v(t + h, \gamma(t + h))$$

our differentiability assumptions then imply

$$v(t, x) \geq L(t, x, \dot{\gamma}(t))h + v(t, x) + \partial_t v(t, x)h + h\nabla_x v(t, x) \cdot \dot{\gamma}(t) + o(h)$$

again diving by h and letting $h \to 0^+$, we find

$$\partial_t v(t, x) \leq -L(t, x, \dot{\gamma}(t)) - \nabla_x v(t, x) \cdot \dot{\gamma}(t) \leq H(t, x, \nabla_x v(t, x)). \qquad \square$$

Assuming that there exists an optimal curve which is smooth is a technical assumption which is not so annoying because it can often be guaranteed by the Euler–Lagrange equations. On the contrary, assuming that the value function is differentiable is unrealistic, as we shall see below, even in simple cases, there are points at which the value function is not differentiable. Nevertheless, thanks to the theory of viscosity solutions of Crandall and

Lions, it is still possible to characterize value functions by an Hamilton–Jacobi equation. The theory of viscosity solutions is beyond the scope of this book, but a detailed exposition can be found in the survey by Crandall, Ishii and Lions [37] and connections with optimal control and variational problems are very clearly explained in the textbooks of Barles [10], Cannarsa and Sinestrari [30], Bardi and Cappuzzo-Dolcetta [9] or Fleming and Soner [50].

To see that the value function need not be differentiable it is instructive to look at the case where the Lagrangian only depends on the velocity variable (and satisfies all the smoothness, convexity and coercivity assumptions of this paragraph). In this case, the value function reads

$$v(t, x) = \inf_{\gamma \in W^{1,1}((t,b), \mathbb{R}^m)), \, \gamma(t)=x} \left\{ \int_t^b L(\dot{\gamma}(s)) ds + \psi(\gamma(b)) \right\}.$$

It follows from Jensen's inequality that

$$\int_t^b L(\dot{\gamma}(s)) ds \geq (b-t) L \left(\frac{1}{b-t} \int_t^b \dot{\gamma}(s) \, ds \right) = (b-t) L \left(\frac{\gamma(b) - \gamma(t)}{b-t} \right).$$

Therefore, among curves joining x at time t to some y at time b, the straight line

$$s \mapsto x + \frac{(s-t)(y-x)}{b-t}$$

is minimizing the integral of the Lagrangian. To solve $v(t, x)$ it is therefore enough to find the terminal condition $\gamma(b) = y$ by minimizing

$$(b-t) L \left(\frac{y-x}{b-t} \right) + \psi(y)$$

and the value function is given by the Lax–Oleinik formula

$$v(t, x) = \inf_{y \in \mathbb{R}^m} \left\{ (b-t) L \left(\frac{y-x}{b-t} \right) + \psi(y) \right\}.$$

If ψ is convex, there exists a unique optimal terminal condition y and v is differentiable thanks to the envelope theorem. However if ψ (even very smooth) is non-convex, there may exist several minimizers which will result in a non-differentiability of v as a consequence of Proposition 5.3.

9.5.3. *A verification theorem*

We end this chapter by a verification theorem which may be viewed as a sufficient optimality condition involving the Hamilton–Jacobi equation.

Theorem 9.19. *Assume that* $w \in C^1([a,b] \times \mathbb{R}^m)$ *solves the Hamilton–Jacobi equation with terminal value* ψ:

$$\partial_t w(s,y) = H(s,y,\nabla_x w(s,y)), \ (s,y) \in (a,b) \times \mathbb{R}^m,$$

$$w(b,y) = \psi(y), \ \forall y \in \mathbb{R}^m.$$

Let $t \in [a,b]$ *and* $x \in \mathbb{R}^m$. *If* $\gamma \in C^1([t,b],\mathbb{R}^m)$ *solves the Cauchy problem*

$$\dot{\gamma}(s) = -\nabla_p H(s,\gamma(s),\nabla_x w(s,\gamma(s))), \ s \in (t,b], \ \gamma(t) = x, \tag{9.93}$$

then γ *is optimal for* $v(t,x)$ *and* $v(t,x) = w(t,x)$.

Proof. By definition of H, the fact that w solves the Hamilton–Jacobi equation may be rewritten as

$$\partial_t w(s,y) \geq -\nabla_x w(s,y) \cdot \xi - L(s,y,\xi), \ \forall (s,y,\xi) \in [a,b] \times \mathbb{R}^m \times \mathbb{R}^m$$

with an equality exactly (recalling (9.84) and (9.85)) when

$$\xi = -\nabla_p H(s,y,\nabla_x w(s,y)).$$

In particular if $\sigma \in W^{1,1}((t,b),\mathbb{R}^m)$, then for a.e. $s \in [t,b]$ one has

$$L(s,\sigma(s),\dot{\sigma}(s)) \geq -\partial_t w(s,\sigma(s)) - \nabla_x w(s,\sigma(s)) \cdot \dot{\sigma}(s) \tag{9.94}$$

and thanks to (9.93), for every $s \in [t,b]$

$$L(s,\gamma(s),\dot{\gamma}(s)) = -\partial_t w(s,\gamma(s)) - \nabla_x w(s,\gamma(s)) \cdot \dot{\gamma}(s). \tag{9.95}$$

Integrating[m] (9.94), we thus get

$$\int_t^b L(s,\sigma(s),\dot{\sigma}(s))ds \geq w(t,\sigma(t)) - w(b,\sigma(b)) = w(t,\sigma(t)) - \psi(\sigma(b)),$$

hence for every $\sigma \in W^{1,1}((t,b),\mathbb{R}^m)$ such that $\sigma(t) = x$ we have

$$\int_t^b L(s,\sigma(s),\dot{\sigma}(s))ds + \psi(\sigma(b)) \geq w(t,x).$$

---

[m]Here, we are using a chain rule for $s \mapsto w(s,\sigma(s))$ as if σ was C^1 whereas it is only $W^{1,1}$, we let the reader check that it can be justified by approximating σ by smooth functions, see Exercise 9.9.

But integrating (9.95) gives

$$\int_t^b L(s,\gamma(s),\dot{\gamma}(s))ds + \psi(\gamma(b)) = w(t,x),$$

which shows that γ is optimal for $v(t,x)$ and $v(t,x) = w(t,x)$. $\qquad\square$

The ODE (9.93) obtained from a (smooth) solution of the Hamilton–Jacobi equation, furnishes an optimal curve in a so-called feedback form, i.e., prescribing the velocity in terms of the current time and current position instead of time only.

9.6. Exercises

Exercise 9.8. *Prove that the functional defined for* $\gamma \in C([0,1], \mathbb{R}^d)$ *by*

$$E(\gamma) := \begin{cases} \int_0^1 |\dot{\gamma}|^2 & \text{if } \gamma \in H^1((a,b), \mathbb{R}^d), \\ +\infty & \text{otherwise} \end{cases}$$

is lsc for the topology of uniform convergence.

Exercise 9.9. *Let* $w \in C^1(\mathbb{R} \times \mathbb{R}^m)$, $\gamma \in W^{1,1}((0,1), \mathbb{R}^m)$ *and define*

$$\theta(t) := w(t, \gamma(t)), \ \forall t \in (0,1).$$

Show (by approximation of γ *by smooth functions) that for every* s *and* t *with* $0 < s < t < 1$ *one has*

$$\theta(t) - \theta(s) = \int_s^t (\partial_t w(\tau, \gamma(\tau)) + \nabla_x w(\tau, \gamma(\tau)) \cdot \dot{\gamma}(\tau))d\tau,$$

deduce that θ *is* $W^{1,1}$ *and compute its weak derivative.*

Exercise 9.10 (The dual of $W_0^{1,p}(\Omega)$). *Let* Ω *be an open bounded subset of* $\mathbb{R}^d$ *with a* C^1 *boundary, let* $p \in [1,+\infty)$, *let* p' *be the conjugate exponent of* p *(with* $p' = \infty$ *if* $p = 1$) *and let* $W^{-1,p'}(\Omega) := (W_0^{1,p}(\Omega))^*$ *be the topological dual of* $W_0^{1,p}(\Omega)$. *For* $\sigma \in L^{p'}(\Omega)^d$, *define* $\text{div}(\sigma)$ *by*

$$\langle \text{div}(\sigma), u \rangle := -\int_\Omega \sigma \cdot \nabla u, \ \forall u \in W_0^{1,p}(\Omega).$$

1) *Show that* $\text{div}(\sigma) \in W^{-1,p'}(\Omega)$ *and that* $\sigma \in L^{p'}(\Omega)^d \mapsto -\text{div}(\sigma) \in W^{-1,p'}(\Omega)$ *is the adjoint of the map* $u \in W_0^{1,p}(\Omega) \mapsto \nabla u \in L^p(\Omega)^d$.

2) *Show that for every $T \in W^{-1,p'}(\Omega)$ there exists $\sigma \in L^{p'}(\Omega)^d$ such that $T = -\operatorname{div}(\sigma)$.*

3) *In the previous question, is there uniqueness of $\sigma \in L^{p'}(\Omega)^d$ such that $T = -\operatorname{div}(\sigma)$?*

Exercise 9.11. *Rigorously solve the following problems:*

$$\inf_{\gamma \in H^1((0,1),\mathbb{R}),\, \gamma(0)=1,\gamma(1)=0} \int_0^1 \left(\frac{1}{2}\dot{\gamma}^2(t) + \frac{1}{2}\gamma^2(t) \right) dt, \tag{9.96}$$

$$\inf_{\gamma \in H^1((0,1),\mathbb{R}),\, \gamma(0)=1} \int_0^1 \left(\frac{1}{2}\dot{\gamma}^2(t) + \frac{1}{2}\gamma^2(t) \right) dt \tag{9.97}$$

and

$$\inf_{\gamma \in H^1((0,1),\mathbb{R})} \int_0^1 \left(\frac{1}{2}\dot{\gamma}^2(t) + \gamma(t) \right) dt + \frac{1}{2}\gamma(1)^2. \tag{9.98}$$

Exercise 9.12. *Rigorously solve the problems*

$$\sup \left\{ \int_0^T e^{-\delta t} \log(-\dot{x}(t)) dt : x(0) = 1,\ x(T) = 0 \right\}$$

(where $\delta > 0$ is a discount rate) and

$$\sup \left\{ \int_0^T e^{-\delta t} \log(-\dot{x}(t)) dt + \log(x(T)) : x(0) = 1 \right\}.$$

Exercise 9.13 (Weak compactness in L^1). *Let Ω be an open bounded subset of $\mathbb{R}^d$. A bounded subset $\mathcal{F}$ of $L^1(\Omega)$ is said to be uniformly integrable if for every $\varepsilon > 0$, there exists $\delta > 0$ such that whenever $A \subset \Omega$ is measurable with $|A| < \delta$ one has:*

$$\int_A |u| \leq \varepsilon,\ \forall u \in \mathcal{F}. \tag{9.99}$$

Let us recall that the dual of $L^1(\Omega)$ is $L^\infty(\Omega)$ hence a sequence u_n in $L^1(\Omega)$ weakly converges to u if $\int_\Omega \varphi u_n \to \int_\Omega \varphi u$ for every $\varphi \in L^\infty(\Omega)$.

1) *Give an example of bounded sequence in $L^1((0,1))$ which does not admit any weakly convergent subsequence in $L^1((0,1))$.*

2) *Let $\mathcal{F}$ be a subset of $L^1(\Omega)$. Show that $\mathcal{F}$ is bounded and uniformly integrable if and only if*

$$\sup_{u \in \mathcal{F}} \int_{|u| \geq M} |u| \to 0 \text{ as } M \to \infty.$$

3) *Show that if $\mathcal{F}$ is a bounded and uniformly integrable subset of $L^1(\Omega)$, then every sequence of elements of $\mathcal{F}$ has a subsequence which converges weakly in $L^1(\Omega)$. (This is part of the Dunford–Pettis theorem which states that there is in fact an equivalence.) (Hint: take a sequence u_n in $\mathcal{F}$, use the fact that both $u_n^{\pm}$ are bounded and uniformly integrable to reduce the problem to the case $u_n \geq 0$, then, for $k \in \mathbb{N}^*$, set $u_n^k := \min(u_n, k)$, use then a diagonal argument to find a subsequence u_{n_j} such that for every k, $u_{n_j}^k$ converges weakly in L^1 to some u^k as $j \to +\infty$, then, use the monotone convergence theorem to show that u^k converges in L^1 to some u and, finally, use uniform integrability to show that u_{n_j} converges weakly to u.)*

4) *Let $g : \mathbb{R}_+ \to \mathbb{R}_+$ be a convex increasing function with $g(0) = 0$ and*

$$\lim_{t \to +\infty} \frac{g(t)}{t} = +\infty,$$

show that if $\mathcal{F}$ is a subset of $L^1(\Omega)$ such that

$$\sup_{u \in \mathcal{F}} \int_\Omega g(|u(x)|) dx < +\infty$$

then $\mathcal{F}$ is bounded and uniformly integrable (De la Vallée–Poussin criterion).

Exercise 9.14. *Let Ω be an open bounded subset of $\mathbb{R}^d$ with a C^1 boundary. Let $F : \Omega \times \mathcal{M}_{m \times d} \to \mathbb{R} \cup \{+\infty\}$ be measurable and such that*

for a.e. $x \in \Omega$, $F(x, .)$ is convex, lsc on $\mathcal{M}_{m \times d}$.

Also assume that for some $\alpha \in L^1(\Omega)$ and some $g : \mathbb{R}_+ \to \mathbb{R}_+$ continuous convex increasing such that $\lim_{t \to +\infty} \frac{g(t)}{t} = +\infty$, one has

for a.e. $x \in \Omega$, $\forall \xi \in \mathcal{M}_{m \times d}$, $F(x, \xi) \geq \alpha(x) + g(|\xi|)$.

Let $G : \Omega \times \mathbb{R}^m \to \mathbb{R}_+ \cup \{+\infty\}$ be measurable and such that $G(x, \cdot)$ is lsc for a.e. $x \in \Omega$. Define for every $u \in W^{1,1}(\Omega, \mathbb{R}^m)$

$$J(u) := \int_\Omega [F(x, Du(x)) + G(x, u(x))] dx$$

and assume that there exists $v \in W^{1,1}(\Omega, \mathbb{R}^m)$ such that $J(v) < +\infty$. Show that J achieves its minimum over $W^{1,1}(\Omega, \mathbb{R}^m)$. (Hint: take a minimizing sequence u_n, show that Du_n is uniformly integrable and then use Rellich–Kondrachov's theorem to deduce that there is a $u \in W^{1,1}(\Omega, \mathbb{R}^m)$ such that $u_n \to u$ in L^1 and $Du_n \rightharpoonup Du$ in L^1 and conclude by a lower semicontinuity argument.)

Exercise 9.15 (A weakly lsc nonconvex functional). *Let Ω be an open bounded subset of $\mathbb{R}^2$ with a C^1 boundary. For every $u = (u_1, u_2) \in W^{1,4}(\Omega)^2$ define*

$$\sigma_u = (-u_2 \partial_2 u_1, u_2 \partial_1 u_1).$$

1) *For $\xi := (\xi_1, \xi_2) \in \mathbb{R}^2 \times \mathbb{R}^2$ define*

$$F_1(\xi) := \det(\xi), \quad F_2(\xi) := \frac{1}{2} \det(\xi)^2.$$

   Show that neither F_1 nor F_2 is convex.

2) *Is the map $\xi \in L^4(\Omega)^2 \mapsto \int_\Omega F_2(\xi(x))dx$ well-defined? Continuous for the strong topology of L^4? Lower semicontinuous for the weak topology of L^4?*

3) *Let $u = (u_1, u_2) \in W^{1,4}(\Omega)^2$. Show that $\det(Du) \in L^2(\Omega)$ and that $\sigma_u \in L^4$, define then $\operatorname{div}(\sigma_u) \in W^{-1,4}(\Omega) := (W_0^{1,\frac{4}{3}}(\Omega))^*$ by*

$$\langle \operatorname{div}(\sigma_u); \varphi \rangle := - \int_\Omega \sigma_u \cdot \nabla \varphi, \quad \forall \varphi \in W_0^{1,\frac{4}{3}}(\Omega).$$

4) *Show that if $u = (u_1, u_2) \in C^2(\mathbb{R}^2, \mathbb{R}^2)$ then $\det(Du) = \operatorname{div}(\sigma_u)$.*

5) *Let $u = (u_1, u_2) \in W^{1,4}(\Omega)^2$ and $u^n := (u_1^n, u_2^n) \in C^2(\mathbb{R}^2, \mathbb{R}^2)$ such that (the restrictions to Ω of) u^n converge to u in $W^{1,4}(\Omega, \mathbb{R}^2)$. Show that $\det(Du_n)$ converges strongly in L^2 to $\det(Du)$ and σ_{u_n} converges strongly in L^4 to σ_u.*

6) *Show that if $u \in W^{1,4}(\Omega)^2$, $\operatorname{div}(\sigma_u) \in L^2$ and $\operatorname{div}(\sigma_u) = \det(Du)$.*

7) *Show that for every $u \in W_0^{1,4}(\Omega)^2$, $J_1(u) := \int_\Omega F_1(Du) = 0$ (in particular, J_1 is weakly lsc!).*

8) *Show that if $u \in W^{1,4}(\Omega)^2$, $u^n \in (W^{1,4}(\Omega)^2)^{\mathbb{N}}$ and $u^n \rightharpoonup u$ in $W^{1,4}(\Omega)^2$ then $\det(Du_n) \rightharpoonup \det(Du)$ in L^2. Deduce that*

$$u \in W^{1,4}(\Omega)^2 \mapsto J_2(u) := \int_\Omega F_2(Du)$$

   is weakly lsc on $W^{1,4}$. Comment on this result.

9) *Show that for any* $v = (v_1, v_2) \in L^4(\Omega)^2$ *the variational problem*

$$\inf_{u \in W^{1,4}(\Omega)^2} J_2(u) + \frac{1}{4} \int_\Omega \left(|\nabla u_1|^4 + |\nabla u_1|^4 + (u_1 - v_1)^4 + (u_2 - v_2)^4 \right)$$

admits at least one solution and write a system of PDEs satisfied by these solutions.

Exercise 9.16. *Let* $F \in C(\mathbb{R}^m, \mathbb{R})$ *be superlinear, we have seen two different ways to define the convex envelope of* F:

- *by taking the Legendre transform of the Legendre transform (see Chapter 3),*
- *by the formula (9.39) from Lemma 9.3.*

Let $\xi \in \mathbb{R}^m$. *Show that these two approaches above to compute* $F^{**}(\xi)$ *are dual to each other in a certain convex duality sense (see Chapter 6).*

Exercise 9.17 (Optimality via duality for an obstacle problem). *Let* Ω *be a non-empty open bounded subset of* $\mathbb{R}^d$ *with a* C^1 *boundary,* $f \in L^2(\Omega)$ *and* $u_0 \in H_0^1(\Omega)$. *We denote by* $H^{-1}(\Omega)$ *the topological dual of* $H_0^1(\Omega)$, *recall (see Exercise 9.10) that if* $\sigma \in L^2(\Omega)^d$, $\operatorname{div}(\sigma)$ *is the element of* $H^{-1}(\Omega)$ *defined by* $\langle \operatorname{div}(\sigma), u \rangle = -\int_\Omega \sigma \cdot \nabla u$ *for every* $u \in H_0^1(\Omega)$. *We are interested in the obstacle problem:*

$$\inf \left\{ \int_\Omega \left(\frac{1}{2} |\nabla u|^2 + fu \right) : u \in H_0^1(\Omega), \ u \geq u_0 \right\}. \tag{9.100}$$

1) *Show that (9.100) admits a unique solution which we denote by* u *and that the infimum of (9.100) coincides with*

$$-\inf \left\{ \int_\Omega \left(\frac{1}{2} |\sigma|^2 + \langle \operatorname{div}(\sigma) - f, u_0 \rangle \right) : \sigma \in L^2(\Omega)^2, \ \operatorname{div}(\sigma) \leq f \right\} \tag{9.101}$$

(where $\operatorname{div}(\sigma) \leq f$ *means that* $\langle \operatorname{div}(\sigma), v \rangle \leq \int_\Omega fv$ *for every* $v \in H_0^1(\Omega)$ *such that* $v \geq 0$).
2) *Show that (9.101) has a unique solution* σ *and that one has the primal-dual and complementary slackness relations:*

$$\sigma = \nabla u, \quad \langle \operatorname{div}(\sigma) - f, u - u_0 \rangle = 0.$$

3) *Show that u is characterized by the conditions*

$$\Delta u \le f, \ u \ge u_0, \ \langle \Delta u - f, u - u_0 \rangle = 0.$$

Exercise 9.18 (Hodge decomposition). *Let Ω be a non-empty open bounded subset of $\mathbb{R}^d$ with a C^1 boundary, $\xi \in L^2(\Omega)^d$, consider*

$$\inf_{u \in H_0^1(\Omega)} \|\nabla u - \xi\|_{L^2(\Omega)}. \tag{9.102}$$

1) *Show that (9.102) has a unique solution u which is characterized by: $u \in H_0^1(\Omega)$ and $\operatorname{div}(\nabla u - \xi) = 0$ in $H^{-1}(\Omega) := (H_0^1(\Omega))^*$.*
2) *Show that for every $\xi \in L^2(\Omega)^d$ there exists a unique pair $(u, q) \in H_0^1(\Omega) \times L^2(\Omega)$ such that*

$$\xi = \nabla u + q, \ \operatorname{div}(q) = 0$$

(this decomposition of the vector field ξ in a gradient part and a divergence-free part is called the Hodge decomposition of ξ).

Exercise 9.19. *Let Ω be a non-empty open bounded subset of $\mathbb{R}^d$ with a C^1 boundary and consider*

$$\lambda(\Omega) := \inf \left\{ \int_\Omega |\nabla u|^2 : u \in H_0^1(\Omega), \int_\Omega u^2 = 1 \right\}. \tag{9.103}$$

1) *Relate $\lambda(\Omega)$ to the best constant in Poincaré's inequality.*
2) *Show that the infimum in (9.103) is attained and that*

$$\lambda(\Omega) := \inf \left\{ \frac{\int_\Omega |\nabla u|^2}{\int_\Omega u^2} : u \in H_0^1(\Omega) \setminus \{0\} \right\}.$$

3) *Show that if $\xi \in \mathbb{R}^d$ and R is a rotation then $\lambda(\xi + R(\Omega)) = \lambda(\Omega)$. Show that if $\alpha > 0$ denoting $\alpha\Omega := \{\alpha x, x \in \Omega\}$ one has*

$$\lambda(\alpha\Omega) = \frac{1}{\alpha^2} \lambda(\Omega).$$

4) *Show that if u is optimal for (9.103) then $-\Delta u = \lambda(\Omega)u$.*
5) *Show that if $d = 1$ and $\Omega = (0, L)$ with $L > 0$ then $\lambda((0, L)) = \frac{\pi^2}{L^2}$.*

Exercise 9.20. *Let $T > 0$, $F \in C^2(\mathbb{R}_+ \times \mathbb{R}^m)$, $G \in C^2(\mathbb{R}_+ \times \mathbb{R}^m)$ be such that for some $C \ge 0$ and some $\Lambda \ge \lambda > 0$, for every $(t, x, v) \in \mathbb{R}_+ \times \mathbb{R}^m \times \mathbb{R}^m$ one has*

$$|D_{xx}^2 F(t, x)| \le C, \ \Lambda \operatorname{id} \ge D_{vv}^2 G(t, v) \ge \lambda \operatorname{id}$$

(where the last inequality has to be understood in the sense of positive semidefinite matrices). For every $\gamma \in H^1((0,T), \mathbb{R}^m)$ then set

$$J(\gamma) := \int_0^T (F(s, \gamma(s)) + G(s, \dot{\gamma}(s)))ds.$$

1) *Show that J is continuous and Gateaux-differentiable over $H^1((0,T), \mathbb{R}^m)$.*

2) *Use a second-order integral Taylor Formula to show that for every γ and θ in $\gamma \in H^1((0,T), \mathbb{R}^m)$ one has*

$$J(\theta) - J(\gamma) - J'(\gamma)(\theta - \gamma) \geq -\frac{C}{2} \int_0^T (\theta - \gamma)^2 + \frac{\lambda}{2} \int_0^T |\dot{\theta} - \dot{\gamma}|^2.$$

3) *Set $T^* := \pi\sqrt{\frac{\lambda}{C}}$, deduce from Exercise 9.19 that whenever $T < T^*$, for any $\gamma_0 \in H^1((0,1), \mathbb{R}^m)$, J is strictly convex over $\gamma_0 + H_0^1((0,T), \mathbb{R}^m)$.*

4) *Show that for $T < T^*$, and $(x,y) \in \mathbb{R}^m \times \mathbb{R}^m$ the problem*

$$\inf_{\gamma \in H^1((0,1), \mathbb{R}^m), \, \gamma(0)=x, \, \gamma(T)=y} J(\gamma)$$

has a unique solution $\gamma^{x,y}$ and that solution is of class C^2.

5) *Show that for $T < T^*$, $(x,y) \in \mathbb{R}^m \times \mathbb{R}^m \mapsto \gamma^{x,y} \in H^1((0,T), \mathbb{R}^m)$ is locally Lipschitz.*

Exercise 9.21 (Isoperimetric inequality by optimal transport).
Let $d \geq 2$, for every Ω, non-empty open bounded subset of $\mathbb{R}^d$ with a C^1 boundary, define

$$I(\Omega) := \frac{\sigma(\partial\Omega)}{|\Omega|^{1-\frac{1}{d}}}.$$

Let B be the unit euclidean ball of $\mathbb{R}^d$, we know from Brenier's theorem (see Section 6.4.3) that there exists u convex such that ∇u transports the uniform measure on Ω to the uniform measure on B (so that $|\nabla u| \leq 1$). Assuming that ∇u is a smooth diffeomorphism between Ω and B, u solves the Monge-Ampère equation

$$\det(D^2 u(x)) = \frac{|B|}{|\Omega|}, \ \forall x \in \Omega. \tag{9.104}$$

1) Use a homogeneity argument to show that $I(B) = d|B|^{\frac{1}{d}}$.
2) Show that $\det(D^2 u)^{\frac{1}{d}} \leq \frac{1}{d}\Delta u$ and deduce that

$$|B|^{\frac{1}{d}}|\Omega|^{1-\frac{1}{d}} \leq \frac{1}{d}\int_\Omega \Delta u.$$

3) Integrate by parts the previous inequality to deduce the isoperimetric inequality: $I(\Omega) \geq I(B)$.

Bibliography

[1] E. Acerbi and N. Fusco. Semicontinuity problems in the calculus of variations. *Arch. Rational Mech. Anal.*, 86(2):125–145, 1984.

[2] R. A. Adams. *Sobolev Spaces*. [Pure and Applied Mathematics, Vol. 65.] Academic Press [A subsidiary of Harcourt Brace Jovanovich, Publishers], New York, 1975.

[3] N. Ahmad, H. K. Kim and R. J. McCann. Optimal transportation, topology and uniqueness. *Bull. Math. Sci.*, 1(1):13–32, 2011.

[4] F. Alvarez. On the minimizing property of a second order dissipative system in Hilbert spaces. *SIAM J. Control Optim.*, 38(4):1102–1119, 2000.

[5] L. Ambrosio, N. Fusco and D. Pallara. *Functions of Bounded Variation and Free Discontinuity Problems*. Oxford Mathematical Monographs. Oxford University Press, New York, 2000.

[6] L. Ambrosio and P. Tilli. *Topics on Analysis in Metric Spaces*, Oxford Lecture Series in Mathematics and its Applications, Vol. 25. Oxford University Press, Oxford, 2004.

[7] H. Attouch and H. Brezis. Duality for the sum of convex functions in general Banach spaces. In *Aspects of Mathematics and Its Applications*, volume 34 of *North-Holland Mathematics Library*, Vol. 34, pp. 125–133. North-Holland, Amsterdam, 1986.

[8] J. M. Ball and V. J. Mizel. One-dimensional variational problems whose minimizers do not satisfy the Euler–Lagrange equation. *Arch. Rational Mech. Anal.*, 90(4):325–388, 1985.

[9] M. Bardi and I. Capuzzo–Dolcetta. *Optimal control and viscosity solutions of Hamilton–Jacobi–Bellman equations*. Systems & Control: Foundations & Applications. Birkhäuser, Boston, MA, 1997.

[10] G. Barles. *Solutions de viscosité des équations de Hamilton–Jacobi*. Mathématiques & Applications (Berlin) [Mathematics & Applications], Vol. 17. Springer-Verlag, Paris, 1994.

[11] H. H. Bauschke and P. L. Combettes. *Convex Analysis and Monotone Operator Theory in Hilbert Spaces*. CMS Books in Mathematics/Ouvrages de Mathématiques de la SMC. Springer, Cham, second edition, 2017.

[12] H. H. Bauschke, R. Goebel, Y. Lucet and X. Wang. The proximal average: basic theory. *SIAM J. Optim.*, 19(2):766–785, 2008.

[13] A. Beck and M. Teboulle. A fast iterative shrinkage-thresholding algorithm for linear inverse problems. *SIAM J. Imaging Sci.*, 2(1):183–202, 2009.

[14] J.-D. Benamou, G. Carlier, M. Cuturi, L. Nenna and G. Peyré. Iterative Bregman projections for regularized transportation problems. *SIAM J. Sci. Comput.* 37(2):A1111–A1138, 2015.

[15] H. Berliocchi and J.-M. Lasry. Intégrandes normales et mesures paramétrées en calcul des variations. *Bull. Soc. Math. France*, 101:129–184, 1973.

[16] D. P. Bertsekas. *Nonlinear Programming.* Athena Scientific Optimization and Computation Series. Athena Scientific, Belmont, MA, second edition, 1999.

[17] J. Frédéric Bonnans, J. Charles Gilbert, Claude Lemaréchal and Claudia A. Sagastizábal. *Numerical Optimization.* Universitext. Springer-Verlag, Berlin, second edition, 2006. Theoretical and practical aspects.

[18] J. Frédéric Bonnans and Alexander Shapiro. *Perturbation Analysis of Optimization Problems.* Springer Series in Operations Research. Springer-Verlag, New York, 2000.

[19] S. Boyd and L. Vandenberghe. *Convex Optimization.* Cambridge University Press, Cambridge, 2004.

[20] C. Boyer, A. Chambolle, Y. De Castro, V. Duval, F. de Gournay and P. Weiss. On representer theorems and convex regularization. *SIAM J. Optim.*, 29(2):1260–1281, 2019.

[21] A. Braides. Γ-*convergence for Beginners*, Oxford Lecture Series in Mathematics and its Applications, Vol. 22. Oxford University Press, Oxford, 2002.

[22] Y. Brenier. Polar factorization and monotone rearrangement of vector-valued functions. *Commun. Pure Appli. Math.*, 44(4):375–417, 1991.

[23] H. Brezis. *Analyse Fonctionnelle.* Collection Mathématiques Appliquées pour la Maîtrise. [Collection of Applied Mathematics for the Master's Degree]. Masson, Paris, 1983. Théorie et applications. [Theory and applications].

[24] H. Brezis. Liquid crystals and energy estimates for S^2-valued maps. In *Theory and Applications of Liquid Crystals* (Minneapolis, 1985), The IMA Volume in Mathematics and Applications, Vol. 5, pp. 31–52. Springer, New York, 1987.

[25] H. Brezis. Remarks on the Monge-Kantorovich problem in the discrete setting. *C. R. Math. Acad. Sci. Paris*, 356(2):207–213, 2018.

[26] G. Buttazzo and M. Belloni. A survey on old and recent results about the gap phenomenon in the calculus of variations. In *Recent Developments in Well-Posed Variational Problems*, Mathematics and its Applications, Vol. 33, pp. 1–27. Kluwer Academic Publishes, Dordrecht, 1995.

[27] G. Buttazzo. *Semicontinuity, Relaxation and Integral Representation in the Calculus of Variations*, Pitman Research Notes in Mathematics Series. Vol. 207. Longman Scientific & Technical, Harlow; copublished in the United States with John Wiley & Sons, Inc., New York, 1989.

[28] G. Buttazzo, M. Giaquinta and S. Hildebrandt. *One-Dimensional Variational Problems*, Oxford Lecture Series in Mathematics and its Applications. Vol. 15. Oxford University Press, New York, 1998. An introduction.

[29] L. A. Caffarelli. Allocation maps with general cost functions. In *Partial Differential Equations and Applications*, Lecture Notes in Pure and Applied Mathematics, Vol. 177, pp. 29–35. Dekker, New York, 1996.

[30] P. Cannarsa and C. Sinestrari. *Semiconcave Functions, Hamilton–Jacobi Equations, and Optimal Control*, Progress in Nonlinear Differential Equations and their Applications, Vol. 58. Birkhäuser, Boston, MA, 2004.

[31] G. Carlier. Duality and existence for a class of mass transportation problems and economic applications. *Adv. in Math. Econ.*, 5:1–21, 2003.

[32] H. Cartan. *Differential Calculus*. Hermann, Paris; Houghton Mifflin Co., Boston, MA, 1971.

[33] L. Cesari. *Optimization—Theory and Applications*, Applications of Mathematics (New York), Vol. 17. Springer-Verlag, New York, 1983. Problems with ordinary differential equations.

[34] S. S. Chen, D. L. Donoho and M. A. Saunders. Atomic decomposition by basis pursuit. *SIAM J. Sci. Comput.*, 20(1):33–61, 1998.

[35] P. L. Combettes. Solving monotone inclusions via compositions of nonexpansive averaged operators. *Optimization*, 53(5–6):475–504, 2004.

[36] R. Cominetti and J. San Martín. Asymptotic analysis of the exponential penalty trajectory in linear programming. *Math. Program.* 67(2, Ser. A):169–187, 1994.

[37] M. G. Crandall, H. Ishii, and P.-L. Lions. User's guide to viscosity solutions of second order partial differential equations. *Bull. Amer. Math. Soc. (N.S.)*, 27(1):1–67, 1992.

[38] M. Cuturi. Sinkhorn distances: Lightspeed computation of optimal transport. In *Advances in Neural Information Processing Systems*, pp. 2292–2300, 2013.

[39] B. Dacorogna. *Direct Methods in the Calculus of Variations*, Applied Mathematical Sciences, Vol. 78. Springer, New York, second edition, 2008.

[40] B. Dacorogna. *Introduction to the Calculus of Variations*. Imperial College Press, London, third edition, 2015.

[41] G. Dal Maso. *An Introduction to Γ-convergence*, Vol. 8. Progress in Nonlinear Differential Equations and their Applications, Birkhäuser Boston, MA, 1993.

[42] J. Douglas, Jr. and H. H. Rachford, Jr. On the numerical solution of heat conduction problems in two and three space variables. *Trans. Amer. Math. Soc.*, 82:421–439, 1956.

[43] J. Eckstein and D. P. Bertsekas. On the Douglas–Rachford splitting method and the proximal point algorithm for maximal monotone operators. *Math. Program.* 55(3, Ser. A):293–318, 1992.

[44] I. Ekeland. On the variational principle. *J. Math. Anal. Appl.*, 47:324–353, 1974.

[45] I. Ekeland. An inverse function theorem in Fréchet spaces. *Ann. Inst. H. Poincaré Anal. Non Linéaire*, 28(1):91–105, 2011.

[46] I. Ekeland and R. Témam. *Convex Analysis and Variational Problems*, Vol. 28. *Classics in Applied Mathematics*. Vol. 28. SIAM, Philadelphia, PA, English edition, 1999.

[47] L. C. Evans and R. F. Gariepy. *Measure Theory and Fine Properties of Functions*. Textbooks in Mathematics. CRC Press, Boca Raton, FL, revised edition, 2015.

[48] A. Figalli. *The Monge–Ampère Equation and its Applications*. Zurich Lectures in Advanced Mathematics. European Mathematical Society (EMS), Zürich, 2017.

[49] D. Firth. Bias reduction of maximum likelihood estimates. *Biometrika*, 80(1):27–38, 1993.

[50] W. H. Fleming and H. Mete Soner. *Controlled Markov Processes and Viscosity Solutions*, Vol. 25 of Stochastic Modelling and Applied Probability. Springer, New York, second edition, 2006.

[51] M. Fortin and R. Glowinski. *Augmented Lagrangian Methods*, Studies in Mathematics and its Applications, Vol. 15. North-Holland Publishing Co., Amsterdam, 1983. Applications to the numerical solution of boundary value problems, Translated from the French by B. Hunt and D. C. Spicer.

[52] D. Gabay and B. Mercier. A dual algorithm for the solution of nonlinear variational problems via finite element methods. *Comput. Math. Appl.* 2:17–40, 1976.

[53] A. Galichon and B. Salanié. Cupid's invisible hand: Social surplus and identification in matching models. *Review of Economic Studies* 2015.

[54] W. Gangbo. An elementary proof of the polar factorization of vector-valued functions. *Arch. Rational Mech. Anal.*, 128(4):381–399, 1994.

[55] W. Gangbo and R. J. McCann. The geometry of optimal transportation. *Acta Math.*, 177(2):113–161, 1996.

[56] M. Giaquinta and S. Hildebrandt. *Calculus of Variations. I*, Grundlehren der Mathematischen Wissenschaften [Fundamental Principles of Mathematical Sciences], Vol. 310. Springer-Verlag, Berlin, 1996. The Lagrangian formalism.

[57] M. Giaquinta and S. Hildebrandt. *Calculus of Variations. II*, Grundlehren der Mathematischen Wissenschaften [Fundamental Principles of Mathematical Sciences], Vol. 311. Springer-Verlag, Berlin, 1996. The Hamiltonian formalism.

[58] D. Gilbarg and N. S. Trudinger. *Elliptic Partial Differential Equations of Second Order*. Classics in Mathematics. Springer-Verlag, Berlin, 2001.

[59] E. Giusti. *Minimal Surfaces and Functions of Bounded Variation*, Monographs in Mathematics, Vol. 80. Birkhäuser, Basel, 1984.

[60] E. Giusti. *Direct Methods in the Calculus of Variations*. World Scientific Publishing Co., Inc., River Edge, NJ, 2003.

[61] A. Hantoute, M. A. López and C. Zalinescu. Subdifferential calculus rules in convex analysis: a unifying approach via pointwise supremum functions. *SIAM J. Optim.*, 19(2):863–882, 2008.

[62] K. Hestir and S. C. Williams. Supports of doubly stochastic measures. *Bernoulli*, 1(3):217–243, 1995.

[63] E. Hewitt and K. Stromberg. *Real and Abstract Analysis. A Modern Treat-ment of the Theory of Functions of a Real Variable.* Springer-Verlag, New York, 1965.

[64] J.-B. Hiriart-Urruty and C. Lemaréchal. *Convex Analysis and Minimiza-tion Algorithms. I*, Grundlehren der Mathematischen Wissenschaften [Fun-damental Principles of Mathematical Sciences], Vol. 305. Springer-Verlag, Berlin, 1993. Fundamentals.

[65] S. G. Krantz and H. R. Parks. *The Implicit Function Theorem.* Birkhäuser, Boston, MA, 2002. History, theory, and applications.

[66] M. Krein and D. Milman. On extreme points of regular convex sets. *Studia Math.*, 9:133–138, 1940.

[67] C. Le Van and R.-A. Dana. Dynamic programming in economics, *Dynamic Modeling and Econometrics in Economics and Finance*, Vol. 5. Kluwer Aca-demic Publishers, Dordrecht, 2003.

[68] C. Léonard. A survey of the Schrödinger problem and some of its connections with optimal transport. *Discrete Contin. Dyn. Syst.*, 34(4):1533–1574, 2014.

[69] V. Levin. Abstract cyclical monotonicity and Monge solutions for the gen-eral Monge–Kantorovich problem. *Set-Valued Anal.*, 7(1):7–32, 1999.

[70] J. Lindenstrauss. A remark on extreme doubly stochastic measures. *Amer. Math. Monthly*, 72:379–382, 1965.

[71] P.-L. Lions and B. Mercier. Splitting algorithms for the sum of two nonlinear operators. *SIAM J. Numer. Anal.*, 16(6):964–979, 1979.

[72] B. Martinet. Régularisation d'inéquations variationnelles par approxima-tions successives. *Rev. Française Informat. Recherche Opérationnelle*, 4(Sér. R-3):154–158, 1970.

[73] B. Martinet. Détermination approchée d'un point fixe d'une application pseudo-contractante. Cas de l'application prox. *C. R. Acad. Sci. Paris Sér. A–B*, 274:A163–A165, 1972.

[74] R. J. McCann. Existence and uniqueness of monotone measure-preserving maps. *Duke Math. J.*, 80(2):309–323, 1995.

[75] J.-J. Moreau. Proximité et dualité dans un espace hilbertien. *Bull. Soc. Math. France*, 93:273–299, 1965.

[76] J. J. Moreau. Fonctionnelles convexes. *Séminaire Jean Leray*, (2):1–108, 1966–1967.

[77] C. B. Morrey, Jr. Multiple Integrals in the Calculus of Variations. *Classics in Mathematics*. Springer-Verlag, Berlin, 2008.

[78] A. S. Nemirovsky and D. B. Yudin. *Problem Complexity and Method Effi-ciency in Optimization.* John Wiley & Sons, Inc., New York, 1983.

[79] Yu. E. Nesterov. A method for solving the convex programming problem with convergence rate $O(1/k^2)$. *Dokl. Akad. Nauk SSSR*, 269(3):543–547, 1983.

[80] J. M. Ortega and W. C. Rheinboldt. *Iterative Solution of Nonlinear Equa-tions in Several Variables.* Academic Press, New York, 1970.

[81] P. Pedregal. Parametrized Measures and Variational Principles, *Progress in Nonlinear Differential Equations and their Applications*, Vol. 30. Birkhäuser Verlag, Basel, 1997.

[82] G. Peyré and M. Cuturi. Computational optimal transport. *Found Trends Mach. Learn.*, 11(5–6):355–607, 2019.

[83] J.-C. Rochet. A necessary and sufficient condition for rationalizability in a quasilinear context. *J. Math. Econ.*, 16(2):191–200, 1987.

[84] R. T. Rockafellar. Characterization of the subdifferentials of convex functions. *Pacific J. Math.*, 17:497–510, 1966.

[85] R. T. Rockafellar. Augmented Lagrangians and applications of the proximal point algorithm in convex programming. *Math. Oper. Res.*, 1(2):97–116, 1976.

[86] R. T. Rockafellar. Monotone operators and the proximal point algorithm. *SIAM J. Control Optim.*, 14(5):877–898, 1976.

[87] R. T. Rockafellar. Convex analysis. *Princeton Landmarks in Mathematics*. Princeton University Press, Princeton, NJ, 1997. Reprint of the 1970 original, Princeton Paperbacks.

[88] W. Rudin. *Real and Complex Analysis*. McGraw-Hill Book Co., New York, third edition, 1987.

[89] F. Santambrogio. Optimal Transport for Applied Mathematicians. *Progress in Nonlinear Differential Equations and their Applications*, Vol. 87. Birkhäuser/Springer, Cham, 2015. Calculus of variations, PDEs, and modeling.

[90] E. Schrödinger. Sur la théorie relativiste de l'électron et l'interprétation de la mécanique quantique. *Ann. Inst. H. Poincaré*, 2(4):269–310, 1932.

[91] L. Schwartz. *Analyse*. Deuxième partie: Topologie générale et analyse fonctionnelle, Collection Enseignement des Sciences, No. 11. Hermann, Paris, 1970.

[92] C. S. Smith and M. Knott. Note on the optimal transportation of distributions. *J. Optim. Theory Appl.*, 52(2):323–329, 1987.

[93] N. L. Stokey and R. E. Lucas, Jr. *Recursive Methods in Economic Dynamics*. Harvard University Press, Cambridge, MA, 1989. With the collaboration of Edward C. Prescott.

[94] M. Valadier. Sous-différentiels d'une borne supérieure et d'une somme continue de fonctions convexes. *C. R. Acad. Sci. Paris Sér. A-B*, 268:A39–A42, 1969.

[95] L. Vandenberghe and S. Boyd. Semidefinite programming. *SIAM Rev.*, 38(1):49–95, 1996.

[96] V. N. Vapnik. *The Nature of Statistical Learning Theory*. Statistics for Engineering and Information Science. Springer-Verlag, New York, second edition, 2000.

[97] C. Villani. *Topics in Optimal Transportation*, Graduate Studies in Mathematics, Vol. 58. American Mathematical Society, Providence, 2003.

[98] C. Villani. *Optimal Transport: Old and New*, Springer Science & Business Media, 2008.

[99] L. C. Young. *Lectures on the Calculus of Variations and Optimal Control Theory*. Foreword by Wendell H. Fleming. W. B. Saunders Co., Philadelphia, 1969.

Index

A

acyclic graph, 229
acyclic matrix, 227
adjoint, 125, 200
alternating direction method of
   multipliers (ADMM), 268
assignment problem, 229
augmented Lagrangian, 240,
   268

B

Baire's theorem, 13
Banach space, 23
Banach–Steinhaus theorem, 29
basis pursuit, 296
Bellman equation, 184
bipolar theorem, 127
Birkhoff–von Neumann theorem,
   230
bistochastic matrix, 230
Brenier's theorem, 225
Brouwer's fixed point theorem, 92

C

c-concave transform, 216
c-cyclical monotonicity, 217
c-superdifferential, 217
Carathéodory's theorem, 102
Cauchy–Schwarz inequality, 31
complementary slackness, 148
conditioning number, 289

conjugate gradient algorithm, 272
contingent cone, 158
convex combination, 33
convex duality, 199
convex envelope, 331
coordinate descent, 270
cyclical monotonicity, 220

D

De la Vallée–Poussin criterion,
   357
divergence formula, 87
Douglas–Rachford algorithm, 264
Du Bois–Reymond's lemma, 315
Dunford–Pettis theorem, 357

E

Ekeland's variational principle, 14
entropic regularization, 231
envelope theorem, 170
epigraph, 11, 106
Euler–Lagrange equations, 338

F

Fenchel reciprocity formula, 113
Fenchel–Moreau theorem, 118
Fenchel–Rockafellar theorem, 202
fast iterative shrinkage thresholding
   algorithm (FISTA), 263
Frank–Wolfe algorithm, 274

G

gauge of a convex set, 94
gradient flow, 257
Green formula, 88

H

Hahn–Banach separation theorem, 95
Hahn–Banach analytic form, 93
Hamilton–Jacobi equation, 351
Hamiltonian systems, 348
Hausdorff distance, 188
heavy ball, 271
Hessian matrix, 70
Hodge decomposition, 360
Hopf–Lax formula, 239

I

implicit function theorem, 75
infimal convolution, 116
inverse function theorem, 73
isoperimetric inequality, 361
iterative shrinkage thresholding algorithm (ISTA), 263

J

Jensen's inequality, 126

K

Kantorovich duality formula, 215
Karush–Kuhn–Tucker multipliers, 147
Krein–Milman theorem, 105

L

Lagrange multipliers, 141
Lagrangian, 143, 210
Lagrangian duality, 211
least absolute shrinkage and selection operator (LASSO), 298
Lax–Oleinik formula, 353
Legendre–Fenchel transform, 113
Leibniz formula, 81
linear quadratic, 112
linear regression, 292

logistic regression, 301
lower semicontinuous, 12

M

Mangasarian–Fromowitz, 150
marginal price of constraints, 212
Minkowski–Farkas theorem, 99
Monge–Ampère equation, 226
Moore–Penrose inverse, 294
Moreau's proximal identity, 276
Moreau–Yosida regularization, 130

N

Nesterov's accelerated gradient descent, 253
Newton's law, 345
Newton's method, 244
no retraction theorem, 92

O

open mapping theorem, 29
Opial's lemma, 259

P

principal component analysis (PCA), 283
penalization, 154
polar cone, 127
projection onto a closed convex set, 33
proximal mapping, 256

Q

quasiconvex, 90

R

Rademacher's theorem, 173
Riesz' representation theorem, 36

S

saddle point, 212
Schur complement, 238

SDP duality, 207
semidefinite programming (SDP), 205
shadow price, 182
Slater condition, 152, 209
soft maximum, 127
strict separation theorem, 96
strong linear programming (LP)
  duality theorem, 204
subdifferential, 120
submodular, 221
superdifferentiability, 172
support function, 98
singular value decomposition (SVD),
  280
support-vector machines (SVM),
  303

T

Tikhonov regularization, 293
topological complements, 139
transversality conditions, 339

U

uniform integrability, 356
Urysohn's lemma, 79
Uzawa's algorithm, 273

W

weak convergence, 38, 46

Y

Young–Fenchel inequality, 114

Printed in the United States
by Baker & Taylor Publisher Services